DATE DUE

GAYLORD			PRINTED IN U.S.A.

SOIL ORGANIC MATTER

SOIL ORGANIC MATTER
BIOLOGICAL AND ECOLOGICAL EFFECTS

ROBERT L. TATE III
Associate Professor
Rutgers University

A WILEY-INTERSCIENCE PUBLICATION
JOHN WILEY & SONS
New York Chichester Brisbane Toronto Singapore

Library of Congress Cataloging in Publication Data:

Tate, Robert L., 1944–
Soil organic matter.

"A Wiley-Interscience publication."
Includes bibliographies and index.
1. Humus. 2. Soil biology. 3. Soil ecology.
I. Title.

S592.8.T38 1987 631.4′17 86-19091
ISBN 0-471-81570-5

Printed in the United States of America

10 9 8 7 6 5 4 3 2 1

TO

ANN, ROBERT, AND GEOFFREY

PREFACE

Soil organic matter—we have tended to ignore its existence, we have abused the processes leading to its function, and we have failed to fully appreciate its importance in total ecosystem function. With increases in societal demands on the soil ecosystem currently occurring and the observation that the pools of organic matter in agricultural soils are declining, we have come to realize that although organic matter is a minor component in most soils, it has an essential function to perform in the ecosystem. As we strive to understand the chemistry of this complex soil component, we must remember its ecosystem impact. In many cases, the total nature and longevity of the aboveground aspects of an ecosystem are controlled by the chemical, biological, and physical properties of the soil organic matter pool. Thus this treatise was prepared with the objective of providing a basic introduction into the biological aspects of the existence of soil organic matter.

Knowledge of this soil fraction is required by a variety of scientists, ranging from ecologists to agriculturalists. Many have strong academic backgrounds in soil, biological, and ecological sciences; many are novices. Thus the topics contained herein were developed with a gradient of complexity. The planks of basic knowledge were first laid before the complete structure of our understanding of each subject was constructed. Frequently, the informational gradient from elementary to complex is steep. At other times, it appears to be rather horizontal. But it is the author's hope that each reader will finish the study of this book with a more complete understanding of the importance of the biological properties of soil organic matter in total ecosystem function, and be better prepared to evaluate the more technical literature available relating to behavior of organic matter in soil.

Concern for soil organic matter is developing in many arenas of our society. Future interest will not only involve the more traditional problems of agriculture and ecosystem studies but it will also be peppered with the added complexities of spills and/or utilization of industrial chemicals. Organic matter associated problems will relate not only to environmental quality and economic issues but also to such difficulties as waste management and soil reclamation. In some cases, new data will be needed, whereas with others application of existing information will lead to more than adequate solutions. This book is offered as an instrument to assist the evaluation of current knowledge and provide a direction in implementation of soil organic matter studies in future land use, ecosystem management, and reclamation plans.

A treatise of this type could not be prepared without considerable input from my friends and associates. I especially thank the many unnamed individuals who unknowingly provided seeds in our discussions that have developed into ideas contained herein. The input from my students through their "suffering" through trial lectures based on this material and their patience when forced to conduct their research while their mentor was preoccupied with writing are greatly appreciated. Finally, the capable assistance of Silvia Taylor in drafting the figures is gratefully acknowledged. A portion of the time devoted to this project was supported by the New Jersey Agricultural Experiment Station, Publication No. A-15187-1-85, supported by state funds.

ROBERT L. TATE III

New Brunswick, New Jersey
January 1987

CONTENTS

ONE

ORGANIC MATTER: A DYNAMIC SOIL COMPONENT

The most obvious aspects of any ecosystem are the aboveground plant and animal communities. They are easily studied and provide the aesthetic aspects that most of us associate with a natural ecosystem. But, these surface communities rely on the existance and function of the complex belowground components of the ecosystem. The subsurface role of plant roots and a variety of micro- and mesofauna are obvious. Perhaps less apparent, but equally important for total ecosystem function, is the soil organic matter fraction. Although it is frequently, from the view of total soil mass, a minor soil component, this organic fraction has a major impact on obligatory ecosystem processes. For example, the living soil organic component catalyzes those biogeochemical reactions basic to providing plant nutrients, whereas all soil organic components interact to contribute to soil structural development.

Division of function between the biotic (living), abiotic (nonliving, chemically derived), and abiontic (nonliving, biologically derived) soil organic matter fractions demonstrates the incongruency associated with frequent references to soil organic matter as if it were a single definable entity. In reality, its composition is quite diverse and ever-changing. Not only do major and minor organic substituents vary between ecosystems, but within a single site, microbial mineralization and spontaneous chemical reactions constantly alter the chemical composition of the organic fraction. The dynamic character of the organic fraction contributes to a variety of essential soil processes including biogeochemical cycling, soil aggregate formation, and mineral solubilization. The mere presence of an organic faction in the soil affects basic soil properties, such as cation exchange capacity and buffer capacity. The historical and chemical aspects of the soil organic fraction have been reviewed a number of times (Kononova, 1966, Stevenson, 1982). Thus the presentation herein will be limited more to the biochemical processes relating to soil organic matter transformations and their impact on total ecosystem function. This chapter contains a general introduction relating to the nature and source of soil organic matter and its role within the ecosystem.

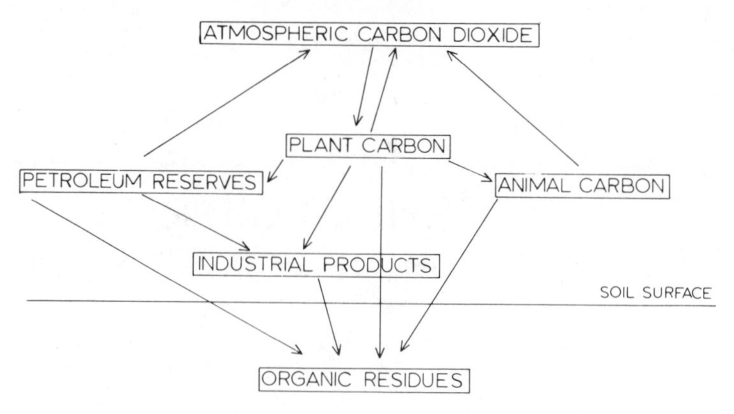

Fig. 1.1.
Aboveground inputs into soil organic matter pools.

1.1. SOURCE OF SOIL ORGANIC MATTER

At one time it could have been said that the ultimate source of all soil organic matter was that carbon fixed through photosynthetic reactions (Fig. 1.1). This would be predominantly through carbon dioxide fixation by plants, but there is also a small input from autotrophic soil bacteria. But, with the development of organic chemical manufacturing processes, synthetically produced biochemicals have become a significant factor affecting the soil organic fraction. Those pesticides commonly added to soil to improve or increase crop production are the most commonly recognized synthetically produced organics entering the soil ecosystem. Other inputs include significant quantities of polycyclic aromatic hydrocarbon produced from burning fossil fuels (Hites et al., 1977; Blumer et al., 1977) as well as a variety of xenobiotics (including plastics) resulting from modern societal development. Anthropogenically produced organic substances generally are insignificant from the view of total mass (although there may be severe localized impacts such as occurs around landfill sites), but their effects on ecosystem function and stability can in many cases be quite dramatic. These organics may be incorporated directly into soil humic substances or they may remain free to interact with soil components or migrate (leach) from the site.

An organic substance entering the soil ecosystem faces three ultimate fates: (1) it may be totally mineralized and thereby returned to the carbon dioxide and mineral nutrient pool (Fig. 1.2); (2) it may be assimilated into microbial biomass; or (3) the organic matter may be incorporated either unchanged or partially modified into the more stable soil humic fraction, that is, it may be humified. The rates of these various conversions depends upon the interactions of the soil microbial, nematode, and microarthropod communities, a variety of chemical reactions, as well as the chemical and physical parameters of the given site.

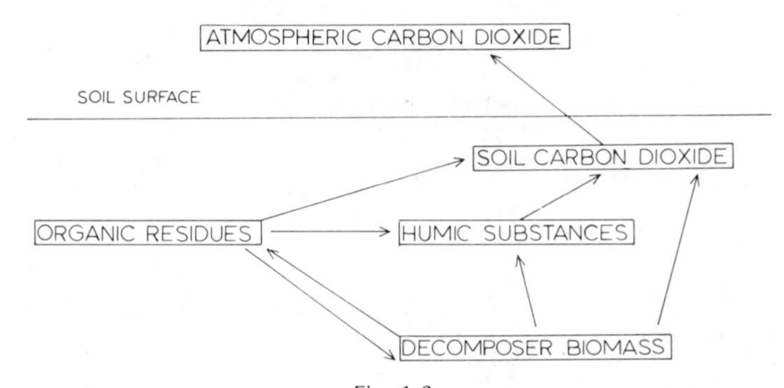

Fig. 1.2.
Biochemical decomposition products of soil organic matter.

Total organic matter accumulated in soils constitutes a major portion of the world's fixed carbon reserves. Bohn (1976) estimated that the soils contain approximately 30 x 10^{14} Kg organic carbon. Distribution of this organic matter among soil types is highly variable and generally not easily predictable from aboveground vegetation types. The quantity of organic material retained within the soil matrix is the difference between total biomass production and decomposition. Therefore, an evaluation of aboveground biomass may provide a deceptive picture of the level of soil organic matter. This is clearly shown when tropical grassland and forest soils are examined. The high biomass production associated with many tropical ecosystems generally does not result in increased soil organic matter levels when compared to comparable temperate ecosystems in which warm moist conditions contribute to plant growth and stimulate soil microbial activity. Along with the aspects of total production and decomposition, a variety of chemical and physical factors of both the ecosystem and the plant material itself control the final concentrations of colloidal organic matter accumulated within a given site. For example, chronically flooded soils generally accumulate greater concentrations of organic matter than do arable soils. The effect of lignification of plant tissues exemplifies plant tissue variability effects on biodecomposition. Woody tissues are much more recalcitrant than succulant green materials.

1.2. FORMS OF ORGANIC MATTER IN SOIL

As is evident from the foregoing discussion, the soil organic fraction is extremely heterogeneous. Variations in the nature and quantity of this component exist not only between divergent ecosystems but also within a single site because of heterogeneity of the plant community. For example, differences in readily decomposable organic components are evident when comparing rhizosphere and nonrhizosphere soil. Root exudates have been shown to contain a variety of

Table 1.1.
Amino Acids in Soy Bean Rhizosphere and Nonrhizosphere Soil (Paul and Schmidt, 1961)[a]

Amino Acid	Nonrhizosphere μg/g	Rhizosphere μg/g
Aspartic acid	0.20	0.59
Threonine	0.20	0.19
Serine	0.16	0.09
Glutamic acid	0.31	0.32
Glycine	0.21	0.32
Alanine	0.18	0.45
Valine	0.20	*[b]
Isoleucine	0.12	0.32
Leucine	0.13	*[b]
Beta Alanine	0.09	0.07
Tyrosine	0.27	0.46
Phenylalanine	0.12	0.18
Gamma amino butyric acid	0.08	*[b]
Lysine	0.52	0.99
Histidine	0.33	0.41

[a]Reproduced from SOIL SCIENCE SOCIETY OF AMERICA PROCEEDINGS, Volume 25, 1961, pp. 359–363 by permission of the Soil Science Society of America.
[b]Added as markers, probably present in concentrations similar to adjacent compounds.

compounds, including amino acids, carbohydrates, organic acids, vitamins, and so on (Alexander, 1977). Paul and Schmidt (1961) demonstrated the quantitative differences in amino acid concentrations between rhizosphere and nonrhizosphere soils (Table 1.1). Individual amino acid concentrations were generally two- to three-fold greater in the rhizosphere soil, although in some cases only slight differences were detected. This concentration gradient resulted from the interaction of the rhizosphere microbial community and the diffusion rate of the organic compounds through the soil. The more active the microbes in decomposing the individual organic component, the lower the gradient with distance from the root. This was dramatically demonstrated in a study of ^{14}C-labeled photosynthetate from wheat roots (*Triticum aestivum* cv Nugaines) grown axenically or inoculated with slight to highly dense bacterial cultures (Fig. 1.3) (Beck and Gilmour, 1983). Other factors contributing to this heterogeneity in the soil organic fraction include spatial variation in aboveground biomass, and differential earthworm and insect activity.

Essentially any compound found in living plant and animal cells can be detected in the soil organic fraction. Concentrations of those materials that are easily metabolized by the microbial community may be quite low and variable with time. These biologically synthesized compounds are supplemented in the

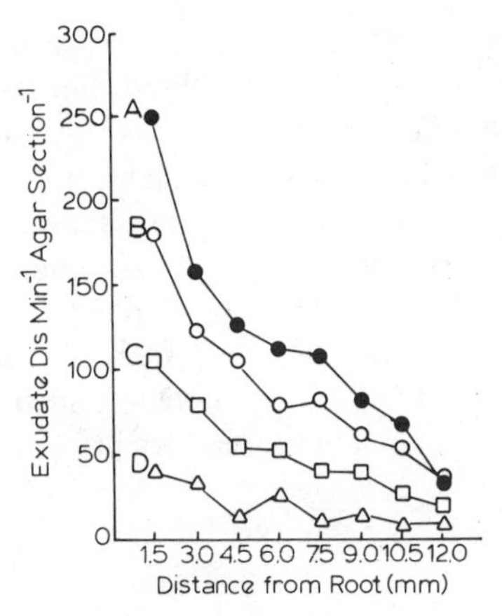

Fig. 1.3.
Organic matter gradients from wheat roots as affected by bacterial culture density (Beck and Gilmour, 1983). Reprinted with permission of Pergamon Press.

soil site by a variety of chemically and enzymatically produced colloidal compounds produced external to the living cell. For this discussion, the soil organic component will be divided into two fractions—a readily decomposable component and a more biodegradation resistant fraction. For a complete discussion of the chemistry of individual soil organic chemicals, see Stevenson (1982).

1.2.1. Readily Decomposable Biochemicals

Readily decomposable biochemicals consist of all of those biochemicals which are rapidly metabolized for cell carbon and energy by the microbial community. (See Chapter 5 for a more detailed discussion of the readily metabolized soil organic matter pool.) Readily decomposable biochemicals may enter the soil system in exogenously supplied biomass, such as plant or animal remains, or be formed *in situ* through the activity of the microbial community. Muramic acids are an example of microbial products commonly isolated from soil, whereas a variety of other amino sugars also found in soil (Stevenson, 1983) may be of indiginous or exogenous origin.

Readily decomposable biochemicals, whether synthesized internally or supplied from outside the soil ecosystem, provide the bulk of the carbon, nitrogen, and energy sources for the microbial community. Two pools of readily decomposable substrates are found in soil. They may be contained within intact cells—living or decaying—or occur free within the soil solution. Half-lives of the compounds located within cell structures, generally protected by more biodegradation resistant external cellular structures, are greater than those in the soil solution. When the cell dies and the protective cell wall is breached, the readily decomposable biochemicals are released into the soil solution where they will be quickly metabolized. Variations in the decomposition rate between various ecosystems result from the physical or chemical limitations to the microbial community. These rates generally correspond to the capability of the soil microbial community to adapt or function in the ecosystem. This microbial activity varies with the soil moisture level, redox potential, pH, presence or absence, or inhibitors, and so on. This physical and chemical control of biodecomposition is exemplified by the accumulation of plant biomass in flooded soils. Under flooded conditions, a major portion of the plant biomass is only partially decayed and metabolic intermediates and fermentation products tend to accumulate, although some catabolism of the organic material occurs in surface sediments.

1.2.2. Humic Substances

Probably the most studied and least understood soil organic components are soil humic substances. These materials consist of three major classes of chemicals, generally catagorized as humic acids, fulvic acids, and humin. They are differentiated by their solubility in alkaline and acid solutions. Humic and fulvic acids are both soluble in alkaline solutions, but humic acids are precipitated in acid. Fulvic acid is soluble in both acidic and alkaline solutions. Humin is soluble in neither acidic nor basic solutions. This differential solubility relates in part to the molecular complexity of the substances; fulvic acids have lower molecular weights and higher oxidation states than humic acids and presumably humin (Stevenson, 1980). The complexity of these humic substances has led to a concentrated effort by organic chemists to elucidate their structure. Earlier studies were designed assuming that humic and fulvic acids possessed a unique chemical structure. Once it became evident that these molecules were in reality random polymers of aromatic and aliphatic substituents, progress could be made in explaining their role in soil. A more complete discussion of the complex chemical structure of humic substances is provided in Chapter 8.

Soil physical and chemical conditions affect the structure of humic substances contained therein. Both temperature and soil pH affect the chemical composition of the humic acids. For example, Schnitzer (1974) noted from the study of alkaline permanganate digests of humic and fulvic acids extracted from acidic and neutral soils that humic acids from acidic soils contained more aliphatic structures than those from neutral soils. But, comparison of the weight ratios of benzene carboxylic to phenolic acids suggested similar structural arrangements of humic and fulvic acids regardless of soil pH during formation. In a similar study of humic substances extracted from arctic soils (Schnitzer and Vendette, 1975), humic acids from surface permafrost soil appeared to be poorly developed. Evidence for this conclusion included their low degree of condensation and aromaticity and a low resistance to mild chemical oxidatants which yielded only small amounts of benzene polycarboxylic acids higher than the di-forms, but relatively large quantities of aliphatic carboxylic acids.

Understanding the biological and chemical properties of these soil components is further limited by the process of humification of otherwise readily biodegradable biochemicals. Humic acids are chemically reactive molecules. Along with a number of carboxy, hydroxyl, and aldehyde groupings with which other soil biochemicals may react, free radicals that can enzymatically or chemically interact with other soil organic components are readily formed in humic acids. Hence, large numbers of proteins, carbohydrates, and so on, which would normally be decomposed easily by the microbial community are, in part, protected from microbial attack by association with the large, complex humic acid complex. Hence, mathematical models developed to describe metabolic turnover of the readily degradable carbon pool must account for that portion of the pool which has been humified (see Chapter 12).

1.2.3. The Living Component

The portion of the living biomass in the soil ecosystem considered part of the soil organic fraction is limited by the basic definition of soil organic matter. Organic debris in which the original structure is still recognizable and can be removed form the soil during processing is not soil organic matter (Jenny, 1980). Thus the living soil organic matter consists essentially of bacterial, fungal, algal, protozoan, actinomycete, and, perhaps, nematode populations. From the view of ecosystem function, the microbial community is a small but highly significant portion of soil organic matter. Essentially all of the soil organic matter transformations that are necessary for a stable, productive ecosystem development are catalyzed, at least in part, by these microorganisms.

This microbial community is the major contributor to one of the most important aspects of the biology of soil organic matter. The presence of organic matter in the soil, while providing some limited improvement of soil physical and chemical properties compared to an organic matterless soil, is of little importance in the absence of an active microbial population. The benefit of organic matter to the soil ecosystem is primarily derived from the active metabolism of the soil organic matter by the soil microbial community (Fig. 1.4). The nutrients contained within plant biomass entering the soil ecosystem are mineralized for plant assimilation by the microbial community. Nutrients derived largely from the soil mineral formations can be solubilized as a result of the direct microbial metabolic activity or indirect action of microbially synthesized extracellular enzymes. Also, microorganisms produce chelators, organic acids, or simply alter the soil pH or redox potential to modify the availability of such important mineral nutrients as iron, manganese, or molybdenum. Similarly toxic metals such as aluminum may be solubilized. Enhancement of soil structure results directly from extracellular products synthesized by the microorganisms as well from the intertwining of fungal mycelia around soil particulates.

The microbial community not only mineralizes organic biomass to produce plant nutrients, but it also constitutes a significant reservoir of such nutrients itself. Anderson and Domsch (1980) in a study of 26 agricultural soils found that

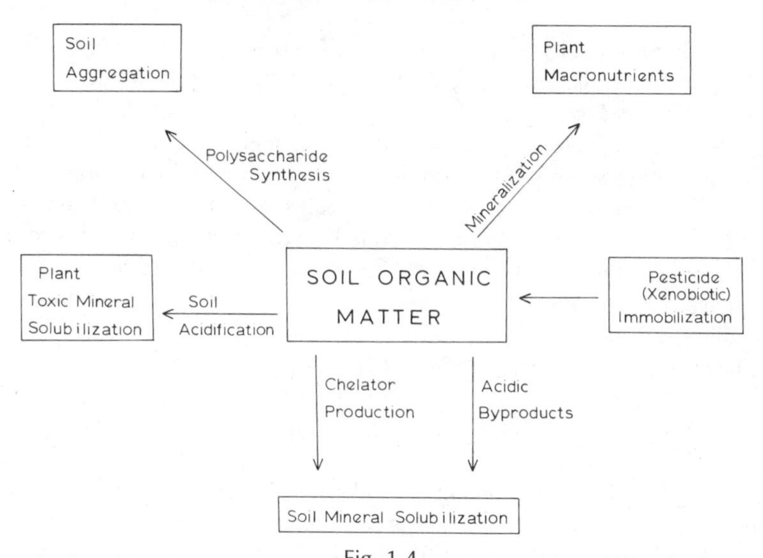

Fig. 1.4.
Major soil processes impacted by microbial transformations of soil organic matter.

the microbial biomass containd between 0.27 and 4.8 percent of the total soil carbon (with a mean of approximately 2.5 percent). Organic nitrogen contents ranged from 0.5 to 15.3 percent of the total soil nitrogen (mean of approximately 5.0 percent). These values suggest an average C to N ratio for the microbial community of their soils of 5:1. They estimated that the surface 12.5 cm of soil contained about 108, 83, 70, and 11 kg per ha nitrogen, phosphorus, potassium, and calcium, respectively, in the microflora.

1.3. DISTRIBUTION OF ORGANIC MATTER IN SOIL

As a result of the particulate nature of soil, the mechanisms of incorporation of organic matter into soil, the variability in plant and animal components of the ecosystem and their activity, and the nature of microbial growth and activity, organic matter and its associated impact on the ecosystem are not distributed uniformly within the soil matrix. For an ecosystem taken as a whole (the macrosite), both horizontal and vertical heterogeneity occurs. Also, the chemical composition and biodegradation susceptibility of the organic matter differs within the microstructure of the soil (microsite). Both the micro and macrosite variations affect ecosystem stability and function in that the interactions directly control such essential soil functions as nutrient availability, moisture holding capacity, soil aggregate formation, and stability.

1.3.1. Macrosite Variation

In undisturbed ecosystems, major mediators of macrosite variation in soil organic matter distribution result from the location and type of primary plant producers and the magnitude of the associated litter and rhizosphere effects. Biochemical transformations in both the litter layer and the rhizosphere are similar in that they are primarily plant controlled rather than soil mediated. Although in many cases, the source of the microbial propagules is soil, whether they contact the plant tissue through direct contact or via insect or other invertebrate vectors, the physical and chemical parameters controlling growth and activity of the microorganisms result from the characteristics of the plant tissue itself. Differences between aboveground inputs and rhizosphere effects relate to the depth of impact of the organic matter in the soil profile. Litter layers initially effect the properties of the surface horizon, whereas roots may extend to great depths within the soil profile. The impact of litter metabolism eventually extends to deeper horizons as the result of leaching of soluble organics from the litter layer. For example, leached organic acids may contribute to podzolization of the soil (Stevenson, 1982). Hence with time, the impact of both organic matter sources will extend throughout the soil profile. In fact, due to the high activity of the microbes of the rhizosphere, the impact of rhizosphere on the total soil profile may be less than is noted with a thick forest litter layer. This is because the distribution of the organics released through the action of plant roots and associated microbial flora into the soil matrix is limited predominantly to the soil in close proximity to the plant root due to the elevated microbial activity occurring therein, as compared with a forest soil where the entire soil surface may be litter covered.

The fate of both the soil microbial and the plant community are tightly coupled. The activity of one portion of the association directly impacts the capability of the other partner to grow and reproduce. For example, in an undisturbed forest system, essentially all of the nitrogen is complexed in either living plant biomass, microbial biomass, or decaying tissues. Thus, nitrogenous nutrients for new plant growth are provided by mineralization of litter and root tissue by the microbial community. In a tightly coupled ecosystem, a direct association of the two populations is needed to assure that little or no nutrients are lost from the ecosystem. Thus, plant roots are frequently concentrated in the layer of decaying plant material and in the immediately adjacent soil layers. This direct physical association between plant nutrient acquiring structures and the microbial community assures long-term stability for the ecosystem.

This interactive nature of plant/microbe associations has also been observed in the rhizosphere. Not only do root exudates control microbial community metabolic activity, but the microbes may stimulate the quantities of nutrients entering the soil from the root. This is exemplified by studies of barley (*Hordeum vulgare* var. Proctor) (Barber and Lynch, 1977). Carbohydrate exudates from axenically grown barley roots and microbial biomass produced in association with the roots were measured. Since the amount of biomass that can be produced per unit of carbohydrate metabolized is well known, the relationship between the quantities of root exudate and microbial productivity was easily determined. Greater microbial biomass was produced in the barley rhizosphere than could be accounted for by root exudate quantities produced in the absence of rhizosphere organisms. Hence, it was concluded that the presence of a microbial community on the plant root resulted in increased quantities of fixed carbon being leached from the root.

Barber and Martin (1976) measured the quantities of photosynthate lost to the rhizosphere microbes by tracing carbon movement through wheat and barley plants incubated under a ^{14}C-labeled carbon dioxide atmosphere. Under axenic conditions, between 5 and 10 percent of the photosynthetically fixed carbon was released from the roots. This compares to losses of 12 to 18 percent of the fixed carbon in the plants with an active rhizosphere population. The latter values were equivalent to 18 to 25 percent of the dry matter biomass of the plants. The authors therefore concluded that the increased carbon dioxide evolution observed in cropped soils as compared to fallow, or noncropped soils, is due largely to the immediate mineralization of plant root exudates.

As a result of this sizeable carbon input, the soil microbial activities are obviously stimulated. Population densities of a number of microbial genera, extracellular enzymes, and microbial metabolic processes, all respond to increases in organic carbon input (Alexander, 1977). The impact of this stimulation is probably best shown by examination of the effect of the rhizosphere association on nitrogen fixation. In this case, the energy contained in the photosynthate is converted to microbial cell mass and used to fix nitrogen. Hence, both soil organic nitrogen and carbon levels are increased. A variety of root-associated nitrogen fixing bacteria have been demonstrated. These include free-living nonsymbiotic bacteria, *Rhizobium*-legume associations, and actinorrhizal associations. The magnitude of carbon transfer through such associations was estimated by Kucey and Paul (1982). They found that in vasicular-arbuscular mycorrhizal and rhizobial infected 4 to 5 week old faba beans (*Vicia faba* L.) about 4 percent of the carbon fixed by the host was consumed by the mycorrhizal fungus and approximately 6 percent of that carbon fixed by nonmycorrhizal hosts and 12 percent of the carbon fixed by mycorrhizal hosts was utilized by the nitrogen-fixing nodules. The mycorrhizal plants fixed more carbon dioxide and their nodules fixed more nitrogen than did the nonmycorrhizal plants.

Along with the input of soluble nutrients as root exudates, the turnover of root mass can be a significant carbon source for the microbial community. In a study of root systems in the surface layers of central Missouri prairie, Dahlman and Kucera (1965) observed that the seasonal biomass in the top 34 inches of the soil profile ranged from 1449 to 1901 g/m^2 from the spring through the end of the growing season. The portion of this mass in the upper 2 inches of the profile varied from 48 to 60 percent in April and July, respectively. Their data suggests approximately 25 percent of this root mass turns over annually. Turnover rates for various parts of the root system were 22.8 and 40.8 percent of the rhizomes and roots in the surface 2 inch zone, respectively. Thus, approximately 475 g/m^2

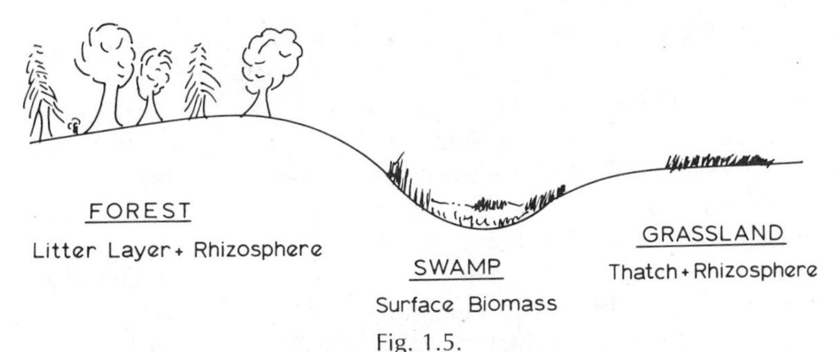

Fig. 1.5.
Plant distribution effects on macrosite variation in soil organic matter.

of root mass was added to the soil organic matter pool annually in this grass land soil.

A third source of organic substrates originating with the plant root are root caps, cortical tissue sheets, tissue fragments and individual cells sloughed from the roots. Griffin et al. (1977) measured the production of such material by roots of peanut (*Arachis hypogaea* L. cv. Argentine) growing in axenic culture solution. Their work indicated approximately 0.26 to 0.73 mg sloughed organic matter per plant per week. Approximately 0.15 percent of the root carbon, nitrogen, and hydrogen were sloughed per week. With the known effects of soil microorganisms on root growth and function, and the increased abrasiveness of soil as compared to nutrient solutions, it is reasonable to suspect that the quantities of sloughed materials in native ecosystems would exceed these values.

As indicated from the data previously cited, there is significant variation within the soil profile due to the presence of plants, and there is an overall variation in the quantity of soil organic matter within a single ecosystem as a result of plant distribution (Fig. 1.5). Sparse plant development in the aboveground community results in extreme heterogeneity of the soil organic matter composition. This results from both the rhizosphere distribution as well as litter incorporation into the soil. A desert ecosystem where organic matter input is highly localized around the widely spaced plants exemplifies this situation. The other extreme is found with dense forest or grassland soils where essentially the entire surface soil layer is in contact with major biomass sources of organic matter. Most undisturbed ecosystems fall between these two extremes. Grass swards, trees, and shrubs are interspersed with barren areas, resulting in wide variation in quantities and types of plant debris incorporated into the soil profile. Similar variation is noted with agricultural systems where major plant growth and development is generally restricted to "well defined" rows.

1.3.2. Microsite Variation

Within the microstructure of soil, organic matter ranges in size from the partially decomposed plant and animal debris which can be easily detected and removed through a variety of colloidal forms to a mixture of water soluble biochemicals. For each of these catagories, the chemical structure may be unmodified from that of the original plant or it may be chemically or biologically altered. Once the plant structure has been lost, the organic matter generally becomes associated with a variety of soil mineral particles. This association can involve attachment to sand and silt particles as microbial cells or capsular material, for example, or clay and other charged particles via electrostatic attraction. Recent studies have shown a dramatic partitioning of colloidal soil organic matter in soil particles of varying densities and partical size. Separation of soil particles by specific gravity in sodium iodide solutions yields a light and a heavy fraction. Sollins et al. (1984) found that the bulk of the organic matter in the light fractions of fine surface soils from sites in Oregon, Washington, and Costa Rica consisted of partially decomposed root fragments plus other plant and microbial remnants. That of the heavy fractions was adsorbed or deposited on mineral surfaces or protected within organo-mineral microaggregates. The heavy fraction nitrogen was more available to microbial activity in that they found more net nitrogen mineralization from this fraction than from whole soils. The nitrogen from the heavy fraction was immobilized by the cells of the light fraction when they were incubated together. The degree of microbial modification of the light fraction organic component was less than that of the heavy fraction since the carbon to nitrogen ratio of the former was wider than that of the latter. This accounts for the immobilization associated with the light fraction, in that with the wide carbon to nitrogen ratio, there would be a nitrogen deficit (Alexander, 1977). Sollins et al. (1984) proposed that as decomposition proceeded, the wide carbon nitrogen compounds of the light fractions, such as polysaccharides, were replaced by humic acids richer in nitrogen but more difficult to decompose. The particulate minerals of the heavy fraction have also provided some protection for the organic material.

Similarly, a difference is found when soil is separated into sand, silt, and clay particle fractions (Young and Spycher, 1979). Organic carbon and nitrogen concentrations were lowest in sand, increased in silt fraction and highest in association with the clay particles. Separation of clay particles into fractions by density again demonstrated two pools of organic carbon. The light particles contained high levels of adsorbed organic carbon, the organic carbon had a wide carbon to nitrogen ratio, and the clay particles contained low amounts of alkali-extractable carbon. The opposite properties were found with the organic matter associated with heavy particles. Young and Spycher explained the existance of the particles in relationship to a soil aggregate formation model. They proposed that light particles developed on the surface of microaggregates and were exposed to a greater extent to the biosphere whereas the heavy particles occupied the intra-aggregate space and were shielded from microbial decomposition.

Anderson et al. (1981) have also shown differences in the nature of the humic substances in soil particles fractionated by size. The course clay and fine silt fractions (5 to 2 μm) contained predominantely conventional humic acids. Studies of their adsorption at 280 nm and their resistance to acid hydrolysis indicated these to be strongly aromatic, reasonable recalcitrant molecules. The organic matter in the fine clay fraction consisted primarily of fulvic acids and a less aromatic humic acid moiety than was detected in the coarser fractions. The carbon to nitrogen to sulfur ratios were highest in the fine silt fraction and reached a minimum value in the fine clay fraction.

Indications that a contributor to the fine clay fractions of the soil organic matter is the metabolism or transformations resulting from microbial activity suggest that this distribution of organic nutrients may be changed by processes affecting microbial populations. This is indeed the situation. One such procedure for stimulating soil microbial activity is cultivation. This disrupts soil particles thereby making occluded organic matter available to the microbial community and aerates the soil. Tiessen and Stewart (1983) examined organic matter distribution versus particle size distribution in a variety of grassland soils that had been cultivated in small grain-fallow rotations for between 4 and 90 years, and compared the separation to similar soils under native praire. Four years of cultivation resulted in a depletion of the organic matter associated with particles greater than 50 μm (43 percent loss of the initial carbon). They suggested that this loss resulted from physical disruption of the particles. Therefore an accumulation of organic material occurred in the finer partile fractions. Over a 60-year period with continued cultivation, there was a net loss of 34 percent of the soil organic carbon and 29 percent of the nitrogen. The organic phosphorus content declined approximately 20 percent. Fine clay associated organic matter declined over the first 60 years but changed little thereafter. A new equilibrium at approximately 50 percent of the original carbon content was reached. Carbon losses in the fine silt and coarse clay fractions was substantially less. The quantities of organic matter in these fractions actually increased with time of cultivation.

1.4. NATIVE LEVELS OF COLLOIDAL ORGANIC MATTER IN SOIL

In a steady state ecosystem, that is, one at equilibrium, the quantities of colloidal organic matter are reasonably constant. With a significant change in the physical, chemical, or biological factors controlling the soil microbial activity, a new equilibrium concentration of soil organic matter is established. These observations result from the principle that the amount of organic matter detected in a soil system is necessarily the difference between input and output, with the primary pathway for exodus of the organic matter from the system being biodecomposition; that is,

$$d[OM]/dt = dS/dt - dD/dt \tag{1}$$

where OM is the organic matter level and S and D represent synthesis and degradation rates, respectively. These two rates are limited by the physical and chemical conditions of the soil. In a steady-state ecosystem, organic matter inputs and losses from the system are balanced, that is,

$$dS/dt = dD/dt \tag{2}$$

Thus were it possible to account for total biomass input into the soil site, this input would necessarily provide a good estimate of biodegradation. (Needless to say, because of the seasonal variability of biomass production and the processes relating to incorporation of this biomass into the soil system, such measurements would have to be conducted over long time periods, perhaps even several growing seasons.) This equality of inputs and losses is obviously not the case in non-equilibrium sites where organic matter levels may be increasing or decreasing. This perhaps is best shown by evaluation of subarctic or antarctic peat soils. In these ecosystems, both biomass production and organic matter degradative processes are limited by soil moisture and temperature. But, since organic matter has accumulated to high levels in the soil, synthesis has occurred at a faster rate than the degradative processes. Otherwise, the accumulation of peat would not have occurred. Other examples include young forests, newly established pastures and so on.

The following discussions provide an evaluation of some of the general principles of the processes controlling organic matter accumulation in soil. Further examples of specific ecosystems are given in Chapter 2. Specific methods of management of soil organic matter levels are discussed in Chapter 13.

The bulk of our understanding of organic matter dynamics in soil is derived from the examination of changes in colloidal organic matter in agricultural soils and from comparison of soils at various stages of recovery from a drastic ecosystem disturbance. An example of the latter situation is provided by work of Sollins and Spycher (1983) in their evaluation of organic matter accumulation in a mudflow chronosequence located at Mount Shasta, California. They measured organic carbon and nitrogen contents in soil fractions of varying density collected from mud flows of differing ages. Soil was separated into a light (specific gravity less than 1.65 g/cm^3) and a heavy fraction. Light fraction organic matter consisted primarily of root fragments and a variety of plant remnants. In the heavy fraction, most of the organic matter was adsorbed on mineral surfaces or occluded with organo-mineral microaggregates. In both the bulk soil and the heavy fraction, carbon and nitrogen levels consistently increased with flow age. Interestingly, the heavy fraction contained 55 to 88 percent of the total soil nitrogen and 37 to 72 percent of the total soil carbon. Thus the heavy fraction constituted the major long-term reservoir of soil nitrogen and was an important sink for carbon. This is of interest because the heavy fraction would be anticipated to consist primarily of microbial and chemical product of organic matter transformations occurring within the soil itself. This implies that a major portion of this heavy fraction organic matter consists of microbial cells which could be expected to be enriched in nitrogen in comparison to the exogenously supplied plant material. The organic matter content of the two soil fractions was found to accumulate in response to different soil factors. Rockiness and tree index accounted for 50 to 80 percent of the variability in the organic matter content of the light fraction but had little effect on this parameter in the heavy fraction. These two factors could be instrumental in controlling amount and location of plant growth. The organic matter in the light fraction apparently correlated with short-term variation in biological activity whereas organic matter in the heavy fraction correlated more with long-term processes such as mineral weathering and soil horizon formation. Analysis of the shape of the accumulation curves suggests that a steady-state level was reached in relatively short periods of times, whereas the heavy fraction organic matter continued to increase for tens of thousands of years. The light fraction therefore would be considered to be more directly associated with short-term processes relating to plant growth, whereas the heavy fraction apparently relates more to those processes involved in synthesis of the more biodegradation resistance soil organic matter fractions.

When grassland soil initially is cultivated, the soil organic matter level declines. This results primarily from the increased availability of the soil organic matter to biological decomposition and the moderating of the physical and chemical restraints on microbial activity. In a virgin prairie, a steady-state balance is achieved where a high level of organic matter is found in the surface horizon of the soil as the result of thatch accumulation and root biomass production (see Chapter 2). Two differences in the soil condition after plowing cause the organic matter decline: (1) the input of plant biomass decreases and (2) the physical structure of the surface horizon soil is disrupted. Thus a portion of the decline in total soil organic matter results from the loss of fixed carbon input. Accompanying this decreased input is the increased metabolism of the plant debris accumulated in the soil. This would include incorporated aboveground plant parts, roots, and thatch. As long as the prairie ecosystem is intact, decomposition of that organic matter entering the surface horizon or thatch layer is controlled by soil aeration, moisture, pH, and so on. Once the prairie is plowed, soil aeration increases due to the mixing of the soil. Any physical protection afforded the organic matter by its association with soil particles is thereby disrupted. Further stimulation of organic matter decomposition may occur should cropping be the reason for development in that nutrient limitations are generally relieved through fertilization and liming of the soil. Thus the organic matter accumulated under the grassland becomes available for microbial decomposition (Jenny, 1980). The soil organic matter level therefore declines to a new steady-state level regulated by the newly established abiotic and biotic parameters of the cropped ecosystem.

Organic matter can also accumulate in cropped soils with some soil management systems. This results primarily from the mixing of soil containing less organic matter than allowed by the steady-state biomass input and abiotic conditions of the ecosystem, alterations in management procedure to encourage biomass acumulation, or through amendment of the soil with fixed carbon in the form of manures, sludges, and so on (see Chapter 13). Smith et al. (1951) report organic matter buildup in agricultural soils cropped to sugarcane. In their situation, subsurface soils, containing low organic matter, developed the high organic matter levels characteristic of the soil surface horizons shortly after they were brought to the surface. This increase of the organic matter content was accelerated by the maintenance of optimal conditions for crop production. Such rapid accumulation of organic matter in soils would not be anticipated without intervention to increase biomass inputs. In this case, the intervention was plowing the soil to bring soil from those horizons not generally receiving high levels of organic matter input into direct contact with the growing sugarcane.

Observations that organic matter levels could be increased in agriculturally developed soils has led to a number of management techniques for conservation of soil organic matter. Some simple procedures have been shown to have rather long-term effects on soil organic matter levels. (See Chapter 13 for a more detailed discussion.) Amendment of soil with manure annually over a 20-year period resulted in greater soil organic carbon and nitrogen levels in the amended as compared to soil not receiving fertilization and similarly cropped over 100 years after the final manure addition (Jenkinson and Rayner, 1977). The effect of management has been shown rather dramatically with a variety of cropping systems on the Morrow plots at the University of Illinois (Stauffer, et al. 1940). Crop patterns included continuous corn, and corn–oats or corn–oats–clover rotations, with and without manure, lime, and rock phosphate. Lowest organic carbon was found in the unfertilized continuous corn plots whereas the highest levels were in the fertilized corn–oat–clover plots. Fertilization decreased carbon losses under all cropping systems. But, it must be noted that none of the cropping procedures tested in this study produced soils with equivalent organic matter status to grass land soils. All of the cropped plots contained less organic carbon than the continuous grass borders surrounding the plots. The studies previously listed demonstrate that soil organic matter levels may be increased (1) by amendment of soils with exogenously produced organic materials and (2) by stimulating indigenous plant biomass production. Accordingly, one rather simple procedure for encouraging accumulation of organic matter in soil would be to maximize aboveground biomass production. Of course, the maximal quantities of organic matter accumulated in the soil are dictated by the equilibrium concentration achievable under the prevailing physical and chemical soil properties.

Organic matter accumulation is a major difficulty in reclamation of mineland soils. Frequently, processes involved in mineral recovery produce mine soils or spoils essentially devoid of native organic matter (Tate, 1985; Cundell, 1977). Yet, soil organic matter is essential for development of a stable, aesthetic ecosystem on the reclaimed minelands (Tate, 1985). Thus prior to reclamation of the sites aside from alleviating another physical or chemical factor limiting biotic development (Visser, 1985), exogenously supplied organic matter is needed to start ecosystem development. This may be accomplished by applying an organic amendment or through encouragement of plant production. The latter alternative may be achieved through fertilization of the soil, if all other physical and chemical limitations to plant growth are mitigated, or to a certain degree through establishment of plants with mycorrhizal associations (Danielson, 1985). This input of organic matter can cause large changes in the activity of the soil microbial populations. Lindemann et al. (1984) found rapid increases in a variety of bacterial populations, dehydrogenase, and fungal populations with amendment of coal mine spoil with sludge or hay. Such stimulation of the microbial populations contribute to nutrient cycling as well as soil structural development.

1.5. IMPORTANCE OF SOIL ORGANIC MATTER

The discussion of the nature and quantities of colloidal organic matter in soil has had the underlying bias that increasing quantities of this soil component benefit the overall ecosystem function. This is a reasonable assumption for mineral soils in general if the physical and chemical conditions are such that this organic matter pool is constantly in transition between mineral and organic forms. Stevenson (1982) listed nine general properties of humic substances and the accompanying effect of these properties on soil. Generally, the associated effects were beneficial to the ecosystem as a whole. Such usually overlooked aspects, as the darkening of soils due to high humus contents contributing to increased soil temperature, were discussed. The effects of colloidal organic matter on basic soil properties can be divided into two general catagories, those that rely on the chemical and biological transformations of the organic component and those that may be increased by biological or chemical transformation but some benefit is gained by the presence of the organic matter. Processes in the former class are reasonably clear. These include the reactions involved in biogeochemical cycling and synthesis of cellular structure and polysaccharides associated with soil granulation (Chapter 11). Process, such as increases in water holding capacity, cation exchange reactions, buffer action, and inactivation of organic pesticides, would all fall into the second catagory.

1.5.1. Physical and Chemical Benefits of Organic Matter

Two major soil properties relating to the physical and chemical status of soil of prime consideration for aboveground productivity are cation exchange capacity and water retention properties. Soil organic matter, albeit modified by the microbial community or not, directly impacts on these properties. Total acidities of isolated soil humus fractions range from 300 to 1400 meq/100 g soil and organic matter can hold up to 20 times its weight in water (Stevenson, 1982).

Cation exchange capacity of soil results from the quantities of charged particulates contained therein. Hence, most of the cation exchange capacity is expected to result from the presence of negatively charged clay and colloidal organic matter particles. This has been shown for the major soil series of New Jersey where it was demonstrated 59 percent of the cation exchange capacity resulted from variations in soil organic matter and clay contents (Drake and

Motto, 1982). The relative contribution of each fraction to the cation exchange capacity and the total cation exchange capacity of the soil depends upon the ratios between the two components and the total quantities of each present. For example, Hallsworth and Wilkinson (1958), in a study of clay and organic matter contribution to cation exchange capacity of soils collected predominantly from Wales, found that the organic matter contribution to cation exchange capacity was approximately nine fold that of the clay.

These studies relate primarily the effect of sodium hydroxide extracted humic substance on soil cation exchange capacity. Although this extract would contain a portion of the recently incorporated and as yet unmodified plant materials, it would consist primarily of biologically and chemically modified colloidal organic components. The question still remains concerning the effect of recently incorporated organic substances on cation exchange capacity. In the vast majority of soil systems this question is immaterial because the cation exchange capacity of the native soil organic matter far exceeds that of any soil amendment. But, some low organic matter and clay containing ecosystems exist where management for reclamation or development is necessary. This may be the case with some mineland and mine spoil reclamation projects. and is most certainly true where reversal of the effects of desertification are desirable. Should significant cation exchange capacity be gained upon soil amendment, nutrient retention within the soil for plant use would immediately be enhanced. Contribution of the soil amendments to cation exchange capacity would depend on the charge of the amendment. At the predominant pH of most soil ecosystems, organic matter is negatively charged. But, as microorganisms are allowed to metabolize it, the total quantity of negative charges of the biological components increases with the oxidation state of the organic matter. Thus the potentiality exists for use of reasonably humified organic materials, such as various sludges, to cause immediate improvement in soil nutrient retention capacity.

The effects of organic matter on soil water retention are a bit more complex. Soil organic matter affects infiltration rates, total quantity of water in the soil, as well as the evaporation from the soil surface. The latter topic which technically does not deal directly with soil organic matter has such an impact on soil moisture that it will be discussed herein. The difficulty relates to the definition of soil organic matter. Evaporation of the water from soil depends upon soil temperature, which is controlled in part by the true soil organic matter and the micrometeorological interactions at the soil surface. These interactions rely upon the type of organic residue on the soil surface and its position. This was recently demonstrated for an agricultural soil in a study of the effects of wheat straw mulch on soil water change of a silt loam soil (Simka, 1983). The straw was placed flat on the soil surface, 3/4 flat and 1/4 standing, and 1/2 flat and 1/2 standing. Soil moisture variation was compared to a bare soil. Soil water losses correlated with wind movement with r^2 values of 0.55, 0.41, 0.41, and 0.32 for bare soil, flat, 3/4 flat-1/4 standing, and 1/2 flat-1/2 standing, respectively. Soil water loss occurred from the bare soil and the soil with flat mulch with winds of 0.08 MS^{-1}, but winds of at least 0.55 MS^{-1} were needed before water loss occurred in those fields with standing mulch. Apparently, the standing mulch functions as a wind break to decrease the evaporative effect, rather than as a wick to increase the soil moisture loss.

As was indicated by Stevenson (1982), soil humic materials can hold up to 20 times their weight in water. Thus it is reasonable to conclude that water retention increases with increased colloidal organic matter content. This was recently demonstrated for coal mine spoils amended with barley (*Hordeum vulgare* L.) straw or pine (*Pinus* sp.) bark (Scholl and Pase, 1984) and with a silt loam amended with sewage sludge (Epstein, 1975). But, the important question is not whether more water is present in the soil, but rather, whether more water is available to the plant in the amended soil. Data suggests that even with the increased soil moisture of the amended soil, the water available to the plant remains unchanged. In the silt loam amended with sewage sludge mentioned above (Epstein, 1975) the amount of water between −0.33 and −15 bars pressure was the same for the amended and unamended soils. Of significance for use of soils for waste water disposal and perhaps for development of soils with low water infiltration rates, is the observation that organic residues do increase water infiltration rates (Johnson, 1957). Since the effect on infiltration rate was maximal under conditions more conducive to microbial mineralization of the incorporated plant material, this effect of soil organic matter most probably results from the improvement of soil structure resulting from microbial metabolism of the organic materials.

Soil organic matter not only complicates decisions involved with the quantities of pesticides used to achieve a desired effect, but also may retard the loss of biodegradable pesticides from the ecosystem. Organic pesticides can interact with soil organic matter through a variety of mechanisms including ionic attraction, van der Waals forces, or covalent bonding to the soil humus component (Stevenson, 1982). Two practical implications of these interactions involve the quantities of pesticides that must be added to the soil to have the desired effect, duration of the effect, and potential for contamination of subsequent crops by pesticides. Because of the expense of pesticides and their environmental impact of usage, it is desirable to use minimal amounts of pesticide where possible. Thus the ideal situation would be where all of the pesticide reaches the target population, perhaps a weed or insect species. But, many of the pesticides are charged molecules containing aromatic rings or other chemically reactive functional groupings which may interact with soil organic matter. Once the pesticide interacts with the soil organic matter, quantities that must be added to the soil to be effective may be greatly increased. For example, N-Serve, a nitrification inhibitor is used in mineral soils at an effective concentration of a few micrograms per g soil. But, in the Histosols of south Florida containing approximately 85 percent organic matter, nitrification was not totally inhibited with as much as 100 micrograms per g soil (Tate, 1977). Similarly Rahman (1976) found that in mineral soils amended with pesticides, from 2 to 4 times more herbicide was needed to be effective in soil containing 22.1 percent organic matter as opposed to an 8-percent organic matter containing soil. The variation in the quantities needed related to the pesticide tested rather than properties of the soil. Pesticides with low water solubility were effected more than those with high water solubility.

An even more serious problem results from binding of the pesticide to soil organic matter. Once a chemical is attached to the complex humic acid structure, its biodecomposition is greatly reduced. Thus although a given pesticide may be rapidly decomposed in many mineral soils, in some soils containing high organic matter concentrations, a biodegradable pesticide may persist long after complete decomposition would be expected. Lichtenstein (1980) reports that although the pesticide may not be extractable from the soil by conventional means, residues could be detected in crops and earthworms growing in the soil. Thus a pesticide used on one crop could appear in subsequent growing season as a residue in another crop.

Much of the positive effects of soil organic matter on ecosystem development and function results from an improved soil structure derived from the microbial transformations of organic matter (see Chapter 11). Products of microbial metabolism are responsible for both reduction of the size of mineral inclusions and cementation of these minerals into water stable aggregates. Acidic by-products, organic and mineral, and chelators produced by the microorganisms are instrumental in mineral solubilization and mobilization. These activities contribute to reduction of soil rockiness and produce mineral nutrients for plants. Another soil structural development process with major impact on plant growth is formation of soil aggregates. McCalla (1942) demonstrated a protective effect of both humic substances and crop residue on the water stable soil structure. When a low organic matter containing soil was mixed with a humus suspension consisting primarily of colloidal organic matter, a water stable structure developed which took considerably more energy from falling waterdrops to destroy than was needed to disrupt the structure of untreated soil. Even more stability was provided when treated soil was protected with straw mulch. Stability of the soil aggregates results from a variety of interactions between soil organic matter and soil mineral components. Microbial and chemical organic matter products implicated in aggregate stabilization are humic substances, polysaccharides, and microbial cells themselves, especially fungal mycelia (Cheshire, et al., 1984; Oades, 1984).

1.5.2. Biological Effects of Soil Organic Matter

A major biological impact of soil organic matter on the total ecosystem results from its role as a source of the plant macronutrients, nitrogen, phosphorus, and sulfur (Chapter 9). As a result of the microbial metabolism of the soil organic fraction, nitrogen, sulfur, and phosphorus compounds which are unavailable to the growing plant in the organic form enter the soil mineral nutrient pool. These nutrients may result from the mineralization of exogenously supplied organic residues, such as organic fertilizers; plant debris, such as leaf litter; or catabolism of indigenously produced organic materials, such as humic substances and microbial cells. Sharpley et al. (1984) found that amendment of a Pullman clay loam with cattle feedlot waste increased the total, inorganic, and organic phosphorus content of the soil. The amounts of phosphorus in the surface soil were highly correlated with the total amount of feedlot waste phosphorus applied and the time since the last application. The increased surface soil inorganic and organic phosphorus with feedlot waste application were attributable primarily to the increased labile fractions of these phosphorus forms. Organic and inorganic phosphorus pools varied independently in that soil organic phosphorus contents decreased to pretreatment levels more rapidly than did inorganic phosphorus contents after treatment ceased. This would be anticipated since availability of mineral phosphate is dependent upon both biological and chemical interactions.

Although they are frequently studied alone, each of the steps of the various biogeochemical cycles depends on levels of activity of the other steps in the process; that is, the biogeochemical cycles do not act independently (Tate, 1985). For the nitrogen cycle to function in an ecosystem, there must be functional carbon and phosphorus cycles. For example, in a forest soil in Wisconsin, Pastor et al. (1984) found that net aboveground production was highly correlated with field measurments of soil nitrogen mineralization. But, soil nitrogen mineralization was positively correlated with litter production, and nitrogen and phosphorus return in litter. Soil nitrogen mineralization was negatively correlated with litter carbon to nitrogen and carbon to phosphorus rations and with phosphorus usage efficiency. Mineralization and nitrification were highly correlated with humus phosphorus contents, whereas differences in nitrification capacity among the soils apparently related to phosphate supply in spring and early summer and to ammonium in mid to late summer. Thus although each portion of the various biogeochemical cycles can be easily demonstrated to affect plant growth and development, a total understanding of the ecosystem requires evaluation of the interaction between the various biogeochemical cycles.

Because the preceding biological effects of the organic matter involve conversion of the organic materials to mineral components, soil humic substances also directly affect a number of plant processes and enzymes. For example, plant invertase, peroxidase, and phosphatase activities may be inhibited or stimulated by humic acids (Malcolm and Vaughan, 1979a, 1979b; Vaughan and Malcolm, 1979) and fulvic acid inhibits enzymatic indoleacetic acid oxidation (Mato et al., 1972). The effect of humic substances on invertase activity varied with plant species and the soil organic matter fraction used. No effect of any soil organic fraction was found on invertase activity of beet root storage tissue, but humic acids and their residues inhibited invertase in carrot storage tissues and wheat roots and coleoptiles and to a lesser extent mungbean hypocotyls. Soil phenolic acids had no effect on any of the plant tissues tested. Interestingly, some stimulation of invertase activity of pea root tissue was observed with water- and acid-soluble humic acid fractions. The inhibitory effect of the humic acids on wheat invertase activity was independent of pH (in the range of 4.5 to 7.0). The Michaelis constant was unaffected by the inhibitor but the maximum velocity was greatly reduced (Malcolm and Vaughan, 1979a).

Anaerobically decomposed wheat straw is also phytotoxic to Barley (*Hordeum distichon*) seedling root growth (Chapman and Lynch, 1983). For comparison, aerobic degradation products of leaves of sweet vernal grass (*Anthroxanthum odoratum*) stimulated root growth. Although some direct effect of the isolated organisms is found, effects of metabolites with microorganisms removed on plant growth suggest that the interactions were predominantly chemical.

Organic matter can be directly incorporated into plant tissue (Vaughan and Ord, 1981). ^{14}C-labeled soil organic matter was accumulated by pea (*Pisum sativum*) roots grown under axenic conditions. More fulvic acids than humic acids were taken up by the tissue. Vaughan and Ord noted that the uptake was the result of both adsorption of the humic substances to the root surfaces and cell walls and a process dependent on plant metabolism, most likely protein synthesis.

1.6. CONCLUSIONS

Although we frequently speak of soil organic matter as if it were a single component, this soil fraction is, in reality, a multifaceted, ever-changing soil component. The organic compounds present depend on the types of organic input into the soil; the physical, chemical, and biological properties of the soil; and time. Although an increasing proportion of the organic substances entering the soil ecosystem are industrially synthesized, the vast majority of soil organic matter originates either directly or indirectly from photosynthetically fixed carbon. Once organic carbon is mixed with soil, it can be mineralized, immobilized within microbial cells, incorporated into the more stable soil organic matter fraction, or in some cases where microbial activity is inhibited, remain unchanged or be modified slightly. An example of the latter would be the accumulation of partially decomposed organic material under swampy conditions. Because of the diversity in the fate of organic materials entering the soil ecosystem, a variation in the types of organic substances in soil also exists. Soil organic components may be divided into easily metabolizable compounds (e.g., amino acids and sugars), biodegradation resistant components (e.g., the soil humic substances), and the living component (e.g., microbial cells). Distribution of the organic component in these various fractions depends on the chemical and physical properties of the soil ecosystem.

Development of the aboveground portion of the ecosystem depends on the quantity and distribution of organic matter in the soil. The distribution of colloidal organic matter varies horizontally and vertically within the soil profile as well as macroscopically and microscopically. Quantities present within a given soil are generally constant, with the absolute amount contained dependent on the prevailing environmental conditions or management procedures. Changes in either generally result in an equilibrium concentration shift. Depending on the amount of environmental modification, the time required to reach the new equilibrium can vary from a few years or decades to several centuries. Interest in distribution and quantities of organic matter in specific soils stems from the generally favorable effect of organic matter on plant growth and development. The soil organic component among other effects moderates soil temperature, affects water content, contributes to the cation exchange capacity, serves as a nutrient reserve, and is a factor in soil structure development.

REFERENCES

Alexander, M. 1977. Introduction to Soil Microbiology. John Wiley & Sons, New York, 377 pp.

Anderson, D. W., S. Saggar, J. R. Bettany, and J. W. B. Stewart, 1981. Particle size fraction and their use in studies of soil organic matter. I. The nature and distribution of forms of carbon, nitrogen, and sulfur. Soil Sci. Soc. Am. J. 45: 767-772.

Anderson, J. P. E., and K. H. Domsch, 1980. Quantities of plant nutrients in the microbial biomass of selected soils. Soil Sci. 130: 211-216.

Barber, D. A., and J. M. Lynch, 1977. Microbial growth in the rhizosphere. Soil Biol. Biochem. 9: 305-308.

Barber, D. A., and J. K. Martin, 1976. The release of organic substances by cereal roots into soil. New Phytol. 76: 69-80.

Beck, S. M., and C. M. Gilmour, 1983. Role of wheat root exudates in associative nitrogen fixation. Soil Biol. Biochem. 15: 33-38.

Blumer, M., W. Blumer, and T. Reich, 1977. Polycyclic aromatic hydrocarbons in soils of a mountain valley: Correlation with highway traffic and cancer incidence. Environ. Sci. Technol. 11: 1082-1084.

Bohn, H. L. 1976. Estimate of organic carbon in world soils. Soil Sci. Soc. Am. J. 40: 468-469.

Chapman, S. J., and J. M. Lynch, 1983. The relative roles of micro-organisms and their metabolites in phytotoxicity of decomposing plant residues. Plant Soil 74: 457-459.

Cheshire, M. V., G. P. Sparling, and C. M. Mundie, 1984. Influence of soil type, crop and air drying on residual carbohydrate content and aggregate stability after treatment with periodate and tetraborate. Plant Soil 76: 339-347.

Cundell, A. M. 1977. The role of microorganisms in the revegetation of strip-mined land in western United States. J. Range Management 30: 299-305.

Dahlman, R. C., and C. L. Kucera, 1965. Root productivity and turnover in native praire. Ecology 46: 84-89.

Danielson, R. M. 1985. Mycorrhizae and reclamation of stressed terrestrial environments. In R. L. Tate III and D. A. Klein (eds.) Soil Reclamation Processes: Microbiological Analyses and Applications. pp. 173-201. Marcel Dekker, New York.

Drake, E. H., and H. L. Motto, 1982. An analysis of the effect of clay and organic matter content on the cation exchange capacity of New Jersey soils. Soil Sci. 133: 281-288.

Epstein, E. 1975. Effect of sewage sludge on some soil physical properties. J. Environ. Qual. 4: 139-142.

Griffin, G. J., M. G. Hale, and F. J. Shay, 1977. Nature and quantity of sloughed organic matter produced by roots of axenic peanut plants. Soil Biol. Biochem. 8: 29-32.

Hallsworth, E. C., and G. T. Wilkinson, 1958. The contribution of clay and organic matter to the cation exchange capacity of the soil. J. Agric. Sci. 51: 1-3.

Hites, R. A., R. E. Laflamme, and J. W. Farrington, 1977. Sedimentary polycyclic aromatic hydrocarbons: The historical record. Science (Washington, D.C.) 198: 829-831.

Jenkinson, D. S., and J. H. Rayner, 1977. The turnover of soil organic matter in some of the Rothamsted classical experiments. Soil Sci. 123: 298–305.

Jenny, H. 1980. The Soil Resource. Spring-Verlag, New York, 377 pp.

Johnson, C. E. 1957. Utilizing the decomposition of organic residues to increase infiltration rates in water spreading. Trans. Am. Geophys. Union 38: 326–332.

Kononova, M. M. 1966. Soil Organic Matter: Its Role in Soil Formation and in Soil Fertility. Pergamon Press, New York, 544 pp.

Kucey, R. M. N., and E. A. Paul. 1982. Carbon flow, photosynthesis, and N_2 fixation in mycorrhizal and nodulated faba beans (*Vacia faba* L.). Soil Biol. Biochem. 14: 407–412.

Lichtenstein, E. P. 1980. "Bound" residues in soils and transfer of soil residues in crops. Residue Rev. 76: 147–153.

Lindemann, W. C., D. L. Lindsey, and P. R. Fresquez, 1984. Amendment of mine spoil to increase the number and activity of microorganisms. Soil Sci. Soc. Am. J. 48: 574–578.

Malcolm, R. E., and D. Vaughan. 1979a. Effects of humic acid fractions on invertase activities in plant tissues. Soil Biol. Biochem. 11: 65–72.

Malcolm, R. E., and D. Vaughan, 1979b. Comparative effects of organic matter fractions on phosphatase activities in wheat roots. Plant Soil 51: 117–126.

Mato, M. C., L. M. Gonzalez-Alonso, and J. Mendez, 1972. Inhibition of enzymatic indoleacetic acid oxidation by soil fulvic acids. Soil Biol. Biochem. 4: 475–478.

McCalla, T. M. 1942. Influence of biological products on soil structure and infiltration. Soil Sci. Soc. Am. Proc. 7: 209–214.

Oades, J. M. 1984. Soil organic matter and structural stability mechanisms and implications for management. Plant Soil 76: 319–337.

Pastor, J., J. D. Aber, C. A. McClaugherty, and J. M. Melillo, 1984. Aboveground production and N and P cycling along a nitrogen mineralization gradient on Blackhawk Island, Wisconsin. Ecology 65: 256–268.

Paul, E. A., and E. L. Schmidt, 1961. Formation of free amino acids in rhizosphere and nonrhizosphere soil. Soil Sci. Soc. Am. Proc. 25: 359–362.

Rahman, A. 1976. Effect of soil organic matter on the phytotoxicity of soil-applied herbicides: Glasshouse studies. N. Z. J. Exp. Agric. 4: 85–88.

Schnitzer, M. 1974. Investigations on the chemical structure of humic substances by gas chromatography-mass spectrometry. Trans. Int. Congr. Soil Sci. 10th. 2: 294–301.

Schnitzer, M., and E. Vendette, 1975. Chemistry of humic substances extracted from an Arctic soil. Can. J. Soil Sci. 55: 93–103.

Scholl, D. G., and C. P. Pase, 1984. Wheatgrass response to organic amendments and contour furrowing on coal mine spoil. J. Environ. Qual. 13: 479–482.

Sharpley, A. N., S. J. Smith, B. A. Stewart, and A. C. Mathers, 1984. Forms of phosphorus in soil receiving cattle feedlot waste. J. Environ. Qual. 13: 211–215.

Smika, D. E. 1983. Soil water change as related to position of wheat straw mulch on the soil surface. Soil Sci. Soc. Am. J. 47: 988–991.

Smith, R. M., G. Samuels, and C. F. Cernuda, 1951. Organic matter and nitrogen build-ups in some Puerto Rican soil profiles. Soil Sci. 72: 409–427.

Sollins, P., and G. Spycher, 1983. Processes of soil organic-matter accretion at a mudflow chronosequence, Mt. Shasta, California. Ecology 64: 1273–1282.

Sollins, P., G. Spycher, and C. A. Glassman, 1984. Net nitrogen mineralization from light- and heavy-fraction forest soil organic matter. Soil Biol. Biochem. 16: 31–37.

Stauffer, R. S., R. J. Muckenhirn, and R. T. Odell, 1940. Organic carbon, pH, and aggregation of the soil of the Morrow plats as affected by type of cropping and manurial addition. J. Am. Soc. Agron. 32: 819–832.

Stevenson, F. J. 1982. Humus Chemistry: Genesis, Composition, Reactions. John Wiley & Sons, New York, 443 pp.

Stevenson, F. J. 1983. Isolation and identification of amino sugars in soil. Soil Sci. Soc. Am. J. 47: 61-65.

Tate, R. L. III. 1977. Nitrification in Histosols: A potential role for the heterotrophic nitrifier. Appl. Environ. Microbiol. 33: 911-914.

Tate, R. L. III. 1985. Microorganisms, ecosystem disturbance, and soil-formation processes. In R. L. Tate III and D. A. Klein (eds.), Soil Reclamation Processes: Microbiological Analyses and Applications. Marcel Dekker, New York, pp. 1-33.

Tiessen, H., and J. W. B. Stewart, 1983. Particle-size fractions and their use in studies of soil organic matter. II. Cultivation effects on organic matter composition in size fractions. Soil Sci. Soc. Am. J. 47: 509-514.

Vaughan, D., and R. E. Malcolm, 1979. Effect of soil organic matter on peroxidase activity of wheat roots. Soil Biol. Biochem. 11: 57-63.

Vaughan, D., and B. G. Ord, 1981. Uptake and incorporation of C-labelled soil organic matter by roots of *Pisum sativum* L. J. Exptl. Bot. 32: 679-687.

Visser, S. 1985. Management of microbial processes in surface-mined land reclamation in western Canada. In R. L. Tate III and D. A. Klein (eds.), Soil Reclamation Processes: Microbiological Analyses and Applications. Marcel Dekker, New York, pp. 203-241.

Young, J. L., and G. Spycher, 1979. Water-dispersible soil organic-mineral particles: I. Carbon and nitrogen distribution. Soil Sci. Soc. Am. J. 43: 324-328.

TWO

ORGANIC MATTER TRANSFORMATIONS: ECOSYSTEM EXAMPLES

The quantities of colloidal organic matter in a specific soil, the structure of that soil, and the capacity of the microbial community to supply the nutrients required by the aboveground plant and animal community are all determined by the interactions of the chemical, physical, and biological properties of the soil ecosystem and the effects of these interactions on soil organic matter accumulation and metabolism. The type of aboveground community that develops at a give site is a direct result of these interactions, whereas this surface aspect of the ecosystem in turn controls the fixed carbon inputs into the soil community. Thus neither the aboveground or the belowground components of the ecosystem function independently. To understand how these interactions differ among major ecosystems, the following discussion is provided to introduce the basic properties of soil organic matter transformations in major ecosystems types (forest, grassland, agricultural, and organic soils). The discussion will be limited to the nature of organic matter inputs, reactions occurring in the soil, and their contributions to the aboveground biomass productivity. Unique properties of each ecosystem will be stressed. More generally applicable principles will be evaluated in subsequent chapters.

2.1. THE FOREST

Although there are a wide variety of forest types with divergent climatic, community, and soil properties, soil organic matter reactions in forests in general can be assessed by considering some common forest properties. Nutrient transfer between the decomposers and plant biomass in forests are tightly coupled. This close association between biotic components for nutrient transfer is demonstrated by the nitrogen values for processes and compartments in a conifer forest. Data compiled by Gosz (1981) (Table 2.1) clearly demonstrate that the forest floor comprises the largest nitrogen pool. Essentially all of the remaining

Table 2.1.
Distribution of Nitrogen and Biomass in Conifer and Deciduous Hardwood Ecosystems (Gosz, 1981).

	Conifer	Deciduous Hardwood
Biomass Nitrogen (kg/ha)	300–1900	250–1600
Foliage biomass (t/ha)	5–20	1–5
Nitrogen (%)	1–5	1.5–3.0
Total litter fall (t/ha)	2–7	5–7
Leaf litter fall (t/ha)	0.3–6	1–5
Leaf persistence (years)	<1–14	<1
Leaf litter N (%)	0.4–1.3	0.5–1.8
Litter fall N (kg/ha/yr)	10–90	45–90
N uptake (kg/ha/yr)	30–75	50–100
N mineralization (kg/ha/yr)	50–100	100–300

nitrogen is contained in the aboveground biomass. Less than 10 percent of the total nitrogen resides in plant roots. With minimum nitrogen inputs in an undisturbed ecosystem, the bulk of the nitrogen mineralized is immobilized into biomass. The efficiency of this association is revealed by the comparison of nitrogen mineralized, plant uptake, and that lost by leaching to stream water. Essentially all nitrogen mineralized is incorporated into plant biomass. This balance may be disturbed by anthropogenic manipulations of the ecosystem, such as those that occur as a result of a variety of management procedures ranging from sewage effluent waste disposal through clear-cutting. For example, the most important nitrogen retention mechanism within the soil in a loblolly pine plantation in North Carolina was immobilization into microbial biomass. Thus any management practices impinging on this reaction directly or through impeding mineral nitrogen from reaching the soil microbial component increased the loss of nitrogen to the aquatic or atmospheric ecosystems (Vitousek and Matson, 1984).

Organic matter enters forest soils predominantly via two cycles: as root exudates and decaying roots, and as forest litter. Obvious differences between forest types relating to the ratio of deciduous to pine trees as well as the species of tree present necessarily controls the amounts of aboveground biomass reaching the soil. In temperate forests and, to a degree, in tropical forests, both root and litter inputs are seasonally pulsed, that is, most biomass enters the soil during the growing season. As will be discussed in the following text, although a smaller portion of the forest nutrient pools are contained in the root biomass, these structures and their decomposition have a major impact on forest nutrition in that they are decomposed at a faster rate than the aboveground biomass, that is, the roots are more readily biodegradable than the leaf tissue. Although litter layer decomposition occurs throughout the year during those times when temperatures and moisture levels are conducive to microbial decomposition, fresh organic matter enters the cycles as leaf litter in deciduous or deciduous/conifer forests primarily during autumn and spring. Johnson and Risser (1974) in their evaluation of primary production of a post oak-blackjack oak (*Quercus stellata–Q. marilandica*) forest in central Oklahoma noted that the yearly mean litter fall of 5400 kg/ha was distinctly bimodal, with peaks in November and March. Aboveground biomass inputs into the litter layer include leaf and needle litter, frass, insect debris, plus any organic matter produced by ferns, mosses, and annuals growing in the system. Litter layer thickness depends on the balance between input and decomposition. The latter rate varies between plant types, with conifer litter being more biodecomposition resistant than deciduous litter (Gosz, 1981), and the physical and chemical properties of the soil. The litter and associated organic decomposition products (muck) layer thickness varies from being imperceptible to several inches or feet thick (Figs. 2.1 and 2.2). This organic layer serves as a reservoir for plant nutrients, effects soil physical properties and subsequent nutrient transfers between the microbial community and plants, and controls plant root and overall plant development.

2.1.1. Nutrient Cycling

The role of litter as a pool of plant nutrients has long been recognized. Hence, the primary procedure for estimation of nutrient cycling in forest ecosystems has been and remains to follow nutrient release and cycling in the litter layer. This is generally accomplished by collecting leaf litter, perhaps sorting it by plant species or litter type (twigs, leaves, etc.), and placing it in mesh bags. Variations in the litter bag mesh size allow for or limit insect activity in the litter, thereby allowing determination of the contribution of micro- and macroarthropods to decomposition. Many excellent studies of nutrient cycling in forest systems have been reported (Bocock, 1963; Johnson and Risser, 1974; Buldegren et al., 1983; Pastor and Bockheim, 1984; Vitousek, 1984). The prime conclusions from each of these studies is that the litter layer provides a major source of nutrients for the growing plant. The proportion of needed mineral nutrients supplied by the litter is less than that found in fresh plant tissues during the growing season in that a major portion of these substituents is translocated to perennial tissues before leaf fall. Pastor and Bockheim (1984) found that translocation was an important mechanism for nitrogen, phosphorus, and potassium conservation in an aspen (*Populus tremuloides*)-mixed hardwood-spodosol sugar maple (*Acer saccharum*) forest. This mechanism was less important for cycling of sulfur, magnesium, and iron and did not occur for calcium and zinc. Of greater concern for the discussion herein is the fate of those nutrients contained in the leaf litter. With the more available nutrients, such as nitrogen, the processes involved in nutrient transformations are primarily biological, although some chemical immobilization into humic substances can occur (Stevenson, 1982). Geological processes partially control the distribution of those nutrients which are likely to be precipitated by soil ions. This includes precipitation of phosphate by divalent ions and sulfide by iron. Wood et al. (1984) found that phosphorus availability

Fig. 2.1.
Forest floor profile of an upland site in the New Jersey Pine Barrens with light litter accumulation.

was regulated in a northern hardwood forest at Hubbard Brook, New Hampshire by this type of combination of biological and geological processes.

A more difficult problem to assess is the distribution of nitrogen by the microbial processes of immobilization and mineralization. This is important for ecosystem function because nutrients immobilized by the microbial community are not immediately available for plant growth, instead, they become part of the soil nutrient pool. As leaf litter decomposes, organic nitrogen is mineralized. But, since the litter contains excess carbon, a major portion of this nitrogen is assimilated by the microbial community into new cell biomass. Hence, this nitrogen becomes unavailable to the plant community until death and decay of the micro-

Fig. 2.2.
Forest floor profile of a transition vegetation site in the New Jersey Pine Barrens with approximately 20 cm of litter and partially decomposed organic matter accumulated over the sand soil surface.

bial cells; because, although these nutrients are unavailable for immediate utilization by the plant and, indirectly, the animal community, they form a pool or reserve of nutrients for future biomass production. Once the litter is decomposed to the point that more mineral nitrogen is produced by decomposition than is required by the microbe, some of the nitrogen becomes directly available to the plant as it is mineralized. A difficulty for those developing models of the forest ecosystem lies in determining when this point of net production occurs. To facilitate the compartmentalization of the nutrient cycle in forest soils, Aber and Melillo (1980) differentiated "soil" and "litter" by their immobilization and miner-

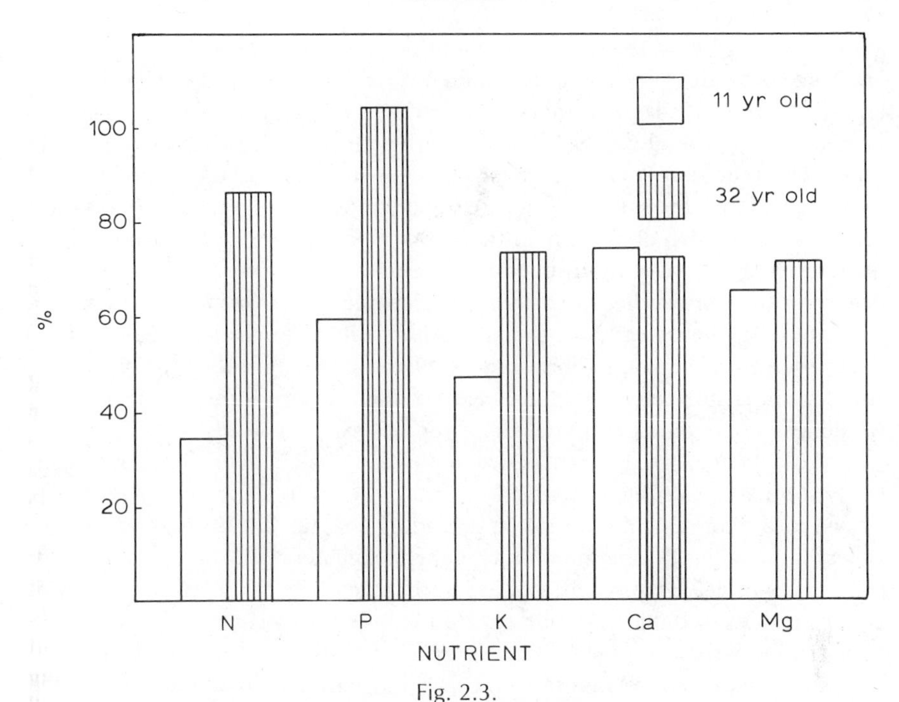

Fig. 2.3.
Nutrient supplying capacity of soil and litter pools of an 11-year-old and a 32-year-old loblolly pine stand (Jorgensen et al., 1980).

alization rates. Litter was defined as that material exhibiting net nitrogen immobilization, whereas soil exhibits net nitrogen mineralization. Thus the point in the forest floor profile where the litter material has been decomposed to soil is that position where nitrogen becomes available to the plant roots. Logically, this would be the point in the profile where the association of roots and litter decomposers would become most prominent.

The proportion of plant nutrient requirements met by the soil and litter nutrient pools varies with age of the forest. Jorgensen et al. (1980) found that in a young stand of loblolly pines (*Pinus traeda* L.), (i.e., 11 years of age), 34, 59, 47, 74, and 65 percent of the nitrogen, phosphorus, potassium, calcium, and magnesium, respectively, incorporated into plant biomass was derived from the forest floor organic layer (Fig. 2.3). In contrast, in a 32-year-old stand of loblolly pines, the forest floor accounted for 86, 104, 73, 72, and 71 percent, respectively of the nitrogen, phosphorus, potassium, calcium, and magnesium. Interestingly, they observed that nitrogen contents of the forest floor did not increase, but remained constant after the first three years. Similarly, Ojeniyi and Agbede, (1980) found coffee (*Coffea robusta*) yields positively correlated with total soil organic matter.

A small portion of the water soluble litter nutrients are lost from the soil by leaching (Berg and Wessen, 1984). The physical structure of the soil ecosystem and the location of the plant roots, considering the quantities of nutrients mobilized by the microbial community, allows these losses to be much smaller. The reduction of the losses occurs because the plant roots are located throughout the surface portion of the soil profile as are soil microbes. Thus nutrients that escape the plants and microorganisms in the upper most surface layers of the soil organic layer will usually be trapped and metabolized by those organisms living deeper in the soil profile. Occasions for accentuated loss of nutrients coincide with unusually heavy raïnfalls that wash nutrients through the profile at rates faster than plants and microbes could trap and metabolize them or during anthropogenic disturbances, such as clear cutting. Surface transport of nutrients would also be accentuated under these conditions.

The biochemical mechanism of decomposition of plant debris has been extensively studied and will be discussed in greater detail in Chapter 3. As was discussed for soil organic matter in the preceding chapter, the decomposition rate of leaf litter varies depending on its biochemical composition. Thus significant differences in decomposition rates are anticipated between leaves with varying concentrations of biodegradation resistant components, such as lignin, as well as divergent carbon:nitrogen ratios. Shanks and Olson (1961) in studying litter decay in Great Smokey Mountain and Oak Ridge forests noted that the greatest variation in first-year weight loss of leaves in nylon net bags resulted from variations in species. Similar results were noted by Gosz et al. (1973) in a study of leaf and branch litter decomposition in the Hubbard Brook Forest, New Hampshire. This variation would result not only from varying contents of lignin and lignin-like compounds, but also from differences in content of easily degraded components and the accessibility of these components to the microbial community. A biodegradable substance becomes biodegradation resistant when protected by lignin. During the first year of decomposition, much of the loss of mass results from biodecomposition of the more readily decomposable components. Anderson (1973) in his study of beech (*Fagus salvatica* L.) and sweet chestnut (*Castanea sativa* Mill.) found that the reasonably constant carbon/hydrogen ratios of the leaf litter through the first year indicated that carbohydrate losses were directly proportional to the weight losses.

Studies of leaf litter decomposition provide a portion of the picture of nutrient cycling in the forest ecosystem, but recent evaluation of fine root turn over rates in forest ecosystems has shown that failure to consider the latter portion of the biomass causes a highly significant under estimation of nutrient cycling rates. McClaugherty et al. (1982) found that fine root production was slightly higher in a hardwood stand than leaf production and about 20 percent less than leaf production in a pine stand. This high synthesis rate and slow decomposition rates indicated that fine roots are a major contributor to the soil organic matter and nitrogen pools. In a study of a young (23 years old) and mature (180 years old) *Abies amabilis* stands, Vogt et al. (1983) found that inclusion of the root turnover into calculations of organic matter residence times in the forest floor reduced the estimate of the nitrogen residence time by 74 percent and for phosphorus by 75 to 86 percent. As with leaf litter, nitrogen initially liberated by mineralization of the fine root tissues and exudates is immobilized into the microbial cells. This is followed by a slower second phase where nitrogen becomes available to the plant community (McClaugherty et al., 1984).

2.1.2. Limitations of Mineralization in Forests

Unique properties of the mineralization cycles in forest ecosystems relate to the nature of the litter layer and the chemical composition of the litter itself. The soil organic layer, which we shall consider to consist of both the litter and the well humified products of microbial catabolism of the litter, can be divided into two rather distinct systems: one controlled initially by leaf properties and the second controlled by the soil microbial community. When a leaf first reaches the litter layer, the microbial decomposition of its structure is primarily controlled by the leaf bacterial flora and the physical and chemical properties of the leaf. Subsequent discussions will indicate that the actual rate of decomposition may relate at least in part to micro- or macrofaunal disruption of the overall leaf structure, but the microbial mediators of the catabolism of the leaf biochemicals are functioning in an ecosystem more reminiscent of the plant than the soil. A property of this portion of the organic layer was indicated previously where it was found that in this material nitrogen liberated through catabolism of leaf components is immobilized into microbial cell biomass. As the leaf components are humified, the product begins to take on the properties of soil humic substances. At this point, the organic component resembles muck and would be considered to be a soil ecosystem or component. Then there would be net mineralization of the nitrogen contained within the leaf. Further data supportive of this separation of the litter layer into two major components was provided by Jorgensen and Wells (1973). In their studies of litter and soil horizons collected from soils in hardwoods and pine forests in the North Carolina Piedmont, they found little variation of microbial respiration in the soil fractions between the various soils types. A significant variation in microbial respiration was found in the litter layer with these differences dependent on the source of the litter. Their data suggest that the differential respiration rate between the litter layer and the underlying soil related to the degree of humification of the plant litter. The microbial community in the partially decomposed plant litter is actively decomposing the readily available carbon sources. This active metabolism varies depending on the quantities of these substances present in the litter as well as the amount of lignin present to occlude carbonaceous substrates from microbial activity. This high rate of microbial activity is the primary cause of nitrogen immobilization. As these readily decomposable substrates are depleted, respiration declines to a rate more characteristic of a soil microbial community slowly decomposing the more biodegradation resistant components reminiscent of soil organic matter. This decline in respiration with increased humification of the plant litter was also noted by Witkamp (1963; 1966) when he observed in studies of forests at

Oak Ridge, Tennessee that microbial respiration was controlled by temperature, bacterial density, moisture, and the number of weeks since leaf drop.

Another factor delimiting the forest litter layer from other sites of active microbial catabolism of plant debris is the role of lignin in the control of microbial catabolic processes. Two major factors that are commonly cited for control of microbial metabolism of organic debris are lignin content and wide carbon:nitrogen ratios. Forest leaf litter provides an example of effects of both variables. In studies of decomposition rates of fir needles, female cones, branches, and bark in several stands of mature vegetations in western Oregon, the decomposition constant was more closely correlated with lignin content than with the carbon:nitrogen ratio (Fogel and Cromack, 1977). This inhibitory effect of lignin can be modified by interactions with the actual evapotranspiration of the site (Meentemeyer, 1978). Under dry conditions, decomposition appeared to be nearly independent of lignin content, but as the moisture limitations were relieved, the lignin effect became more obvious. This provides an example of a major complication in generalizing biological decomposition principles in any ecosystem. Although well documented relationships are known, the impact of the variable may not be observed or be of minor significance when other chemical or physical limitations exist in the ecosystem. In the preceding example, moisture availability overcame the effect of lignin content.

2.2. THE GRASSLAND

Soil organic matter transformations of grasslands are differentiated from those of other ecosystems by the type and quantities of organic matter entering the soil, the proportion of rhizosphere effect in the soil ecosystem, and the temporal distribution of organic matter inputs. During the growing season, soil microorganisms experience a continuous input of easily degraded organic compounds. Organic matter accumulation may be seasonally pulsed, such as is observed in temperate climates where soils are frozen a significant portion of the year or reasonably constant throughout the year as is typically found in tropical regions where seasonal variation involves primarily the magnitude of carbon input. In a study of a tropical grassland soil, Upadhyaya and Singh (1981) recorded litter and root biomass production by *Iseilema anthephoroides* Hack for the 1976–1977 seasons. Maximum litter production occurred during the winter and summer months (December through June) with approximately 33 to 50 percent of that maximum production occurring during the rainy season (June through October). Litter biomass was approximately 20 to 33 percent of the root biomass production (Table 2.2). Similarly, with a temperate grass, Blue Grama [*Bouteloua gracilis* (H. B. K.) Lag.], in a growth chamber study, 78 percent of the ^{14}C-carbon dioxide fixed during the final 31 days of the 55 day growth period between germination and seed set was translocated to the root–soil system (Dormaar and Sauerbeck, 1983). Not all of this fixed carbon was retained within the roots. A distribution of 33 and 45 percent of the total photosynthate carbon

Table 2.2.
Litter Accumulation and Decomposition, Carbon Dioxide Evoluation and Root Biomass Pools in a Tropical Grassland Soil (1976–1977)
(Upadhyaya and Singh, 1981)

Sampling Dates	Litter Accumulation (g/m^2)	Rate of CO_2 Evolution ($mg/m^2/hr$)	Root Biomass (g/m^2)	Litter Decomposition (% Wt. Loss)
June 4	103	377	597	8.00
October 3	111	345	887	23.33
January 2	215	147	645	3.33
March 2	243	136	612	2.00

to the root and soil system, respectively, was detected. This mass transfer of energy to the root soil system is of particular interest in arid soils containing relatively low organic matter contents because they represent a formidable nutrient pool for both the microbial and plant community. Of significance from the view of evaluating the rate of occurrence and temporal variation in organic matter transformations in the soil ecosystem is the fact that this organic matter input in grasslands occurs throughout the growing season, that is, roots and shoots are continuously being produced. Hence, the root system represents a total age gradient from mature to nascent tissue at any given point in time. This root mass in grasslands can be quite large. Total mass of the root systems of *Andropogon gerardi* and *A. scoparius* in a central Missouri prairie ranged from 1449 g/m^2 prior to inception of growth in the spring to 1901 g/m^2 at the end of the growing season (Dahlman and Kucera, 1965). Between 48 and 60 percent of these underground plant parts were found in the upper 2 inches of the soil profile while greater than 80 percent of the root mass was detected in the A_1 horizon or the surface 10 inches of the soil profile. Significant quantities of roots were found in the soil profile to the B_2 horizon (18 to 30 inches).

This continuous input of organic carbon and the dense distribution of the plant roots also has a major effect on the nature of the microbial activity in the soil surface. As a result of the high plant density and the active fixed carbon translocation to the root-soil system, the majority of the surface soil may be loosely defined as rhizosphere soil. Thus the microbial activity of this soil is significantly greater than nonrhizosphere soil and the soil microbial floral activity is primarily controlled by the properties of the plant. This results because the physical presence of the plant alters the chemical and physical parameters of the soil and the plant root serves as the primary nutrient reservoir for the microbial community.

The high activity of this soil microbial community is readily appreciated when we observe that in many grassland soils 80 percent or more of the organic matter entering the soil is mineralized within the first year. The balance between fixed carbon input into the soil and carbon dioxide evolution has been evaluated by a variety of different methods ranging from analysis of ^{14}C-labeled carbon dioxide production from specifically labeled plant components to simple quantification of biomass production and carbon dioxide evolution in the field. An example of the latter procedure is provide by Coleman (1973) where they found that carbon dioxide evolution from a broomsedge (*Andropogon virginicus* L. and *A. ternarius* Michx.) old field approximated the organic matter inputs from roots and litter. Evaluation of the annual turnover of specific plant parts by assessment of quantities within the soil profile reveals a differential mineralization of various belowground components. Dahlman and Kucera (1965) noted a turnover rate of 22.8 percent and 40.8 percent for rhizomes and roots, respectively, in the 0- to 2-inch zone in a central Missouri prairie. A number of workers have amended soil with ^{14}C-labeled grass components to evaluate their metabolic rates (Jenkinson, 1965; Nyhan, 1975; Dormaar, 1975; Klein, 1977). In each of these studies, as much as 90 percent of the plant organic matter was mineralized during the initial six months of incubation. That portion of the plant residue remaining in the soil after the first 6 to 12 months is generally very stable. For example, Jenkinson (1965) found about 20 percent of the ryegrass amended to soil remained after four years incubation. Such studies support the conclusion that a major portion of the root and shoot debris reaching the soil microbial community is rapidly decomposed. The remains are composed of the lignified plant components which are more resistant to microbial attack. Hence, variation in the decomposition properties of various grass species relates to the quantities of lignin contained therein. High lignin containing components would be expected to be more stable than the more easily metabolized components. Obviously, as with the forest system previously discussed, this generalization is tempered by other limiting factors, such as soil moisture or nitrogen or phosphorus availability, which may override any effect of plant chemical composition. For example, Tesarova and Gloser (1976) concluded that soil temperature and moisture were among the most significant factors controlling carbon dioxide evolution from grassland soils.

As a result of the high biomass input and the accompanying microbial activity, grassland soils characteristically have high organic matter contents. The effect of the highly productive plant biomass portion of the ecosystem on the steady-state levels of soil organic matter is demonstrated by the effect of development of grasslands for agricultural production. Loss of soil carbon, nitrogen, and phosphorus in a variety of prairie soils was examined by Tiessen et al. (1982). In clay and silt loam soils, after 60 to 70 years of cultivation, approximately 35 percent reduction in soil carbon was observed. Variations in soil nitrogen necessarily depended upon cropping history. Growth of legumes resulted in less nitrogen depletion than occurred with grain crops. Nitrogen losses in the prairie silt loams and clay soils ranged from 18 to 34 percent. Phosphorus losses were approximately 12 percent with the total loss accounted for by changes in the organic phosphorus pool. Greater declines in the organic nutrient pools of lighter texture soils were reported. In a sandy loam over a similar cultivation period, 46, 46, and 29 percent declines in carbon, nitrogen, and phosphorus concentrations, respectively, were detected. Interestingly, with these large nutrient losses, the soil had not achieved a steady-state condition because a decrease in the rates of decline was not noted in a silt loam soil that had been cultivated for 90 years. This reinforces the observation that achievement of a steady-state condition is a slow process. Practical implication of this observation relates to the time course of improving agricultural soils through modification of management practices to improve organic matter accretion rates. Patience is necessary when evaluating the benefits of changes in management practices. Whitehead et al. (1975) studied organic matter components in soils that had been (1) under continuous cultivation, (2) 17 years under grass after prolonged cultivation, and (3) old pasture. Organic carbon contents of the three soils were 0.9, 1.7, and 4.8 percent, respectively, Thus even after nearly two decades of grass, the grass field still contained less than half of the organic carbon of the old pasture. In a similar study, White et al. (1976) found that with their system organic matter and total nitrogen increased about 0.02 and 0.001 percent per year.

These studies emphasize the unique properties of grassland soils contributing to organic matter accumulation and the fact that accumulation of this organic matter is a long-term process. The contribution of the intimate association of the highly productive grass species with the microbial community in a reasonably undisturbed state is seen by the dramatic declines in soil organic matter levels accompanying grassland development. Both the physical and biological properties of the ecosystem contribute to organic matter accumulation.

2.3 AGRICULTURAL SOILS

Although cropped soils actually comprise a minor portion of worlds soils, the impact of these soils on world hunger and societal development is highly significant. Thus any changes in organic matter that may impact crop yields have far reaching consequences. The most revealing predictor of organic matter problems in agricultural soils is the changes in native organic matter levels as these soils are brought into cultivation. Most probably a few examples of soils exist where minor changes in organic matter contents are observed on initial cultivation of virgin soils, but generally rapid reductions of organic matter occur over the first few years following development with more gradual declines over subsequent years. The amount of loss as well as its rate depend on the cropping system employed. This decline in colloidal soil organic matter can be quite dramatic. For example, 70 years of cultivation of Blackland clay soils of Texas resulted in loss of more than 50 percent of the surface horizon organic matter (Smith et al., 1954). Cultivation of a variety of forest soils in Georgia (U.S.A.) resulted in losses of 57 percent of the native soil organic matter. A major portion of the organic matter decline occurs shortly after development in that in six of the forested Piedmont soils under simulated cultivation for two years, an average 20 percent reduction in the soil carbon was noted (Giddens, 1957). This rapid initial loss followed by a gradual but dramatic decline in soil organic matter levels over several decades is common to all mineral soils developed for agriculture. The main variable is the extent of organic matter loss that depends on the cropping system as well as the physical, biological, and chemical traits of the ecosystem. The rapid initial decline in organic matter is an immediate response to the loss of biomass input as well as the reduction of pools of easily metabolizable organic matter. Although agricultural soils can be and generally are highly productive, due to the necessary loss of organic matter with harvest of the crop, less organic matter enters the soil than occurred under predevelopment conditions. Also, when generally compared to a productive grassland soil, considerably less organic matter enters the soil through the plant roots. This decline in input combined with the continued mineralization of that readily decomposable organic matter remaining in the soil immediately following cultivation accounts for the initially rapid loss of soil organic matter. This accentuated microbial activity is further stimulated by reduction of the physical and chemical factors formerly controlling microbial activity. Changes in physical conditions include the increased aeration due to plowing, whereas chemical transformations relate in part to increased availability of occluded organic matter. Prior to development of the virgin soil, a portion of the soil organic matter is protected from mineralization by the physical structure of the soil; for example, it may be contained within soil granules or be located within complexes of clay particles. After the more readily decomposable organic matter pools are depleted, the microbial community is limited to metabolizing the more resistant plant components and soil organic fractions plus the organic matter entering the soil as a result of aboveground biomass production. Hence, the organic matter mineralization rate is greatly reduced. The soil organic matter level continues to decline until a new equilibrium is established between input and mineralization. As has been indicated previously, this usually takes decades, especially if soils are managed for maximal crop production with little organic matter return to the soil. Achievement of the new steady-state equilibrium may be further delayed if the management procedures vary with time. A generalized prediction of the final soil organic matter levels resulting from various cropping systems is now difficult, if not impossible to make.

If managed properly, a major benefit to the aboveground biomass during this adjustment period of soil organic matter levels is the increased availability of mineral nitrogen. Cultivation leads to significant decreases in total soil nitrogen, nonhydrolyzable nitrogen and hydrolyzable nitrogen (total, ammonium, hexosamine, amino acid, and unidentified nitrogen) (Keeney and Bremner, 1964). Cultivation has little effect on the percent distribution of the nitrogen in these various fractions. In their study of effect of cultivation on 10 virgin soils, Keeney and Bremner found an average loss of 36.2 percent of the total soil nitrogen. Benefits of this nitrogen mineralization relate to crop production, but if the soil is not properly managed, increased leaching of mineral nitrogen into regional streams and lakes could augment eutrophication rates (Reinhorn and Avnimelech, 1974). In a comparison of nutrient budgets for conventional-tillage, no-tillage, and old-field systems, Stinner et al. (1984) found that the largest nutrient flow in each system was into the crops, weeds, and old-field plants. Leaching losses were small compared to fertilizer inputs, but leaching losses were least in the old-field system and greatest under conventional till.

Of more practical importance from the view of environmental conservation is the effect of the loss of soil organic matter on soil physical and chemical properties. Decreases in soil organic matter result directly in declines in soil air space, aggregation, and water holding capacity (e.g., see Laws and Evans, 1949). As the aggregate structure of the soil is lost, compaction (bulk density) increases, as does runoff of surface waters, and thereby, soil erosion (Moldenhauer et al., 1967). This loss of soil has not only an effect on soil fertility in that the soil surface horizons are most conducive for cropping but a regional impact on streams and lakes is also observed. Also, the increased compaction or bulk density of the soil has a detrimental effect on crop root development. Root development is improved in less compact soils, whereas reductions in root growth are found in the more compact soils (Schuurman, 1965).

Cultivation of grassland soils results in a dramatic change in the distribution of organic matter with soil particle size (Fig. 2.4) (Tiessen and Stewart, 1983). After four years of cultivation dramatic losses of organic matter associated with particles greater than 50 μm in a Blaine Lake silt loam were observed. This accounted for a 43 percent loss of the initial carbon. This loss was considered by Tiessen and Stewart to be a result of a physical disintegration that caused an accumulation of the organic materials in finer particle-size fractions. The shift of organic matter towards smaller particles was associated with a net loss of 34 and 29 percent of the carbon and nitrogen, respectively, over 60 years of cultivation. Fine clay (0.2 μm) associated organic matter rapidly declined over the first 60 years but little change was observed with subsequent cultivation. Losses of organic matter in the fine silt (5 to 2 μm) and coarse clay (2 to 0.2 μm) fractions were substantially less. The proportion of the total soil organic matter found in the latter fractions increased with time of cultivation. The increased occurrence of the more humified materials with time indicates a reduced capability of the soil organic matter pool to supply plant nutrients via mineralization.

The natural ^{15}N abundance in the soil particle size fractions may provide a measure the biodegradation resistance and, hence, nitrogen supplying capability of the organic matter fractions (Tiessen et al., 1984). In their evaluation of ^{15}N abundance with particle size in two cultivated soils and two native grassland soils, nitrogen associated with the coarse soil fractions showed a low enrichment typical of recently deposited plant materials. The ^{15}N enrichments were similar for the silt-size fractions whereas the clays had a higher ^{15}N abundance which is characteristic of organic matter resulting from microbial metabolism. Thus reduction of the ^{15}N abundance in the large particle fractions apparently directly indicates of a reduction in fresh plant biomass incorporation.

Soil management procedures have a significant influence on the final levels of soil organic matter achieved in the cultivated soils. As will be discussed in Chapter 13, greatest accumulations of organic matter are found under no-till systems and greatest depletions are found in soils with maximum removal of grain crops.

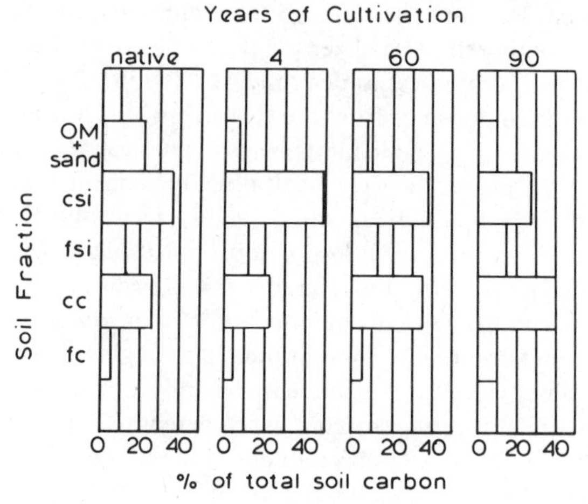

Fig. 2.4.
Carbon distribution in partical size fractions of Blaine Lake silt loam under native grassland, or cropped to small grain-fallow rotations for 4, 60, or 90 years. (OM = organic matter; csi = coarse silt; fsi = fine silt; cc = coarse clay; fc = fine clay) (Tiessen and Stewart, 1983) Reproduced from SOIL SCIENCE OF AMERICAL JOURNAL, Volume 47, 1983, pp. 509–514 by permission of the Soil Science Society of America.

Residue management and fertilization have a significant effect on the soil organic matter levels. Larson et al. (1972) evaluated the effect of amendment of Marshall silty clay loam with cornstalks (*Zea mays* L.), sawdust, oatstraw (*Avena sativa* L.), and bromegrass (*Bromus inermis* Lyeyss). Changes in the soil organic matter levels were followed over an 11 year period. At amendment rates of 16 tons/ha/yr, the soil carbon, nitrogen, sulfur, and phosphorus increased 47, 37, 45, and 14 percent, respectively, over the control plots. At an amendment rate of 8 tons/ha/yr, no influence of amendment type was noted on the carbon or phosphorus contents, but nitrogen and sulfur and nitrogen contents were lower in soils receiving sawdust as compared to those receiving residues. These workers calculated that with their management system and soil about six tons of cornstalk residue would have to be added per hectare per year to maintain the soil organic carbon levels. Barber (1979) found that after 12 years of amendment of a Raub silt loam with cornstalk residue in a fertility experiment, about 8 percent of the residues were synthesized into new soil organic matter. Corn roots appeared to contribute at least 18 percent of their carbon to new soil organic matter. Depletions of organic matter in high organic matter containing mineral soils may be less severe in that Bloom et al. (1982) found that for high organic matter containing mollisols of Minnesota, changes of organic matter due to cropping to continuous corn were slight. As a result of these small differences in organic matter levels due to cropping, their study period of 13 years was insufficient to develop a precise estimate of the changes in soil carbon levels.

In many cases, the nature of the biological activity of agricultural soils is rather similar to that of the grass soils. These similarities relate to the biodegradation rate of organic matter entering the ecosystem. Activity of the microbial community is in large part determined by the quantities of easily metabolizable organic matter entering the ecosystem. As with grasslands and, to a certain degree, in contrast to forest soils, most of the plant debris entering agricultural soils is easily metabolizable. Hence, a burst of microbial activity occurs during the rapid mineralization of this organic matter, followed by low levels of activity as allowed by the decomposition of more biodegradation resistant carbon compounds. In some ways, the physical and chemical ecosystem of cropped soils is usually less complex than is found in previously discussed ecosystems because agricultural soils are amended and manipulated to provide optimal growth conditions for the crop. This would include liming to raise the pH of acid soils and amendment with trace minerals to overcome any deficiencies in these nutrients. Thus plant debris is metabolized rather rapidly and extensively in cropped mineral soils. Shields and Paul (1973) found that over a four-year period, 80 percent of wheat straw amended to the plow layer of a cropped soil was decomposed. The decomposition was controlled by soil moisture, temperature (especially in the winter months), and degradation resistance of the straw. The readily decomposable straw components had a half life of 48 growing-months in the cropped soil, whereas the degradation resistant components exhibited a half-life of 96 growing-months. In soils of low nitrogen availability, nitrogen would be expected to limit decomposition process, especially for the more rapid decomposition phases. During the rapid decomposition period, net nitrogen immobilization is anticipated, whereas subsequent metabolism results in nitrogen mineralization. The nitrogen immobilized during the rapid growth period becomes available to the plant community during subsequent periods. This nitrogen limitation can be masked (i.e., overcome for cropping) by nitrogen fertilization.

The common procedures of planting crops in rows and the cycling of organic matter production between planting and harvest cause the considerable variation in organic matter distribution and decomposer activity characteristic of agricultural soils. Microbial activity varies in cropped soils from being essentially equivalent to uncropped or fallow soils to being equivalent or somewhere between this activity and that of grassland soils. This was demonstrated in Pahokee muck where the carbon mineralization capacity of the microbial community was compared between a sugar cane (*Saccharum* sp. L.) field soil, a St. Augustinegrass (*Stenatopharum secundatum* (Walt) Kuntz) soil, and a fallow field soil (Tate, 1979). Succinate and salicylic acid metabolism were greater in the grassland soils than in the cane or fallow field soils whereas, no significant difference was found with acetate metabolism. The differences in the microbial activity of the various soils depended on crop growth in that activities in sugar cane field soils approximated activities of uncropped soils when the crop was young but the activities increased to values intermediate between the bare and the grass field soils under mature cane. Practical consequences of such observations relate to sampling procedures in cropped soils. Assessments of residual nutrients in cropped soils and microbial activity depend on the maturity of the crop as well as whether the samples were taken from between rows or within rows.

A further point of importance in evaluating soil organic matter properties and conservation in cropped soils relates to development of practices to encourage accretion of organic matter or to at least minimize losses. This important topic will be discussed in Chapter 13.

2.4. HISTOSOLS

Histosols, organic soils, constitute a small portion of the worlds soils; yet once drained, they can prove to be highly productive agricultural soils. Soil organic matter transformations are accentuated in Histosols in that by definition they must contain at least 12 percent organic matter (more if the soils contain clay) and organic matter may comprise as much as 100 percent of the soil mass (Lucas, 1982). This enrichment of organic matter, which resembles that organic matter of mineral soils, frequently provides an unique opportunity for study of soil organic matter transformations in the absence of complications associated with the interactions of organic and mineral soil substituents. Because of the high organic matter contents and the nature of the ecosystem leading to their genesis, some specialized management problems are associated with organic soil development; including water management and soil subsidence. Both are directly related to the biochemical mineralization of the soil organic matter. Deposits of organic soils originate as swampy or water-logged soils where biomass production is greater than decomposition. As the plant and animal remains fall to the soil surface, some biodecomposition occurs. But, partially decomposed organic matter accumulates, in part, as a result of the protective effect of the water saturated conditions. Other factors implicated as contributors to the organic matter accumulation are soil acidity, oxygen depletion or low redox potential, and the inhibitory effect of protonated organic acids (Kilham and Alexander, 1984). As long as the protective mechanism(s) remain in tact and aboveground biomass production continues, organic matter can accumulate at substantial rates. Accumulation rates have been estimated to range anywhere from 0.6 to 10 cm annually, with the bulk of the estimates at the lower end of the range (Broadbent, 1960; Bohn, 1978; Dawson, 1956; Richardson and Allen, 1979).

Once the flooded soils are drained, the balance between organic matter synthesis and decomposition, favoring accumulation, is shifted towards decomposition. Organic matter mineralization replaces accumulation. Chemical properties of the organic matter and the biochemical reactions involved in its decomposition vary with the length of time since drainage. Initially, biochemical decomposition is not too different from that observed in mineralization of partially decomposed plant debris or even compost. A significant portion of easily metabolizable organic substituents that comprised the living plant remain as carbon and energy sources for the microbial community. As these carbon sources are depleted, the microbes, as with general metabolism of plant debris, are left with oxidation of the more highly resistant organic substituents. At this point the organic substrate has lost the biochemical and structural traits that link the biochemical properties of its decomposition to those of plant tissue. The substrate for mineralization now resembles soil humic substances. Hence, the mineralization rate slows and the soil microbial community and enzymes reflect those activities associated with general humic substance decomposition.

2.4.1. Management of Organic Matter in Histosols

The most severe management problem associated with utilization of drained Histosols is soil subsidence. Because the unique physical and biochemical properties of a soil that contains high quantities of organic matter, the surface elevation of the field begins to subside essentially as soon as drainage is started (Figs. 2.5 and 2.6). Causes of this subsidence include loss of the buoyant effect of the water, shrinkage due to drying, wind and water erosion, compaction, and mineralization of the organic matter. Little can be done to control the problems of buoyancy loss and drying effects, but effective management can be used to control the others. The impact of wind and/or water erosion depends on the size of the deposit being managed. In extensive areas such as are found in the Florida Everglades Agricultural Area, erosion difficulties are not of major practical importance in that they merely involve movement of the organic soil from one field to another. Water erosion is considered primarily from the view of maintaining drainage canals. In small deposits of organic soils, a number of techniques have been developed to minimize soil erosion. A good review of best management procedures for organic soils has been prepared by Lucas (1982).

Biochemical oxidation of the soil organic matter and the physical and chemical conditions limiting the process are of greater concern from the view of controlling subsidence because with this process, organic matter is lost from the ecosystem by its conversion to gaseous and water soluble products. To date, the sole practical method to reduce or prevent this oxidation has been to return the soil to the moisture conditions existent prior to development. This entails maintaining as large a portion of the soil profile under water saturated conditions. It has been known for many decades that the soil subsidence rate is indirectly and linearly proportional to the depth of the water table. Thus the lower the water table is in the soil, the higher the soil subsidence rate (Neller, 1944; Allison, 1956; Stephens, 1969; Zimenko, 1972). Therefore, a doubling of the depth to the water table doubles the subsidence rate. Thus proper management of Histosols involves maintaining the highest water table conducive to crop growth. Obviously, production of crops, such as rice, which may be flooded for a major portion of the year is more logical on these soils than is the growth of dry-land crops.

With the development of the chemical industry, the question has often been raised concerning the feasibility of developing a chemical means of controlling the biological oxidation of the soil organic matter in Histosols. The major assumption behind such proposals is that there exists a specific reaction primarily

Fig. 2.5.
Subsidence pole, Agricultural Research and Education Center, Belle Glade, Florida, April, 1979. The top of the pole was at ground level on April 3, 1924.

instrumental in the decomposition of this organic matter and that a specific inhibitor can be found that would inhibit this reaction. Although the truth of this assumption would be welcomed, examination of the biochemical data as well as the general reactions involved with oxidative processes in Histosols indicate that development of a specific inhibitor is highly unlikely. Recent biochemical data involving the study of soil enzymes active in soil organic matter decomposition in Pahokee muck of the Everglades Agricultural Area indicate that the enzymes associated with the subsidence are those involved with organic matter decomposition in general (Tate, 1984). This would have been suspected *a priori*. Thus an inhibitor of biochemical oxidation in Histosols must be a general inhibitor of microbial metabolism. Mathur et al. (1979), Mathur and Rayment (1977), and Mathur and Sanderson (1980) have demonstrated that several soil enzymatic activities and endogenous respiration of organic soils of southern Canada are inhibited by copper amendments. High correlations were found between rates of copper amendment and soil subsidence rates in some Canadian peatlands.

Institution of these expensive procedures for controlling subsidence of organic soils is justified through an examination of subsidence rates of world Histosols and the environmental impact of the oxidation of this reserve of organic matter. Soils in the Everglades Agricultural Area are subsiding an average of 3 cm/yr (Tate, 1980), with the exact rate depending on the management proce-

Fig. 2.6.
House located at the Agricultural Research and Education Center, Belle Glade, Florida. Note the subsidence from around the foundation and from above the septic tank.

dures. Subsidence of those soils cropped to rice is necessarily less than the vegetable or sugarcane field soils. For comparison, rates of 2.07 cm/yr in southern Quebec (Millette, 1976), 1.75 cm/yr in the western Netherlands (Schothorst, 1977), and 7.6 cm/yr in the California delta (Weir, 1950) have been reported.

Allowing some of the biochemical oxidation to occur is justified from the view of supplying a portion of the nutrients necessary for crop growth, whereas some difficulties have been encountered when these nutrients are leached from the soils into regional waters. Based on calculations from the known soil composition and biochemical oxidation rates, it is estimated that a subsidence rate of 3 cm/yr in the Everglades Agricultural Area results in the production of from 1200 to 1400 kg mineral N/ha/yr (Tate, 1977; Terry, 1980). This nitrogen can accumulate in the soil to substantial concentrations (Avnimelech, 1971; Terry and Tate, 1980). Avnimelech reported concentrations of approximately 100 to 2000 μg nitrate-nitrogen per g dry soil in samples from the Hula Valley of Israel. Terry and Tate detected 38 to 311 μg nitrate-nitrogen per g dry muck in Everglades Histosols. This nitrogen is removed from the ecosystem in the crop, via denitrification, or in drainage waters. The latter option only accounts from 20 to 40 kg/ha/yr (Florida Sugar Cane League, 1978). Loss of some of the mineral nitrogen as nitrous oxide accounted for 50 to 165 kg/ha/yr (Duxbury et al., 1982). Needless to say, with this quantity of nitrogen produced in the agricultural soils from mineralization of native soil organic matter, nitrogen fertilization is generally not necessary.

2.4.2. Biochemical Properties of Histosols

The granular structure, the mass transfer of air throughout the soil profile involved with water table depth maintenance, and the generally high organic matter content, all contribute to the elevated decomposer activity found in Histosols when compared to mineral soils. Not only is the general microbial activity stimulated, but significant organic matter decomposition occurs throughout that portion of the soil profile above the water table as compared to the limitation of the activity to the top few centimeters of most mineral soils. This is attributed in part to the movement of air throughout the soil profile as the soils are drained and the water table depth maintained and the easement of air diffusion limitations by the open granular soil structure. These properties help to create an environment conducive to an active decomposer population.

Two major types of biochemical ecosystems can be discerned in these soils; one in the newly drained soils, and a second that predominates after the more readily decomposable organic matter pools are depleted. When a Histosol is first drained, the organic matter resembles the parent plant material. Thus the material contains high levels of polysaccharides and retains a major portion of the plant cellular structure. Much of the initial microbial activity is associated with the decomposition of these structural polymers. As decomposition of the more easily metabolized components progresses the cellular structure is lost. Reduction of the plant structure results in a decreased particle size which causes an increased soil bulk density. Since microbial activity is highest at the soil surface and declines with profile depth, greatest decomposition of the organic matter occurs at the soil surface, whereas soils at the base of the profile are more reflective of predrained conditions. Thus in drained Histosols, bulk density declines with profile depth. For example, in the Everglades Agricultural Area, the bulk density at the 0- to 5-cm depth is 0.35 g/cm^3 whereas the bulk density at a 60- to 65-cm depth is 0.16 g/cm^3 (Fig. 2.7) (Duxbury and Tate, 1981). Two other physical changes in the soil resulting from organic matter decomposition are a reduction of the carbon/nitrogen ratio and an increased ash content. During the initial microbial activity, the microbes metabolize carbon rich polysaccharides and immobilizing mineral nitrogen. As the available carbon pools are depleted, this net immobilization becomes a net mineralization. Ash accumulation results from conversion of organic compounds to gaseous and water soluble compounds. Associated mineral matter, such as silicates, salts, and clays, is retained within the soil profile. Thus the relative proportion of these soil components increases. Ash content was used by Broadbent (1960) as an indicator of the amount of microbial oxidation of soil organic matter and the extent of this oxidation throughout the soil profile. Microbial oxidation of the soil organic matter

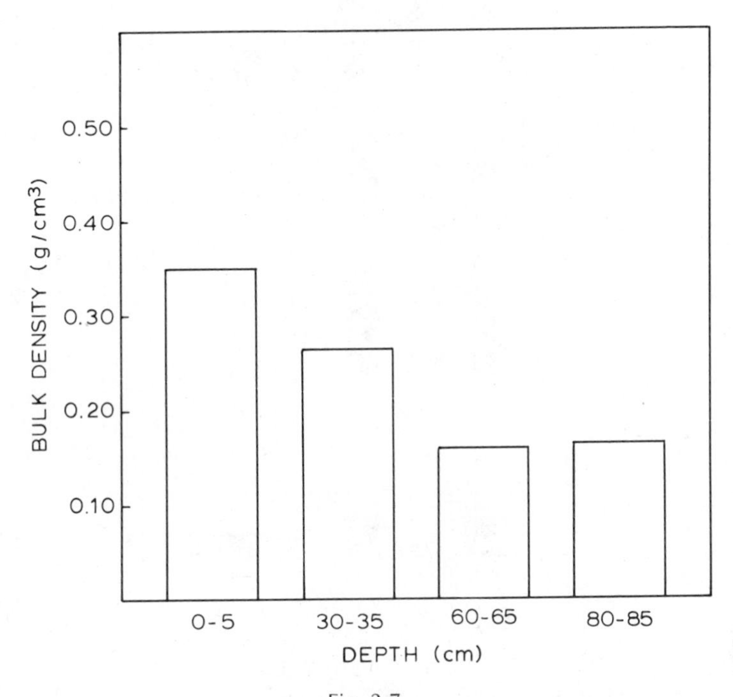

Fig. 2.7.
Change in the bulk density of Pahokee muck profile at the Agricultural Research and Education Center, Belle Glade, Florida.

was shown to occur throughout the soil profile to water table surface which was maintained at approximately 60-cm depth.

Although the nature of the carbon source metabolized by the soil microorganisms changes in mature drained organic soil, this occurrence of microbial decomposition throughout the soil profile above the water table still occurs. In drained organic soils, extensive decomposition has been found at all depths up to the 60- to 70-cm samples (Fig. 2.8) (Tate, 1980). As with any soil, the maximum depth of aerobic, carbon decomposition is determined by the balance between oxygen diffusion and microbial activity. Necessarily, if the soils were drained to greater depths, a point would be reached where aerobic activity would cease and anaerobic activity predominate. A gradual transition between predominantly aerobic and anaerobic growth patterns is suggested by the fact that although aerobic carbon oxidation was occurring at the 60- to 70-cm depths in the previous studies, facultatively anaerobic bacterial populations were increasing. This suggests that the occurrence of anaerobic microsites also increased with soil depth. Thus a gradient of decreasing aerobic activity and increasing anaerobic activity exists in these soils.

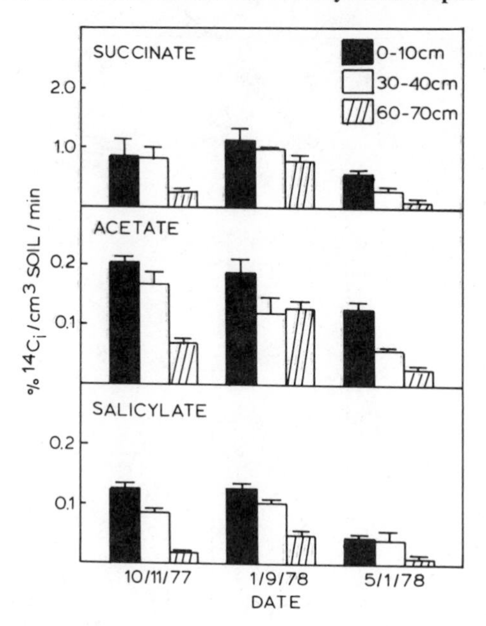

Fig. 2.8.
Microbial activity in fallow Pahokee muck versus depth of soil. (% $^{14}C_i$ = percentage of ^{14}C added at zero time evolved as $^{14}CO_2$) (Tate, 1980).

As the microbes continue to oxidize the native soil organic matter, its degree of humification increases. With increased humification of a carbonaceous substrate, the ability of the microorganisms to mineralize it declines. Thus as the bulk of the soil organic matter reaches a highly humified state, the microbial community becomes carbon limited. This situation exists in the Pahokee muck of the Everglades Agricultural Area. Pahokee muck contains greater than 80 percent organic matter (Terry and Tate, 1980), but most of it, because of the degree of humification, is unavailable or of limited availability to the microbial community. This is supported by the observation that inputs of carbon associated with crop growth increase microbial activity in these soils (Tate, 1979).

Limiting factors for microbial activity alluded to thus far are soil moisture/aeration and degree of humification of the carbon source. Since the organic matter of drained, humified organic soils is rather similar to that of mineral soils and the biochemical transformations of both indigenous and exogenous organic sources are the same, factors known to control or limit organic matter decomposition in other soil types would also be operative in Histosols. These would include pH, temperature, and presence of inhibitors in the decaying plant material. Ivarson (1977) found a 2.4- and 3.2-fold stimulation of carbon dioxide evolution of surface and subsurface peat soils, respectively, collected in Canada by adjusting the soil pH with lime from pH 3.8 to 7.9. A two-fold stimulation was found by adjusting the soil pH of both soils to 6.0. Stephens and Stewart (1976) used temperature relationships to explain the differences in microbial contributions to Histosol subsidence between temperate and subarctic soils and more soils from more tropical regions. They postulated that the microbial decomposition of the organic matter of these soils followed a typical Q_{10} relationship. Inhibitors can be quite effective in retarding microbial decomposition of peat-forming organic matter (Kavanagh and Herlihy, 1975). Inhibitors, such as the sphagnols of *Sphagnum*, greatly retard microbial decomposition of the organic matter.

2.5. CONCLUSIONS

Four widely divergent ecosystem types were evaluated in this analysis. Each has a significant organic matter input and characteristic equilibrium concentrations of colloidal soil organic matter. The magnitude and type of decomposer activity within each soil depends on environmental conditions that may be constant or seasonally variable, soil physical and chemical factors, ratio of easily decomposable to more degradation resistant carbonaceous substrates present in the plant debris, the C:N ratio of the plant debris, and its content of metabolic inhibitors. Although obvious differences are noted between the surface plant communities of these ecosystems both in composition and biomass productivity, an overall conclusion derived from this analysis is that the same physical and chemical properties control the rate of organic matter decomposition and accumulation. Variations in these rates relate to the specific combinations of limiting factors operative in a particular site. Thus by gaining a firm understanding of overall biochemical transformations of plant and animal remains and the native soil organic matter, including the effects of various chemical and physical factors on the rates of reaction and the basic ecological properties of the system of interest, a general idea or model of the organic matter transformations can be developed. The resultant general organic matter decomposition model may be "fine tuned" through incorporation of the particular chemical and physical values descriptive of the site of interest.

REFERENCES

Aber, J. D., and J. M. Melillo, 1980. Litter decomposition: Measuring relative contributions of organic matter and nitrogen to forest soils. Can. J. Bot. 58: 416–421.

Allison, R. V. 1956. The influence of drainage and cultivation on subsidence of organic soils under conditions of Everglades reclamation. Proc. Soil Crop Soc. Fl. 16: 21–31.

Anderson, J. M. 1973. The breakdown and decomposition of sweet chestnut (*Castanea sativa* Mill.) and beech (*Fagus sylvatica* L.) leaf litter in two deciduous woodland soils. II. Changes in the carbon, hydrogen, nitrogen and polyphenol content. Oecologia (Berlin) 12: 275–288.

Avnimelech, Y. 1971. Nitrate transformation in peat. Soil Sci. 111: 113–118.

Barber, S. A. 1979. Corn residue management and soil organic matter. Agron. J. 71: 625–627.

Berg, B., and B. Wessen, 1984. Changes in organic-chemical components and in growth of fungal mycelium in decomposing birch leaf litter compared to pine needles. Pedobiologica 26: 285–298.

Bloom, P. R., W. M. Shcuh, G. L. Malzer, W. W. Nelson, and S. D. Evans, 1982. Effect of N fertilizer and corn residue management on organic matter in Minnesota mollisols. Agron. J. 74: 161–163.

Bocock, K. L. 1963. Changes in the amount of nitrogen in decomposing leaf litter in sessile oak (*Quercus petraea*). J. Ecol. 51: 555–566.

Bohn, H. 1978. On organic soil carbon and CO_2. Tellus 30: 472–475.

Broadbent, F. E. 1960. Factors influencing the decomposition of organic soils of the California Delta. Hilgardia 29: 587–612.

Buldgren, P., D. Dubois, and J. Remacle, 1983. Principal component analysis applied to nutrient balances in organic layers of beech and spruce forests. Soil Biol. Biochem. 15: 511–518.

Coleman, D. C. 1973. Soil carbon balance in a successional grassland. Oikos 24: 195–199.

Dahlman, R. C., and C. L. Kucera, 1965. Root productivity and turnover in native prairie. Ecology 46: 84–89.

Dawson, J. E. 1956. Organic Soils. Adv. Agron. 8: 377–401.

Dormaar, J. F. 1975. Susceptibility of organic matter of chernozemic Ah horizons to biological decomposition. Can. J. Soil Sci. 55: 473–480.

Dormaar, J. R., and D. R. Sauerbeck, 1983. Seasonal effects on photoassimilated carbon-14 in the root system of blue grama and associated soil organic matter. Soil Biol. Biochem. 15: 475–479.

Duxbury, J. M., D. R. Bouldin, R. E. Terry, and R. L. Tate III. 1982. Emissions of nitrous oxide from soils. Nature (London) 298: 462–464.

Duxbury, J. M., and R. L. Tate III, 1981. The effect of soil depth and crop cover on enzymatic activities in Pahokee muck. Soil Sci. Soc. Am. J. 45: 322–328.

Florida Sugar Cane League, 1978. Water quality studies in the Everglades Agricultural Area of Florida. The Florida Sugar Cane League. Clewiston, FL.

Fogel, R., and K. Cromack, Jr, 1977. Effect of habitat and substrate quality on Douglas fir litter decomposition in western Oregon. Can. J. Bot. 55: 1632–1640.

Giddons, J. 1957. Rate of loss of carbon from Georgia soils. Soil Sci. Soc. Am. Proc. 21: 513–515.

Gosz, J. R. 1981. Nitrogen cycling in coniferous ecosystems. In F. E. Clark and T. Rosswall (eds.), Terrestrial Nitrogen Cycles. Ecol. Bull. (Stockholm) 33: 405–426.

Gosz, J. R., G. E. Likens, and F. H. Bormann, 1973. Nutrient release from decomposing leaf and branch litter in the Hubbard Brook Forest, New Hampshire. Ecol. Monogr. 43: 173–191.

Ivarson, K. C. 1977. Changes in decomposition rates in microbial population and carbohydrate content of an acid peat-bog after liming and reclamation. Can. J. Soil Sci. 57: 129–137.

Jenkinson, D. S. 1965. Studies on the decomposition of plant material in soil. I. Losses of carbon and C labelled ryegrass incubated with soil in the field. J. Soil Sci. 16: 104–115.

Johnson, F. L., and P. G. Risser, 1974. Biomass, annual net production and dynamics of six mineral elements in a post oak-blackjack oak forest. Ecology 55: 1246–1258.

Jorgensen, J. R., and C. G. Wells, 1973. The relationship in organic and mineral soil layers to soil chemical properties. Plant Soil 39: 373–387.

Jorgensen, J. R., C. G. Wells, and L. J. Metz, 1980. Nutrient changes in decomposing loblolly pine forest floor. Soil Sci. Soc. Am. J. 44: 1307–1314.

Kavanagh, T., and M. Herlihy, 1975. Microbiological aspects. Appl. Bot. 3: 39–49.

Keeney, D. R., and J. M. Bremner, 1964. Effect of cultivation on the nitrogen distribution in soils. Soil Sci. Soc. Am. Proc. 28: 653–656.

Kilham, O. W., and M. Alexander, 1984. A basis for organic matter accumulation in soils under anaerobiosis. Soil Sci. 137: 419–427.

Klein, D. A. 1977. Seasonal carbon flow and decomposer parameter relationships in a semiarid grassland soil. Ecology 58: 184–190.

Larson, W. E., C. E. Clapp, W. H. Pierre, and Y. B. Morachan, 1972. Effects of organic residues on continuous corn: II. Organic carbon, nitrogen, phosphorus, and sulfur. Agron. J. 64: 204–208.

Laws, W. D., and D. D. Evans, 1949. The effects of long-time cultivation on some physical and chemical properties of two Rendzina soils. Soil Sci. Soc. Am. Proc. 14: 15–19.

Lucas, R. E. 1982. Organic soils (Histosols): Formation, distribution and chemical properties and management for crop production. Research Report 435. Michigan State University Agricultural Experiment Station and Cooperative Extension Service. East Lansing, Michigan.

Mathur, S. P., H. A. Hamilton, and M. P. Levesque, 1979. The mitigating effect of fertilizer copper on the decomposition of an organic soil *in situ*. Soil Sci. Soc. Am. J. 43: 200–203.

Mathur, S. P., and A. F. Rayment, 1977. The influence of trace element fertilization on the decomposition rate and phosphatase activity of a mesic fibrisol. Can. J. Soil Sci. 57: 397–408.

Mathur, S. P., and R. B. Sanderson, 1980. The partial inactivation of degradative soil enzymes by residual fertilizer copper in Histosols. Soil Sci. Soc. Am. J. 44: 750–755.

McClaugherty, C. A., J. D. Aber, and J. M. Melillo, 1982. The role of fine roots in the organic matter and nitrogen budgets of two forested ecosystems. Ecology 63: 1481–1490.

McClaugherty, C. A., J. D. Aber, and J. M. Melillo, 1984. Decomposition dynamics of fine roots in forested ecosystems. Oikos 42: 378–386.

Meentemeyer, V. 1978. Macroclimate and lignin control of litter decomposition rates. Ecology 59: 465–472.

Millette, J. A. 1976. Subsidence of an organic soil in southwestern Quebec. Can. J. Soil Sci. 56: 499–500.

Moldenhauer, W. C., W. H. Wischmeier, D. T. Parker, 1967. The influence of crop management on runoff, erosion, and soil properties of a Marshall silty clay loam. Soil Sci. Soc. Am. J. 31: 541–546.

Neller. J. R. 1944. Oxidation loss of low moor peat in fields with different water tables. Soil Sci. 58: 195–204.

Nyhan, J. W. 1975. Decomposition of carbon-14 labelled plant materials in a grassland soil under field conditions. Soil Sci. Soc. Am. J. 39: 643–648.

Ojeniyi, S. O., and O. O. Agbede, 1980. Soil organic matter and yield of forest and tree crops. Plant Soil 57: 61–67.

Pastor, J., and J. G. Bockheim, 1984. Distribution and cycling of nutrients in an Aspen-mixed-hardwood-spodosol ecosystem in northern Wisconsin. Ecology 65: 339–353.

Reinhorn, T., and Y. Avnimelech, 1974. Nitrogen release associated with the decrease in soil organic matter in newly cultivated soils. J. Environ. Qual. 3: 118–121.

Schthorst, C. J. 1977. Subsidence of low moor peat soils in the western Netherlands. Geoderma 17: 265–291.

Schuurman, J. J. 1965. Influence of soil density on root development and growth or oats. Plant Soil 22: 352–374.

Shanks, R. E., and J. S. Olson, 1961. First year breakdown of leaf litter in southern Appalachian forests. Science (Washington, D. C.) 134: 194–195.

Shields, J. A., and E. A. Paul, 1973. Decomposition of ^{14}C-labelled plant material under field conditions. Can. J. Soil Sci. 53: 297–306.

Smith, R. M., D. O. Thompson, J. W. Collier, and R. J. Hervey, 1954. Soil organic matter, crop yields, and land use in the Texas Blackland. Soil Sci. 77: 377–388.

Stephens, J. C. 1969. Peat and muck drainage problems. J. Irr. Drainage Div. Am. Soc. Civil Eng. 95: 285–305.

Stephens, J. C., and E. H. Stewart, 1976. Effect of climate on organic soil subsidence. In Proceedings of the Anaheim Symposium. December 1976. Int. Assoc. Hydrological Sci. Publ. No. 121. Washington, D.C., pp. 647-655.

Stevenson, F. J. 1982. Humus chemistry. John Wiley & Sons, New York, 443 pp.

Stinner, B. R., D. A. Crossley, Jr., E. P. Odum, and R. L. Todd, 1984. Nutrient budgets and internal cycling of N, P, K, Ca, and Mg in conventional tillage, no-tillage, and old-field ecosystems on the Georgia Piedmont. Ecology 65: 354-369.

Tate, R. L., III 1977. Nitrification in Histosols: a potential role for the heterotrophic nitrifier. Appl. Environ. Microbiol. 33: 911-914.

Tate, R. L., III 1979. Microbial activity in organic soils as affected by soil depth and crop. Appl. Environ. Microbiol. 37: 1085-1090.

Tate, R. L., III 1980. Microbial oxidation of organic matter of Histosols. Adv. Microbial Ecol. 4:169-201.

Tate, R. L., III 1984. Function of protease and phosphatase activities in subsidence of Pahokee muck. Soil Sci. 138: 271-278

Terry, R. E. 1980. Nitrogen mineralization in Histosols. Soil Sci. Soc. Am. J. 44: 747-750.

Terry, R. E., and R. L. Tate III, 1980. Denitrification as a pathway for nitrate removal from organic soils. Soil Sci. 129: 162-166.

Tesarova, M., and J. Gloser, 1976. Total CO_2 output from alluvial soils with two types of grassland communities. Pedobiologia 16: 364-372.

Tiessen, J., R. E. Karamanos, J. W. B. Stewart, and F. Selles, 1984. Natural nitrogen-15 abundance as an indicator of soil organic matter transformations in native and cultivated soils. Soil Sci. Soc. Am. J. 48: 312-315.

Tiessen, H., and J. W. B. Stewart, 1983. Particle-size fractions and their use in studies of soil organic matter. II. Cultivation effects on organic matter composition in size fractions. Soil Sci. Soc. Am. J. 47: 509-514.

Tiessen, H., J. W. B. Stewart, and J. R. Bettany, 1982. Cultivation effects on the amounts and concentrations of carbon, nitrogen, and phosphorus in grassland soils. Agron. J. 74: 831-835.

Upadhyaya, S. D., and V. P. Singh, 1981. Microbial turnover of organic matter in a tropical grassland soil. Pedobiologia 21: 100-109.

Vitousek, P. M. 1984. Litterfall, nutrient cycling, and nutrient limitations in tropical forests. Ecology. 65: 285-298.

Vitousek, P. M., and P. A. Matson, 1984. Mechanisms of nitrogen retention in forest ecosystems: A field experiment. Science (Washington, D.C.) 225: 51-52.

Vogt, K. A., C. C. Grier, C. E. Meier, and M. R. Keyes, 1983. Organic matter and nutrient dynamics in forest floors of young and mature *Abies amabilis* stands in western Washington, as affected by fine-root input. Ecol. Monogr. 52: 139-157.

Weir, W. W. 1950. Subsidence of peatlands of the Sacramento-San Juaquin Delta, California. Hilgardia 20: 37-56.

White, E. M., C. R. Krueger, and R. A. Moore, 1976. Changes in total N, organic matter, available P, and bulk densities of a cultivated soil 8 years after tame pastures were established. Agron. J. 68: 581-583.

Whitehead, D. C., H. Buchan, and R. D. Hartley. 1975. Components of soil organic matter under grass and arable cropping. Soil Biol. Biochem. 7: 65-71.

Whitkamp, M. 1963. Microbial population of leaf litter in relation to environmental conditions and decomposition. Ecology 44: 370-376.

Witkamp, M. 1966. Decomposition of leaf litter in relation to environment, microflora, and microbial respiration. Ecology 47: 194-201.

Wood, T., F. H. Bormann, and G. K. Voigt, 1984. Phosphorus cycling in a northern hardwood forest: Biological and chemical control. Science (Washington, D.C.) 223: 391-393.

Zimenko, R. G. 1972. Activity of the microorganisms and the mineralization of the organic matter in peat soils with varying levels of subsoil water. Izv. Akad. Nauk Kaz. SSR Ser. Biol. 1972: 846-854 (in Russian).

THREE

BIOLOGICAL MEDIATORS OF SOIL ORGANIC MATTER TRANSFORMATIONS

Those biochemical reactions leading to the conversion of exogenous plant materials to humic substances, the mineralization of organic matter, and the modifications of soil structural elements as a result of these organic matter reactions are all the product of soil biotic activity. Although our attention is generally attracted by the microbial community because this portion of the biota is primarily responsible for the biochemical modification of organic materials, the pattern of organic matter metabolism characteristic of each ecosystem is the result of interactions of a variety of plant, microbial and animal species, and their products. Plants produce enzymes, such as peroxidase, which is involved in the humification process; microorganisms are the primary mediators of the biochemical mineralization and humification reactions; and the soil fauna affect decomposition rates not only directly through their metabolic processes but also indirectly through mixing of the plant debris throughout the soil A horizon, providing a source of fungal and bacterial inoculum, and physically disintegrating the organic debris. The emphasis of the following discussion will relate to the major biological groupings involved in soil organic matter metabolism and the problems associated with assessment of their activity *in situ*.

3.1. SOIL FAUNA

The actions of the soil fauna may have a dramatic effect on mineralization of plant debris and litter entering the soil ecosystem. Specifically identified interactions include (1) the physical mixing of the organic matter within the soil profile (this would include harvesting of organic debris by ants), (2) inoculation of the plant litter with decomposer populations, (3) adjustment of the soil physical properties to levels more conducive for organic matter decomposition, (4) physical disintegration of organic matter, such as is involved in the processing of organic matter within ant colonies, (5) direct metabolism of the organic components, and (6) stimulation of decomposer populations through interactions which increase or decrease activity at various trophic levels.

Ants and earthworms participate in all of these functions. Ant colonization thereby may lead to localized increases in native soil organic matter levels and fertility. Ofer et al. (1982) found that the soil mixing by harvester ants resulted in increased humic and fulvic acid contents in ant nests as well as greater soil fertility in the mounds. The latter effect was exemplified by the increased growth of pasture plants in the area of the ant mounds. Earthworms form long burrows and feeding caverns within the soil surface layer. Along with the effects of burrows and caverns on soil structure and water movement, soil particulate structure is modified through feeding by the worm (Martin, 1982). The species of earthworm active in a given soil, the extent of its activity, and the growth rate of the worm vary with soil organic matter content (Martin, 1982). Martin found that *Allolobophor trapezoides* (Duges) was capable of growth in soils with lower organic matter than was *Lumbricus rubellus* (Hoffmeister). Interestingly, the pattern of burrow formation varied with the quantities of soil organic matter, but the earthworm growth rate was unchanged. This suggests a feeding compensation by the earthworm that maintains its growth rate in soils with reduced organic matter contents.

The impact of soil invertebrates on surface litter decomposition is generally quantified by measurement of litter decomposition in litter bags either containing chemical inhibitors of animal activity, or designed to exclude or include animals from the litter decomposer population. Bocock (1964), using coarse mesh nylon net bags, found that litter decomposition by large invertebrates, such as eathworms and millipedes, was more rapid in a soil with a mull humus than in a soil with moder humus. During the initial five months of decomposition, these animals accounted for 40 percent decomposition of ash (*Fraxinus excelsio*) litter on the mull site, but only 10 percent on the moder site. Seastedt and Crossley (1983) used the inhibitor naphthalene to exclude arthropods from decomposition of dogwood (*Cornus florida* L.) litter. Their study demonstrated that the soil fauna not only stimulated the litter decomposition rate but their activity also resulted in increased nutrient concentrations of the decomposing litter. Thus soil fauna do have a major impact on litter decomposition rates in temperate sites. Somewhat contrasting conditions apparently exist in tropical forests in that although litter fall and decomposition rates are greater in tropical than in temperate forests, animal populations tend to be lower (Lee, 1974). In these latter ecosystems, microbial decomposition combined with leaching account for most of the litter breakdown. This is not a general property of tropical ecosystems because termites have a significant impact on litter decomposition rates in tropical savannas. Lee concluded that mixing of soil profiles by burrowing animals and competition for plant biomass between termites and grazers are significant factors in energy and nutrient cycling in tropical savannas.

Soil and litter fauna may accelerate litter decomposition either directly by digesting litter components themselves or indirectly through feeding on populations of litter decomposers or populations of organisms that limit activity of litter decomposers. The woodlouse, *Porcellio scaber*, provides an example of direct involvement in litter decomposition. This organism feeds on decaying pine needles. In fact, it prefers more decayed pine needles. This isopod consumed fewer brown needles and more yellowish decayed needles, with dark brown and blackish needle consumption occurring at an intermediate rate (Soma and Saito, 1983). The structural toughness of the needles appeared to be the major needle property determining feeding value. A variety of soil invertebrates including an isopod (*Oniscus asellus*), a millipede (*Pseudopolydesmus serratus*), a slug (*Deroceras reticulatum*), a snail (*Oxychilus drapanaldi*), and an earthworm (*Eisenia foetida*) metabolize cinnamic acid and vanillin, both decomposition products and precursors of lignin, to carbon dioxide (Neuhauser et al., 1978). This is not a universal capability of these invertebrate groups in that other isopods were incapable of this activity. A further chemical modification of plant products by animals or their products relates to synthesis of humic substances through the action of invertebrate synthesized peroxidases. These enzymes are involved, in part, in the polymerization of aromatic compounds associated with humification (Neuhauser and Hartenstein, 1978).

The energy derived from the decomposition of soil organic matter supports the activity of a number of trophic levels of soil organisms. Major activity relating to biochemical transformations are conducted by the soil bacterial, fungal, and actinomycete populations. These populations are fed on by, among other secondary feeders, protozoa and nematodes, which serve as food sources for soil invertebrates, such as mites. Hence, changes in the feeding pattern of these latter populations indirectly control the biochemical catabolism of soil organic matter as well as productivity of the plant community. The latter effect is most pronounced in ecosystems where aboveground biomass production is closely linked to organic matter mineralization rates. Also, this interaction of soil biota may stimulate plant inputs to the soil ecosystem not only through those effects associated with total biomass production but through stimulation of root processes. Acid phosphatase synthesis by blue gramagrass [*Bouteloua gracilis* (H. B. K.) lag. ex Steud] growing in hydroponic cultures is stimulated by the presence of root bacteria (*Pseudomonas* sp.) and a combination of bacteria and amoebae (*Acanthamoeba* sp.) (Fig. 3.1) (Gould et al., 1979). Since proteolytic activity was detected in the solutions containing the bacteria, the actual phosphatase activity produced may have been underestimated. Bacterial populations increased in association with the grass roots but amoebal populations declined. This activity did not result in a direct benefit to the plants in the hydroponic cultures because no significant effect of the bacterial and/or amoebal populations on the plant weight was observed. The induction of the new phosphatase activity did not result from microbially catalyzed increases in phosphate concentrations since the bacteria and amoebal populations did not affect the solution phosphate concentrations. The effect of the microbes is apparently a direct stimulation of plant enzyme synthesis. Gould et al. noted that bacterial populations stimulated phosphatase synthesis, but generally enzyme synthesis was greatest in those cultures containing the plant, bacteria, and amoebae. Sim-

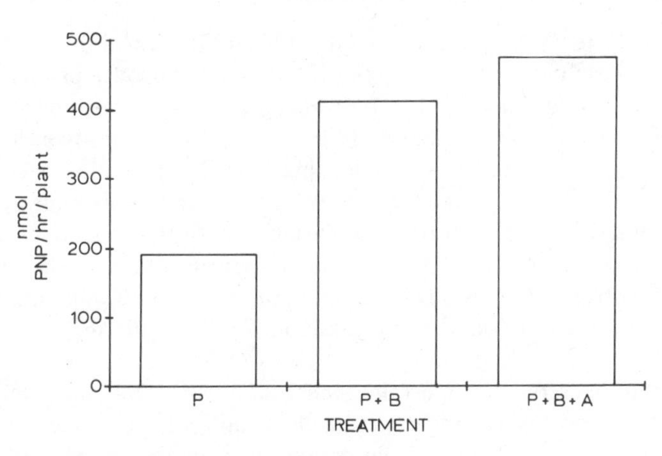

Fig. 3.1.
Alkaline phosphatase activities in root culture solutions in the presence in the absence of bacteria or amoebae (P), or in the presence of bacteria (P + B) or bacteria plus amoebae (P + B) (Gould et al., 1979).

ilar effects of soil microbes are associated with elevated organic carbon exudation in the presence of soil microbes (Barber and Lynch, 1977).

As primary decomposers grow and reproduce in soil—mainly bacterial, actinomycete, and fungal populations—nutrients, which exist in the mineral form and thus are available to the plant community, are immobilized into microbial cells. These nutrients therefore become unavailable to the plant until the microorganism dies and its cell substance is mineralized. Cycling plant nutrients through microbial biomass provides a role in nutrient cycling for those populations feeding upon the bacteria. Anderson et al. (1981) directly demonstrated this in their study of the effect of nematodes (*Acrobeloides* sp. and *Mesodiplogaster lheritieri*) on substrate utilization and nitrogen and phosphorus mineralization in soil. Both of the nematode species increased carbon metabolism by soil microbes (as measured by ^{14}C-labeled glucose conversion to ^{14}C-carbon dioxide) and nitrogen and phosphorus mineralization. But, total carbon metabolized and total quantities of inorganic phosphorus and nitrogen formed over the 65 day incubation period were no different in the presence or absence of nematodes. Thus the grazing nematodes, while not affecting the total decomposition of the ecosystem, did affect the kinetics of the decomposition; that is, they acted to maintain the dynamic nature of the nutrient cycles. Total carbon mineralization is controlled more by the quantities of organic matter present and its susceptibility to biological decomposition. Thus in an ecosystem receiving a single input of organic matter, the interactions between the two trophic levels of biota would primarily affect the time necessary for the ecosystem to return to an equilibrium state. In soils receiving steady inputs of organic matter, the metabolic rate can be anticipated to be higher when exhibiting the more complex trophic interactions.

A tri-part system was studied by Santos et al. (1981) and Santos and Whitford (1981). They evaluated the interaction of microarthropods, primarily tydeid mites, bacteriophagic nematodes, and bacteria in decomposition of creosote bush litter in a Chihuahuan desert. Elimination of the mites from the system increased the populations of nematodes that were feeding on the bacteria. This in turn caused a decline in the number of bacteria and a resultant decrease in litter decomposition. In contrast, elimination of both the nematodes and mites directly increased bacterial numbers. The 40-percent reduction in decomposition resulting from elimination of the mite populations indicates that the mites were affective in controlling litter decomposition by regulating the size of the bacterial grazing populations.

These selected reports support the general impression that, although primary emphasis has been given to the direct biochemical processes associated with organic matter decomposition and the microorganisms involved in the catalysis, soil animals are major contributors to the dynamics of the ecosystem. Thus any model or conceptualization of the overall ecosystem must include an assessment of the animal component and its direct and/or indirect interactions with the decomposer populations and soil organic matter.

3.2. SOIL MICROBES

The biochemistry of microbial interactions in soil is well studied, and the major role of these organisms in colloidal organic matter metabolism is unquestioned. The biochemical mechanisms of the organic matter reactions and the types of reactions will be discussed in greater detail in subsequent chapters. But first, because of difficulties in collecting data representative of microbial activities within a soil ecosystem, some analysis of the common problems in soil sample collection and storage for biochemical analyses is necessary. The objectives at this point relate more to the special problems associated with the measurement of these activities in soil and determination of what to measure; that is, how to decide which microbial parameters are indicative of the kinetics and properties of the soil organic matter reactions of interest. Most of the difficulties associated with soil microbiology result from the relatively small size of the microbes and their metabolic diversity. Hence, discussions in this section will have the objective of answering the following questions: (1) How can soil samples that are representative of the processes occurring in the overall ecosystem be collected, and (2) How can the microbial activity in these samples be quantified?

3.2.1. Soil Sampling for Microbial Activity

Once experimental objectives have been established so that the type of soil sample required for implementation of the project is known, the most critical decision in any study of soil microbial activity relates to the actual sample collection, storage, and analysis procedures. The problems associated with soil sampling are derived from the uneven distribution of microbial propagules and colonies, and the variation in microbial nutrient sources within the soil, and the fact that microbial activities may change as a result of sample collection and/or storage procedure. Each of these difficulties has to be addressed for each soil and biological parameter to be measured.

The heterogeneity of the soil ecosystem was discussed in Chapter 1. Choice of soil sampling procedures for soil biochemical analyses and the inherent problems are controlled in part by the breadth of the study proposed. Less concern for sample size and degree of representation of the overall ecosystem is required if microsite properties are to be studied than if an overall estimate of ecosystem function is desired. The problems can be enormous for ecosystem activity estimates. The size of a representative sample must be found. For more homogeneous soils this could consist of a few grams of soil per sample, whereas a site-descriptive sample size may not in reality exist for a heterogeneous soil system. Sample collection may involve selection of a series of composited samples from each distinct ecosystem type with in the site followed by replicated assays of biochemical activities within each composite.

Once truly representative samples are collected they must be treated in a manner to preserve the field levels of activity. Many studies of soil microbial activity have been invalidated by improper sample handling procedures. It is relatively easy to repeatedly assay an activity until the standard deviation is reduced to an acceptable level. More time is required to assure that the values recorded in reality represent the *in situ* activity. Of course, one means of avoiding this difficulty is to conduct assays *in situ* at an undisturbed site. This procedure is amenable to collection of higher plant or animal population densities or estimation of gaseous eminations from soils, but unfortunately it is impractical for many microbiological assays. Thus the only option available for the assessment of most microbial activities is to collect soil samples for laboratory analysis. Data inaccuracies can result from the collection of nonrepresentative soil samples as described previously plus inappropriate sample storage and processing techniques. Major difficulties encountered involve storage temperature, air-drying verses field moist conditions, and grinding of samples to decrease heterogeneity. Each of these procedures has been repeatedly discussed in the literature. Examples of studies of each are provided in the following text as an indication of the types problems that can be encountered.

Storage temperature variation has an unpredictable effect on microbial activity parameters. Activities may increase, decrease, or remain unchanged. The magnitude as well as the nature of the impact of storage temperature varies with both the microbial activity measured and soil type. The same microbial activity may be affected differently by storage temperature depending upon the soil type. The activity may decrease due to inability of the microorganisms to survive under the storage conditions or as a result of instability of the enzymes themselves. Increases in biochemical parameters frequently relate to organic matter availability. Disruption of the soil structure during sampling increases the availability of organic matter to microbial attack. Should the microbial activity assessed be one that is proportional to growth of the microbial populations stimulated by sampling, then immediate assays are necessary or the samples must be stored under conditions that retard this microbial growth. If microbial growth were the only cause of artifacts due to storage, then the development of appropriate storage conditions would be rather simple; that is, all samples would be stored at temperatures inhibitory to microbial growth, for instance 4°C. Unfortunately some microbial populations, especially fungi, are active at 4°C. Freezing of the sample would preclude microbial growth but soil structure would be disrupted and occluded organic matter liberated. Availability of organic matter is also increased by the death of microbial cells due to freezing. As is clear at this point in the discussion, there is no set rule for sample storage. Optimum storage temperatures or even the possibility of soil storage must be determined for each enzymatic or microbial activity to be studied.

Selection of the optimum sample handling procedure must not only relate to the specific activity to be measured but also the assay procedure itself must be considered. Ross et al. (1980) evaluated the effect of 28- and 56-day storage periods at 25, 4, and −20°C on four soils with three procedures for estimating microbial biomass (biomass carbon, adenosine triphosphate (ATP) concentrations, and mineralizable nitrogen). An *a priori* conclusion would be that all biomass measures would respond similarly to storage temperature. Biomass carbon estimates changed at all storage temperatures, but the change was least at −21°C. Storage under frozen conditions also was optimal for estimates of biomass using ATP analysis, but mineral-N values were most consistant at 4°C. Ross et al. concluded that no storage condition was adequate for all procedures but, for short periods of time, 4°C was best. Storage at 25°C was totally unsuitable for preservation of field microbial biomass levels. This variant effect on the three biomass estimates is understandable when it is realized that the indicator of biomass used for each of the assays is independently controlled within the cell. Since a variety of enzymes is involved in regulating the concentrations of each measure of microbial biomass and these enzymes have their own characteristic temperature optima, the parameters respond differently to environmental perturbations. Thus although one true biomass value is representative of the level in the field, laboratory estimates of it could vary considerably due to sample manipulation and assay procedure selected. Soil microbial biomass is so sensitive to changes in soil conditions that some have chosen to avoid sample storage all together (Tate and Terry, 1980).

A standard procedure in many soils laboratories is air-drying and sieving of soil samples prior to storage. This is acceptable for many chemical and physical analyses, but it is totally unacceptable for most, if not all, microbiological analyses. Bartlett and James (1980) note that drying and storage tended to push a soil that was metastable towards increased surface acidity, reduced manganese, and increased solubility and oxidizability of the soil organic matter. Reformation of metastable soil is a slow process involving the interaction of the abiotic and biotic soil components. Basically these data indicate that the soil ecosystem is in a dynamic state defined by the interactions of the abiotic and biotic ecosystem components. Air-drying disrupts both the living and the nonliving soil components, and thereby causes a complete change in the soil ecosystem. Although the air-dried–remoistened sample represents a soil ecosystem, as a result of the changes which occurred during drying it may or may not resemble the original soil sample. In the air-dried ecosystem, microbial activity is controlled by the organic matter liberated during drying and the newly established physical and chemical conditions, not necessarily those previously existing in the field. Hence, the data collected may not even resemble the situation existant under field conditions. The current consensus among soil microbiologists is that field moist soil be stored at temperatures determined to result in minimal changes in microbial activity, and that the soil not be air-dried except in cases where air-drying is unavoidable. For example, some degree of drying is frequently necessary for the manipulation of sediment or flooded soil samples.

Associated with the air-drying of soil samples is the effects of grinding on microbial and nonmicrobial organic matter in soil. Powlson (1980) measured the effect of grinding of native soils and of chloroform-fumigated soils. The increase in activity associated with grinding was less in the fumigated soils than was found with the native soils. The data suggest that the flush of decomposition resulting from grinding was derived from both killed microorganisms and non-biomass portions of the soil organic matter. Grinding killed about 25 percent of the microbial biomass.

These studies adequately demonstrate the care needed in sampling soil ecosystems for microbial activity. In many cases, *in situ* analysis is preferable, if possible. In others, laboratory analysis of field samples is possible, but for best estimates of field activity, the assays must be performed immediately. Frequently, data interpretation must also involve consideration that changes in activity due to sampling were unavoidable. Each study reported as well as each experiment attempted must be planned with these limitations in mind.

3.2.2. Indicators of Soil Microbial Activity

Microbial parameters used to estimate the microbial activity associated with soil organic matter transformations range from a variety of general assays, such as, population densities, respiration, and biomass, to assays of specific enzymatic activity. The primary prerequisite of any of these procedures is that the data collected reflect actual field conditions and not potential activity which may or may not be expressed under the limitations generally existant in the field. This is particularly significant in the more general techniques in that much of the microbial population may be inactive or in a resting state *in situ*. Thus the questions associated with evaluating microbial activity in soil involve deciding which measurement is most representative of those reactions or processes of interest and developing techniques for determining the deviation in the level of activity detected from the actual activity *in situ*. The difficulty in reaching a valid solution to these problems is clearly demonstrated with the procedures used to estimate total soil microbial activity. The first impulse in selecting a procedure to quantify general microbial activity is simply to count the number of bacteria, fungi, or other convenient soil microbial population. An assumption behind this procedure is that the microbial activity exhibited within the ecosystem must be proportional to the number of individuals present. This would most probably be valid in studies of plant or animal populations, but it is rarely valid in studies of soil microorganisms. Situations where the assumption may hold, at least to some degree, relate to drastically disturbed soil sites or soils where new microbial populations are being developed. In this latter case, conditions could dictate that essentially all of the microbes present would be active. In steady-state ecosystems, the microbial population consists of both living and resting, or inactive microbial cells. Assuming that it were possible to extract and enumerate all of the microorganisms in a soil sample, which again is improbable, the data would still be complicated by the inability of simple population counts to separate active from resting microbes. Several studies have involved the evaluation of population densities of varying physiological groups of bacteria in soils (e.g., see Kauri, 1983). Since these data are subject to similar limitations as isolation of general microbial populations, care must also be used in their interpretation. Changes in population density are of greater value in situations where the population growth is linked specifically to a given process. For example, autotrophic nitrifiers can only attain their energy from the oxidation of ammonium or nitrite. Hence, presence of these microbial populations is reasonable evidence for the occurence of nitrification. Unfortunately, because of limitations in extracting the nitrifiers from soil and the difficulties imposed by the nitrifer enrichment medium, population densities frequently do not relate directly to the amount of nitrification detected (Dommergues et al., 1978).

Similar or equally troublesome problems are frequently associated with techniques for estimating biochemical activities in soil. The remainder of this discussion will involve presentation of a variety of techniques used to estimate microbial activity in soil. In each case, the capability of overcoming the limitations associated with simple population counts will be discussed.

Probably the simplest technique for quickly assessing the overall activity of the soil community is to measure carbon dioxide evolution rates. Since carbon dioxide is a terminal oxidation product of aerobic respiration, measurement of carbon dioxide flux rates provides an indication of aerobic carbon mineralization. The procedure is limited first by those factors associated with estimation of gaseous eminations from soil and mathematical manipulations involved with extrapolation of the data to the total ecosystem and second by the problem of separating microbially respired carbon dioxide from that produced by other soil sources. Such measurements may be conducted *in situ* by measuring the rate of carbon dioxide accumulation under soil aovers (Fig. 3.2) or discrete soil samples can be collected and analyzed. With the former, respiration of the total biotic community is measured, whereas in the latter samples, root and animal contributions can be excluded. Data collected with discrete soil samples may overestimate microbial respiration, in that occluded organic matter becomes available to the microbial community during sample collection and preparation thereby resulting in increased microbial respiration rates. The sensitivity of the procedure may also be limited by the technique for quantifying the carbon dioxide, titration methods being less sensitive than infrared gas analysis or gas chromatographic separation of the soil gases. A variety of modern modifications of the procedure for application to specific soil ecosystems have been developed (Coleman, 1973; Dejong and Schappert, 1972; Edwards, 1982). The values collected provide an indication of the total respiration of the soil site. Correlation of these data with other measures of microbial activity within the same site provide a picture of the rate of function of the biological community and the factors limiting this function. Vogt et al. (1980) related monthly field carbon dioxide levels and forest floor ATP concentrations with seasonal variation in litter weight loss in four forest ecosystems in western Washington. Soil ATP levels provides an estimate of total soil biomass (Ausmus, 1971). Vogt et al. found that soil biomass levels peaked in the winter and spring months in all four ecosystems, whereas carbon dioxide evolution maxima were found in summer and autumn—times when temperatures may have been more conducive to maximizing microbial and root respiration. Since ATP of both active and resting microorganisms would have been measured, it is logical that the two peaks of activities did not coincide.

The ATP measurements discussed in the preceding text provided an indirect estimate of the soil microbial biomass. This type of procedure has the advantage that essentially all of the ATP in the soil sample can be accounted for whereas with viable counts, 90 percent or more of the microorganisms may not be extracted from the soil sample. Indirect estimates of microbial biomass commonly used are the ATP procedure (Ausmus, 1971) and the chloroform fumigation procedure (Jenkinson and Powlson, 1980). Both procedures rely on analysis of a specific microbial cell component (total cell carbon, nitrogen, or phosphorus for the fumigation procedure) and conversion of the data to values representative of the total biomass. Both techniques are limited by the conversion factor to be used; that is, actual measurements must be corrected for recovery efficiencies. Also a primary assumption for each procedure is that the component measured has a constant concentration in all living microbial cells. But, cellular ATP concentration varies depending upon the nutrient state of the cell and the fumigation procedure is limited by the availability of the cell carbon to microbial metabolism following fumigation. The availability of the carbon (if nitrogen or phosphate are measured, the extractability) varies with the propensity of the component to be adsorbed by soil clays or organic matter. Vorney and Paul (1984) found that using a correction factor of 0.41 for cell carbon and no correction for an unfumigated control provided the best estimate of the microbial biomass with the chloroform procedure. Jenkinson and Powlson (1980) originally proposed use of an unfumigated control correction of the data. Since microbial biomass levels can change rapidly and are limited by a number of factors that can be relieved during sample collection, preparation and storage, both measures of microbial biomass can vary considerable with minor changes in sample handling procedure. Ross and Tate (1984) found effects of incubation time, soil

Fig. 3.2.
Measurement of carbon dioxide fluxes from a forest soil site.

moisture and mesh size and preincubation procedures on the biomass estimate with the fumigation procedure. Thus they indicate that experimental procedure must be considered in interpreting and/or comparing biochemical indices of microbial biomass and their ratios attained with different or modified experimental procedures.

A variety of direct observation methods have been developed for analysis of microbial populations in soil (see review of Foster and Martin, 1981). Due to the rather undistinctive morphology of most microbial species, except where specialized techniques have been developed to allow idenitification of individual microbial species, direct observation of microbial cells is primarily of value in the observation of spatial variation (Baath and Sonderstrom, 1982) or in estimation of microbial biomass (Witkamp, 1974; Van Veen and Paul, 1979). Immunofluorescence procedures have been useful in the identification and enumeration of a variety of microbial species in soil. The utility of immunofluorescent techniques in autecological studies of soil microorganisms has been recently reviewed (Bohlool and Schmidt, 1980). This method has been used to study *Rhizobium* population changes in soil (Bohlool and Schmidt, 1973), nitrifiers in soils (Fliermans et al., 1974; Fliermans and Schmidt, 1975) and a variety of other bacteria and fungi in soil and water (Bohlool and Schmidt, 1980). Limitations of this useful procedure relate to its high specificity and the limited view of the soil ecosystem provided by analysis of a number of individual microscopic fields. Once the population density of either the total microbial populations or individual species, as allowed by the fluorescent antibody techniques, has been determined, the biovolume is calculated in a step in developing an estimate of the microbial biomass. Conversion of the biovolume measurements to biomass estimates requires knowledge of the specific gravity of the microorganisms. van Veen and Paul (1979) found the specific gravity of microorganisms to be quite variable depending on the moisture stress conditions. For example, with moisture held at 34, 330 and 1390 kPa suction, the specific gravity of fungi ranged from 0.11 to 0.41 g/cm^3. Their data indicated that moisture stress, which would be common within the soil ecosystem, affected the carbon, nitrogen, and phosphorus contents of the microorganisms as well as the ash contents. Their data indicated that existing data cannot be used to develop factors for the conversion of biovolume to biomass.

Most of the aforementioned procedures provide a measure of the relative contribution of bacterial and fungal populations to total microbial biomass, but this does not necessarily relate to the relative contributions of the respective populations to overall soil respiration. Use of specific metabolic inhibitors has proven useful in differentiating fungal and bacterial respiration in soil samples. Most commonly, streptomycin and actidione are used to inhibit bacterial and fungal populations, respectively. In their studies of a variety of soils, Anderson and Domsch (1972, 1975) noted through the use of these inhibitors that fungal populations provided for the majority of the soil respiratory activity. In contrast, Nakas and Klein (1980) found that bacteria contributed 40 to 80 percent of the glucose mineralization capacity from soils from the rhizosphere-rhizoplane of a semiarid grassland.

The aforementioned studies indicate the difficulties associated with analysis of a variety of microbial activities involved with soil organic matter transformations and the basis for the caution that must be exercised in analysis or comparison of soil microbiology data from divergent studies. The major concern in evaluating any of the procedures or data collected with any of the procedures relates to the potential of introduction of artifacts as a result of sample preparation and handling. Populations that are not active *in situ* may be stimulated during sample preparation and storage, whereas active populations may senesce. The populations analyzed may in reality represent a minor portion of the active biomass. Also, inhibitors may work quite well in one soil type and be totally inactivited in a second due to adsorption onto soil mineral or organic components. The researcher continually must question whether the parameters measured are truly representative of the *in situ* activity and are in reality measuring the activity of interest.

3.3. CONCLUSIONS

The biotic interactions controlling modification and mineralization of soil organic matter involve several trophic levels, including microbial, plant, and animal populations. The degree of participation of any single population is dependent on the physical and chemical properties of the ecosystem limiting the microbial activities. Greatest difficulties in the study and quantification of organic matter transformations at the population and biochemical level relate to selection of a biotic parameter whose activity is proportional to the transformation of interest and collection of a representative sample. The problems stem from the extreme heterogeneity of most soil systems and the minute size of microbial populations; that is, neither the substrates for microbial growth nor the microbial colonies themselves are evenly distributed. Thus great care is necessary in collecting sufficient quantities of a representative soil sample to produce statistically significant data. Collection of representative samples is particularly important if extrapolation of the activity measured to the total ecosystem is desired. Minor variations in precision at the subsample level make ecosystem-wide analyses difficult. Because of the general limitation of microbial activity *in situ* as a result of carbonaceous nutrient availability and the changes in the availability of this nutrient during sampling and storage, significant differences in microbial populations and their activities are found in stored soils, especially if the soils are air-dried and/or ground. Each of these factors must be considered in evaluating experimental data. These problems are of even greater significance when comparing activities between ecosystems because effects of variation in sampling handling procedures varies with soil types.

REFERENCES

Anderson, R. V., D. C. Coleman, C. V. Cole, and E. T. Elliott, 1981. Effect of the nematodes *Acrobeloides* sp. and *Mesodiplogaster lheritieri* on substrate utilization and nitrogen and phosphorus mineralization in soil. Ecology 62: 549–555.

Anderson, J. P. E., and K. H. Domsch, 1973. Quantification of bacterial and fungal contributions to soil respiration. Arch. Microbiol. 93: 113–127.

Anderson, J. P. E., and K. H. Domsch, 1975. Measurement of bacterial and fungal contributions to respiration of selected agricultural and forest soils. Can. J. Microbiol. 21: 314–322.

Ausmus, B. S. 1971. Adenosine triphosphate—A measure of active microbial biomass. Biologie Sol 14: 8–9.

Baath, E. and B. Soderstrom, 1982. Seasonal and spatial variation in fungal biomass in a forest soil. Soil Biol. Biochem. 14: 353–358.

Barber, D. A., and J. M. Lynch, 1977. Microbial growth in the rhizosphere. Soil Biol. Biochem. 9: 305–308.

Bartlett, R., and R. James, 1980. Studying dried, stored soil samples—some pitfalls. Soil Sci. Soc. Am. J. 44: 721–724.

Bocock, K. L. 1964. Changes in the amounts of dry matter, nitrogen, carbon and energy in decomposing woodland leaf litter in relation to the activities of the soil fauna. J. Ecol. 52: 273–284.

Bohlool, B. B., and E. L. Schmidt. 1973, Persistence and competition aspects of *Rhizobium japonicum* observed in soil by immunofluorescence microscopy. Soil Sci. Soc. Am. Proc. 37: 561–564.

Bohlool, B. B., and E. L. Schmidt, 1980. The immunofluorescence approach in microbial ecology. In M. Alexander (ed.), Adv. Microbial Ecol. 4: 203–241.

Coleman, D. C. 1973. Compartmental analysis of "total soil respiration": An exploratory study. Oikos 24: 361–366.

De Jong, E., and H. J. V. Schappert, 1972. Calculation of soil respiration and activity from CO_2 profiles in soil. Soil Sci. 113: 328–333.

Dommergues, Y. R., L. W. Belser, and E. L. Schmidt, 1978. Limiting factors for microbial growth and activity in soil. In M. Alexander (ed.), Adv. Microbial Ecol. 2: 49–104.

Edwards, N. T. 1982. A time saving technique for measuring respiration rates in incubated soil samples. Soil Sci. Soc. Am. J. 46: 1114–1116.

Fliermans, C. B., and E. L. Schmidt, 1975. Autoradiography and immunofluorescence combined for autecological study of single cell activity with *Nitrobacter* as a model system. Appl. Microbiol. 30: 676–684.

Fliermans, C. B., B. B. Bohlool, and E. L. Schmidt, 1974. Autecological study of the chemoautotroph *Nitrobacter* by immunofluorescence. Appl. Microbiol. 27: 124–129.

Foster, R. C., and J. K. Martin, 1981. *In situ* analysis of soil components of biological origin. In E. A. Paul and J. N. Ladd (eds.), Soil Biochem. 5: 75–111.

Gould, W. D., D. C. Coleman, and A. J. Rubink, 1979. Effect of bacteria and amoebae on rhizosphere phosphatase activity. Appl. Environ. Microbiol. 37: 943–946.

Jenkinson, D. S., and D. S. Powlson, 1980. Measurement of microbial biomass in in tact soil cores and in sieved soil. Soil Biol. Biochem. 12: 579–581.

Kauri, T. 1983. Fluctuations in physiological groups of bacteria in the horizons of a beech forest soil. Soil Biol. Biochem. 15: 45–50.

Lee, K. E. 1974. The significance of soil animals in organic matter decomposition and mineral cycling in tropical forest and savanna ecosystems. Trans. Int. Congr. Soil Sci. 10th. 3: 43–51.

Martin, N. A. 1982. The interaction between organic matter in soil and the burrowing activity of three species of earthworms. Pedobiologia 24: 185–190.

Nakas, J. P., and D. A. Klein, 1980. Mineralization capacity of bacteria and fungi from the rhizosphere-rhizoplane of a semiarid grassland. Appl. Environ. Microbiol. 39: 113–117.

Neuhauser, E. F., and R. Hartenstein, 1978. Reactivity of soil microinvertebrate peroxidases with lignins and lignin model compounds. Soil Biol. Biochem. 10: 341–342.

Neuhauser, E. F., R. Hartenstein, and W. J. Connors, 1978. Soil invertebrates and the degradation of vanillin, cinnamic acid, and lignins. Soil Biol. Biochem. 10: 431–435.

Ofer, J., R. Ikan, and O. Haber, 1982. Nitrogenous constituents in nest soils of harvester ants *Messor-Ebeninus* and their influence on plant growth. Commun. Soil Sci. Plant Anal. 13: 737–748.

Powlson, D. S. 1980. The effects of grinding on microbial and non-microbial organic matter in soil. J. Soil Sci. 31: 77–85.

Ross, D. G., and K. R. Tate, 1984. Microbial biomass in soil: Effects of some experimental variables on biochemical estimations. Soil Biol. Biochem. 16: 161–167.

Ross, D. J., K. R. Tate, A. Cairns, and K. F. Meyrick, 1980. Influence of storage on soil microbial biomass estimated by three biochemical procedures. Soil Biol. Biochem. 12: 369–374.

Santos, P. F., J. Phillips, and W. G. Whitford, 1981. The role of mites and nematodes in early stages of buried litter decomposition in a desert. Ecology 62: 664-669.

Santos, P. F., and W. G. Whitford, 1981. The effects of microarthropods on litter decomposition in a Chihuahuan desert ecosystem. Ecology 62: 654-663.

Seastedt. T. R., and D. A. Crossley, Jr, 1983. Nutrients in forest litter treated with naphthalene and simulated throughfall: A field microcosm study. Soil Biol. Biochem. 15: 159-165.

Soma, K., and T. Saito, 1983. Ecological studies of soil organisms with reference to the decomposition of pine needles. II. Litter feeding and breakdown by the woodlouse, *Porcellio scaber*. Plant Soil 75: 139-151.

Tate, R. L. III, and R. E. Terry, 1980. Variation in microbial activity in Histosols and its relationship to soil moisture. Appl. Environ. Microbiol. 40: 313-317.

Van Veen, J. A., and E. A. Paul, 1979. Conversion of biovolume measurements of soil microorganisms, grown under various moisture tensions, to biomass and their nutrient content. Appl. Environ. Microbiol. 37: 686-692.

Vogt, K. A., R. L. Edmonds, G. C. Antos, and D. J. Vogt, 1980. Relationship between CO_2 evolution, ATP concentrations and decomposition in four forest ecosystems in western Washington. Oikos 35: 72-79.

Vorney, R. P., and E. A. Paul, 1984. Determination of K_C and K_N *in situ* for calibration of the chloroform fumigation-incubation method. Soil Biol Biochem. 16: 9-14.

Witkamp, M. 1974. Direct and indirect counts of fungi and bacteria as indexes of microbial mass and productivity. Soil Sci. 118: 150-155.

FOUR

SOIL ENZYMES AND ORGANIC MATTER TRANSFORMATIONS

The preceding discussions related to those environmental factors controlling or delimiting overall decomposer population dynamics and interactions within the soil ecosystem. These interactions necessarily include physical as well as chemical aspects. But, a primary concern within this ecosystem is the chemical modifications of organic matter, including humification as well as decomposition, once it has entered the soil ecosystem. Although such transformations may be biologically or chemically mediated, the primary emphasis of the following discussion will involve the biological processes. (For elucidation of the chemical reactions, see Stevenson (1982) and Chapter 8). The biological catalysts of soil organic matter reactions are soil enzymes. An appreciation of the primary importance of soil enzymes in soil organic matter transformations and the impact of these reactions on the ecosystem as a whole is grained by examining the practical implications of some of the more recent studies of soil enzymes. Enzymatic activity has been used to evaluate the chemical components of humic acids (Mathur, 1971). Phenolase and peroxidase have been demonstrated to be primary mediators of humic acid and humic acid-like polymer synthesis (Martin and Haider, 1980; Sulfita and Bollag, 1980; Sjoblad and Bollag, 1981). Amidases convert soil amides to fertilizer nitrogen (Cantarella and Tabatabai, 1983), whereas a number of enzymatic activities are associated with pesticide decomposition in soil (Burns and Edwards, 1980), including peroxidase (Bordeleau and Bartha, 1972).

As we examine these soil enzyme studies, it becomes clear that the basic philosophy behind the experimental design differs from that previously encountered. Two views of soil dominate the assumptions behind most soil biological research. First, soil samples may be reduced to the individual components and the properties of each isolated component evaluated. Data so collected are then reassembled with the goal of reconstructing a picture of the total ecosystem. These types of studies are more commonly associated with biological population studies. Second, the soil may be studied as a whole. For these experiments, the soil is considered to be a single tissue with the individual microbial populations and enzymes as components of that tissue. Because of the difficulty involved with isolation of distinct enzymes from soil, this is the predominant nature of soil enzyme work. Some limitations are obviously associated with this type of research. First, since purified enzymatic activities are not isolated, it is not possible to clearly determine the number of isozymes contributing to the activity in question, although, as will be discussed in the following text, enzyme kinetic studies have provided some measure of enzyme diversity *in situ*. Associated with this inability to truly disassociate most individual isozyme activities from the soil matrix is an enhancement of the complexity of interpretation of enzyme reaction rate data due to interference with enzymatic activities by associated microenvironmental variation; that is, substrate concentrations, pH, ionic strength, or inhibitor levels measured in the bulk soil generally differ from the conditions in the proximity of the individual enzyme molecules. Evaluation of the impact of soil enzymatic activity on soil organic matter transformations is quite complex and, in many cases, rather speculative. Thus the objective of this chapter is to evaluate the basic properties of soil enzymes and how these relate to soil organic matter transformations *in situ*. For a more complete review of general soil enzyme research, see reviews by Skujins (1967; 1976), Kiss et al. (1975), or Burns (1978).

4.1. BASIC PROPERTIES OF SOIL ENZYMES

The validity, interpretation, and generalizations that can be made from any soil enzymatic analysis depend upon a basic appreciation for general properties of soil enzymes. Problems in data analysis relate to the laboratory methods used, enzyme source within the soil ecosystem, environmental (especially in relationship to the microenvironment) influences, stability of the enzymatic activity *in situ*, interactions of the enzyme molecules with clays and soil organic matter, and the resultant effect of each of these processes on overall enzyme kinetics. The objective of this section is to evaluate the importance of each of these factors on basic soil organic matter research.

4.1.1. Laboratory Analyses of Soil Enzymes

As a result of the molecular size of enzymes and the heterogeneity of their location in soil, those difficulties associated with collection of representative soil samples for microbial analysis discussed in Chapter 3 also affect evaluation of soil enzymatic activity, especially should an ecosystem-wide estimate of the activity be desired. Microsite enzyme activity levels vary with the type and location of organic matter activity of the microbial community and the physical composition of the soil sample.

As with sample collection for microbial population enumeration, a major difficulty in soil enzymatic analysis involves sample storage. Some enzymatic activities are reasonably stable under most common storage conditions, whereas others are not. Urease activity provides an example of one of the more stable soil enzymatic activities. Urease activity in 10 soils with a wide range of chemical and physical properties was found to be affected minimally by freeze-drying (−60°C) for 60 hours, air-drying for 48 hours, oven-drying (55°C) for 24 hours, or storage at 20, 5, −10, or −20°C for periods up to three months. Activity in air-dried soils was stable up to one year in soils stored at 21 to 23°C (Zantua and Bremner, 1975a). Caution is still recommended in selection of storage conditions in that others have noted increases and decreases in this enzymatic activity in air-dried soils (Skujins, 1967). Since variation is noted in urease stability in various soils, even with these data as justification for most any storage procedure chosen, a control experiment should be conducted to verify urease stability in the specific soil type under study.

Urease activity generally is associated with extracellular activity. Hence, as discussed in the following text, reaction of urease activity with soil organic matter results in a highly stabilized activity. This stability contrasts to that associated with a cellular enzymatic activity, dehydrogenase. Dehydrogenase activity, since it relies on an intact cellular electron transport system, is not found outside the microbial cell. This activity therefore is quite sensitive to any changes in the soil environmental conditions which would stimulate or retard microbial metabolic rates. Any factor relieving limitations to microbial growth results in increased dehydrogenase activity. For example, increases in organic matter availability necessarily cause increased dehydrogenase activity. Air-drying of samples prior to assay of dehydrogenase activity drastically affects the quantities of this activity present in the soil sample. Ross (1970) found that over half of the dehydrogenase was lost during air-drying of New Zealand top soils collected from pastures of grasses and clovers. Storage of the air-dried soils at room temperature increased this loss of activity. Although air-drying of the soil would be expected to increase the availabiltiy of native soil organic matter (thus microbial activity would be stimulated in the remoistened soil), the decline in dehydrogenase activity most likely results from death of microbial populations due to desication. Simple storage of fieldmoist samples at room temperature for periods of less than 77 days proved to be quite satisfactory, although the results were variable with this storage procedure. Freezing or storage at 4°C appeared to be most satisfactory for sample storage. Validity of these "apparently" acceptable storage procedures is limited by the changes in microbial activity occurring during or following storage. Field moist soils containing pools of newly available organic matter experience a burst of microbial activity. Similarly, the organic matter made available from the death of a portion of the microbial population during a freeze/thaw cycle stimulates microbial growth and metabolism during any subsequent incubation period. Thus under either of these storage conditions, dehydrogenase activity could be overestimated. Tate and Terry (1980a) found that in the Histosol, Pahokee muck, immediate assay of dehydrogenase activity was necessary for estimation of field activities.

Many of the difficulties associated with accurate assays of enzymatic activity in soil samples relate to the inability to free the enzyme from its association with soil particulate material. Both extracellular enzymatic activity and cellular activity generally is linked to various soil colloidal fractions, such as humic substances and clays. Soil organic matter fractions retaining various enzymatic activities have been isolated, but in most cases these enzymes are still bound to the soil humic materials. Urease, phosphatase, and protease activities have been extracted from soils with a variety of techniques which solubolize soil organic matter (McLaren et al., 1975; Lloyd, 1975; Nannipieri et al., 1975, Nannipieri et al., 1980). Although some enzymatic activities are removed from soil with water, the best procedure appears to be the use of sodium pyrophosphate as an extractant. Ceccanti et al. (1978) found that the most effective purification of soil urease involved exhaustive ultrafiltration of the sodium pyrophosphate extract against 0.1 M sodium pyrophosphate (pH 7.1) followed by gel chromatography. Urease specific activities were increased 6.9 to 18 fold over that of the initial soil extract. Similarly, although Ladd (1972) found protease activity from a highly organic Mount Gambier soil to be water extractable, increased extraction efficiency (50 to 75 percent of the activities of unextracted soil) was attained with tris-borate buffer. Peroxidase, an enzymatic activity implicated in soil humic substance synthesis, is analyzed routinely in water or buffer soil extracts (Bartha and Bordeleau, 1969).

Each of those extraction procedures involving purification of those enzymatic activities bound to humic substances is limited by the ability to separate the enzyme from its associated humic component. This was recognized by Chalvignac and Mayaudon (1971) when they noted the association of the brown humic substance color with their sodium carbonate extracted, ammonium sulfate purified tryptophane metabolizing enzymes. This intimate linkage of the enzymatic activity with various fractions of soil organic matter was also noted by Cacco and Maggioni (1976) when examining multiple forms of acetyl-naphthyl esterase activity in soil extracts. Polyacrylamide gel isoelectric focusing of enzyme extracts demonstrated several peaks of enzymatic activity, each associated with humic isoelectric bands.

Even with optimum sample selection and storage procedures, data interpretation and applicability may be limited by the assay procedure. Continued debate occurs relating to activity optimization and the development and use of uniform techniques. Malcolm (1983) stresses the necessity of adoption of procedures "obeying the rules governing simple enzyme kinetics" and urges the adoption of more uniform assay techniques. The first suggestion is obligatory for sound study of soil enzymes, whereas utilization of uniform assay procedures is unlikely. Greater complexity and diversity of soil systems than is usually found in cellular extracts necessitates a certain degree of adaptation of the procedures to the soils to be studied. Thus although an optimum pH may have been shown for tissue enzymatic activity, simple use of a buffer of that pH in the soil enzyme assay will not assure that the selected pH value occurs at the site of enzyme activity. Clay particles and organic matter interactions frequently alter the soil microsite pH. Similarly, assay substrate concentrations must be adjusted so that the enzyme is saturated but not inhibited by excess substrate. The reaction rate must be proportional to enzyme concentration and not the substrate. This requirement is met reasonably easily with tissue extracts but in complex soil systems, charged or aromatic substrates may adsorb or even covalently bind to various soil components. This necessitates use of substrate levels higher than are generally required in more defined systems. Thus optimization of the assays in simple soils, such as sands, would require less substrate than in heavy clay soils where adsorption would be more of a problem.

Of perhaps greater significance once it is determined that the procedures selected do not violate basic enzyme assay principles, is whether an estimation of potential enzyme activity is desired or if some measure of the actual field enzyme activity is the objective. With the former goal, the assay procedure should be optimized to provide maximum, or total, enzyme activity present, whereas for the latter objective, the assay conditions would be designed to mimic the actual *in situ* soil conditions as closely as possible. The basis for existence of these two "activities" is the harsh, heterogeneous soil microenvironment. Enzymes retained within cellular structures operate under more favorable conditions than those outside of the cell wall. This was demonstrated when Zantua and Bremner (1975b) compared urease activity in 16 soils in buffered and unbuffered assay mixtures. They found that buffering of the assay mixture yielded significantly higher activities (i.e., potential activities) than was detected with unbuffered mixtures (i.e., actual field rates). Greater assay precision was found with the unbuffered solutions. This would be expected because optimization of the assay pH value would allow for a more uniform environment for the enzymatic activity to be expressed. They felt that the unbuffered assay provided a better estimate of urea hydrolysis under native conditions in that the assay mixture pH approximated that of the soil sample.

When field levels of enzymes are to be estimated, duration of the assay is also a problem for those procedures requiring long incubation times. For most enzyme assays, a metabolizable substrate is added to the soil under conditions conducive for microbial growth. Thus for assays extending over periods of hours or days, the enzymatic activity measured may represent that initially present in the soil plus that synthesized by the growing microbial community. A number of procedures have been developed to overcome this incubation artifact. Perhaps the most common technique is to add toluene to the reaction mixture (Skujins, 1976). This is necessarily precluded for cell associated enzymatic activities because disruption of the cell membrane by the toluene interferes with the assay. Problems also are associated with resistance of a number of fungi and bacteria to toluene at concentrations of 0.05 to 0.1 percent (Kaplan and Hartenstein, 1979). Thus higher inhibitor concentrations must be maintained. Alternate methods include radiation sterilization of the soil sample and use of antibiotics to inhibit microbial growth (Skujins, 1976). Since in many cases the extended incubation period has been necessitated by the inability to detect small quantities of the reaction product, development of more sensitive assay procedures has allowed reduction of the incubation periods. Incorporation of liquid or gas chromatographic procedures into enzymatic assays have proven to be quite productive.

4.1.2. Source and *In Situ* Location of Soil Enzymatic Activity

The types of enzymatic activities present within a given soil ecosystem are dependent on the metabolic capabilities of those organisms responsible for enzyme synthesis, the presence of chemical compounds for the induction of enzyme synthesis, and existence of the environmental physical and chemical conditions necessary for the expression of the enzyme activity. Soil enzymes are produced predominantly by the microbial community, but significant contributions may be made by plant and soil animal communities. The metabolic potential of the microbial community is selected to a certain degree by the environmental conditions; that is, a microbial species that relies on a certain metabolic process for its existence. For example, nitrification and the autotrophic nitrifiers would most probably not be present in a system where that process was prohibited. Exceptions would involve a highly variable site where compatible conditions for the process previously existed. In that situation, the microbial cells could exist in a resting state until conditions conducive for full activity again are existant. Hankin et al. (1974) examined the relationship of degradative enzyme synthesis by soil bacterial populations and soil use in a number of Connecticut soils. Soils were collected from cultivated lands, pasture, orchards, forests, forest litter, tidal marsh, and swamp ecosystems. Bacterial activities measured were cellulolytic, pectolytic, amylolytic, lipolytic and proteolytic activities. Protein and starch hydrolytic activity appeared to be related to the current soil use, whereas no influence was found on cellulose or pectin degradation ability by land use. Variations in cropping practices had no significant effect on the bacterial diversity in Pahokee muck, a drained, cultivated organic soil of South Florida (Tate and Mills, 1983). Similar bacterial diversity indices were detected for soils collected from fallow, sugarcane (*Saccharum* sp.), and St. Augustinegrass [*Stenatophrum secundatum* (Walt) Kuntz] fields. Comparison of the metabolic capabilities of the isolated bacteria strains with actual enzymatic activities of the soil samples from which they originated indicated that similar enzyme potentials existed in all three ecosystems and that the actual chemical and physical conditions *in situ* determined which of the enzymatic activities would be induced. This conclusion was verified by a study of the induction of new glucose degradative activity in Pahokee muck in comparison with microbial growth rate variation (Tate, 1985). New glucose respiratory activity was induced in the absence of significant increases in microbial growth rates suggesting enzyme synthesis in previously existing microbial cells.

Macroinvertebrates may also be significant sources of soil enzymatic activities (Hartenstein, 1982). Soil isopods, diplopods, molluscs, and oligochaetes were found to be sources of catalase, cellulase, and in some cases, peroxidase, aldehyde oxidase, and polyphenol oxidase. Greatest specific activity of peroxidase was found in the earthworms than in any other group which exhibited similar levels of this enzyme. This suggested a role for earthworms in the humification process. Similarly, earthworms and ryegrass (*Lolium perenne* L.) significantly increased respiratory and enzyme activities of subsoil of Judgeford silt loam collected from a pasture (Ross and Cairns, 1982). The presence of earthworms in the soil resulted in increased oxygen respiration, cellulase, and sulphatase activities. Generally, ryegrass resulted in augmentation of all biochemical activities, but the combination of both earthworms and ryegrass provided maximal invertase, amylase, urease, and phosphatase activities. The authors concluded that earthworms stimulated biochemical activities and nutrient cycling in the pastures and thereby enhanced pasture recovery after top soil removal.

Because of the multitude of sources of enzymatic activity and the uneven distribution of these activities within soil, the microsite location of these enzymes is also extremely variable. Enzyme molecules are associated with a variety of biotic and abiotic soil components. The latter components include colloidal humic substances and clay particles. Biotic components include active as well as resting cells and cellular debris [For further discussion of these various pools of soil enzymatic activity and their interaction, see Burns (1982)]. Hence, any measure of soil enzymatic activity is a composite of the reservoirs. Since those kinetic parameters which are descriptive of the basic enzymatic activity are directly affected by the enzyme microenvironment, soil enzyme kinetic constants in reality represent an average of the activities of a variety of enzyme forms rather than the value associated with any specific individual enzymatic activity. (These variable forms include isoenzymes as well as cellular, extracellular, organic matter bound, and clay associated forms.) For example, by examining changes in the urease activity with time and amount of microbial activity with multiple regression analysis of the resultant data, Paulson and Kurtz (1969) were able to separate cell associated urease activity from extracellular activity. They estimated by their procedure that 79 to 89 percent of the steady state urease activity was extracellular and associated with soil colloids. Through combination of this procedure with kinetic analysis, they demonstrated that the Michaelis constant for that activity associated with the microbial cells was 0.057 M, whereas as that associated with soil colloids was 0.252 (Paulson and Kurtz, 1970). These data demonstrate a decrease in efficiency of this enzyme with excretion from the cell and adsorption to soil colloids.

4.1.3. Environmental Influences on Enzyme Activity

As is clear from the aforementioned study of Paulson and Kurtz, the physical and chemical properties of the microenvironment around an enzyme can have major effects on both the amount of enzymatic activity expressed and the reaction kinetics. Factors affecting the enzymatic activity include binding of the enzyme to soil colloids, pH effects, interactions with metals, and effects of interstitial water ionic strength (salinity). The latter topics will be analyzed in this section. Effects of binding to soil colloids will be reviewed in the next section.

The complexity of the effects of soil microenvironment on enzymatic activity was demonstrated by a study of phosphatase activity in woodland soils (Harrison, 1983). Significant interactions between soil enzymatic activity and soil organic matter, moisture, clay plus silt contents, total nitrogen, isotopically exchangeable phosphorus, extractable magnesium contents, and soil pH were found. When all the data from soils of both 0 to 5 and 10 to 14 cm depths were analyzed, 66 percent of the total variation in phosphatase activity was explained by variation in these soil properties. Importance of each individual property on enzymatic activity varied with soil depth, soil type, season, vegetation type, and underlying rock type. These ecosystem properties accounted for 99 percent of the variation in phosphatase activity. Similarly, Galstian (1974) found that enzyme activity in Armenian soils depended upon soil acidity, calcareousness, particle-size distribution, temperature, adsorption, and seasonal dynamics.

Each enzymatic activity within a soil sample has a characteristic pH optimum. In a study of the effect of acidity on a variety of enzymatic activities of sod-podzolic soils, urease, protease, dehydrogenase, polyphenol oxidase, and catalase were shown to have pH optima between pH 6.3 and 7.2. Phosphatase and invertase activity were optimal in a pH range of 4.2 to 4.5 (Chunderova, 1970). Frankenberger and Johanson (1982) found that small changes in soil pH resulted in declines in urease, acid phosphatase, alkaline phosphatase, and phosphodiesterase activity by a reversible reaction involving ionization or deionization of acid or basic groups in the active center of the enzyme protein. With greater changes in the soil pH, irreversible denaturation of the enzyme protein occurred. The pH stability of the various enzymatic activities was shown to be highly dependent on soil type. This variation between soils was attributed to the diversity of vegetation, microorganisms, and soil fauna which served as enzyme sources as well as variation in the protective interactions within the soil matrix itself.

As a result of the use of trace minerals in fertilization of agricultural soils and the interest in using soils in waste disposal plans, considerable data has been accumulated that is descriptive of the effect of a variety of trace metals on soil enzymatic activities. In nearly every situation, these metals are inhibitory to soil enzymatic activity. This would be expected in that the activity of enzymes is dependent upon its tertiary structure. Any interaction that disrupts this structure either inhibits or totally inactivates the enzyme. In most cases this interaction of the metal with the enzyme molecule in a manner which alters the tertiary structure results in total loss of enzymatic function. Thus observation of a reduction of enzymatic activity in a particular soil sample treated with a metal mixture would not indicate partial inhibition of a specific enzyme molecule. Because of the soil heterogeneity, some of the enzyme molecules may be totally inactivated by the metal molecules, whereas other enzyme proteins do not even encounter the inhibitor and therefore are unaffected. This inactivation of enzymatic activity, in many cases, results from chemical binding the metal atom with the sulfhydral groups which are instrumental in both structural integrity and, in some reactions, the enzyme active site itself. Other inhibition effects include precipitation of cofactors (such as, precipitation of phosphate), substitution of an active metal ion with an inactive ion in the enzyme structure, and simple salt or ionic effects. In contrast with the effect of tertiary structure modification, these latter mechanisms may not inactivate totally the individual enzyme proteins, but merely modify the apparent Michaelis Constant (K_{app}) or maximum velocity

(V_{max}). Most trace elements have been shown to be inhibitory to enzyme activity to some degree. Juma and Tabatabai (1977) found greatest inhibition of acid phosphatase activity by Hg(II), As(V), W(VI), and Mo(VI) with average inhibition greater than 50 percent and that less than 10 percent inhibition was observed with Ba(II), Co(II), and As(III). Other trace minerals which also inhibited the enzymatic activity were Cu(I), Ag(I), Cd(II), Cu(II), Zn(II), Mn(II), Sn(II), Ni(II), Pb(II), Fe(II), Cr(III), Fe(III), B(III), Al(III), V(IV), and Se(IV). The inhibitory effect varied with soil type. Many of these metals are inhibitory in high concentrations but stimulating to phosphatase activity at lower concentrations. Dick and Tabatabai (1983) removed exchangeable and soluble metals from three soils by leaching with 1 N ammonium acetate (pH 8). They found that amendment of these soils with low concentrations of Ba(II), Ca(II), Co(II), Mg(II), Mn(II), Ni(II), or Zn(II) promoted phosphatase activity, whereas potassium and sodium amendment had no effect. Fe(II) and Cu(II) were still inhibitory to phosphatase activity. Due to physical and chemical differences between the soils, the concentration of metal ions for optimum activity varied between soil types.

A variety of enzymatic activities have been demonstrated to be inhibited by low levels of copper amendment (Schinner et al. 1980; Mathur and Sanderson, 1980). Mathur and Sanderson (1980) observed that in Histosols developed for agricultural usage in southern Canada, with copper contents ranging from 18 to 275 μg per g, C_1-cellulase, C_x-cellulase, cellobiase, xylanase, amylase, inulase, lichenase, and lipase enzymatic activities in air-dried and stored soil samples correlated negatively with soil copper content. This relationship was verified in freshly collected soil samples. The authors propose amendment of developed Histosols with low levels of copper as a means of controlling the soil subsidence resulting from the microbial oxidation of the soil organic matter.

Of related interest, especially in relationship to soil salinity problems, is the effect of saline or high ionic strength conditions on soil enzymes. Frankenberger and Bingham (1982) found that activities of amidase, urease, acid phosphatase, alkaline phosphatase, phosphodiesterase, inorganic pyrophosphatase, arylsulfatase, rhodonase, α-glucosidase, γ-galactosidase, dehydrogenase, and catalase activities declined as the soil salinity increased. The degree of inhibition varied among the enzymes assayed and the nature and amounts of the salts added. Dehydrogenase activity was severely inhibited by salinity, whereas the hydrolases were inhibited to a much lesser degree. The authors propose three mechanisms of inhibition of the enzymatic activity, (1) osmotic desiccation of microbial cells releasing intracellular enzymes which are vulnerable to proteolytic attack, (2) "salting out" effect resulting in a modification of the enzyme active site, and (3) specific ionic toxicities.

4.1.4. Protein Interactions with Colloidal Organic Matter and Clay

Binding of enzymes to clay or colloidal organic matter, specifically humic acids, may have essentially any conceivable affect on the enzymatic activity; that is, enzymatic activity may be increased, decreased, or left unchanged. The specific consequences of the binding of the enzyme protein depends on a variety of factors, including the degree of change in enzyme tertiary structure, the capability to maintain active site integrity, and changes in substrate availability. Three major mechanisms may explain changes in enzymatic activity following binding to clays or organic matter. These are classed as structural modifications, kinetic effects, and environmental effects.

Two means of structural modification may enhance or inhibit enzyme-substrate association of bound enzymes. Attachment of the protein to humic acids may occur through a covalent or hydrogen bonding mechanism that disrupts those associations stabilizing the protein tertiary structure. This modification of tertiary structure causes a change in active site three dimensional structure which in turn alters the capability of the substrate and/or product molecules to associate freely with the enzyme. Alternatively, the enzyme-colloidal particle association may be directly at or in the vicinity of the active site. This physical occlusion of the active site may retard or totally inactivate the enzyme. Binding near the active site may provide some degree of steric hindrance to the substrate–enzyme association while still allowing the reaction to occur, albeit at a reduced rate. Actual kinetic expression of this effect may be noted by changes in the Michaelis (K_m) and/or the V_{max}.

Kinetic effects of the binding of the enzyme protein to the colloidal material may result from the physical modification of the protein tertiary structure as indicated previously or by modification of substrate availability. For a reaction to occur at maximum efficiency, the approach of the substrate to the enzyme as well as the quantities of substrate available to the enzyme must be maintained at optimal concentrations. Enzyme activity may be reduced by an interaction with colloidal particles that impede approach of the substrate to the active site. Steric hindrance, as a mechanism of such impedance, has already been introduced. A potentially overlooked mechanism relates to the binding of the enzyme to the colloidal particle preventing the enzyme from diffusing to the site of substrate availability. Soil is a heterogeneous ecosystem. An extra-cellular enzyme must be capable of reaching sites of high substrate concentrations. Physical attachment of the enzyme to colloidal substances may preclude this mobility. This physical attachment may also increase the probability of enzyme-substrate encounter. This can be anticipated to occur with substrates which may bind to the clay or organic matter structures. Should these substrates be present in insufficient concentrations for efficient catalysis to occur, concentration of the substrates on clay or organic matter surfaces may create pools of substrates where enhanced enzyme reactions may occur.

The presence of charged clay and/or humic substances in a microenvironment may alter the chemical environment in a manner which increases or decrease enzyme reaction rates. For example, each enzyme has a pH range where the reaction rate is maximized. Thus a given reaction may not be expected to occur because of adverse soil pH, yet when this activity is assayed, significant levels of activity are detected. This anomalous observation is explainable by the

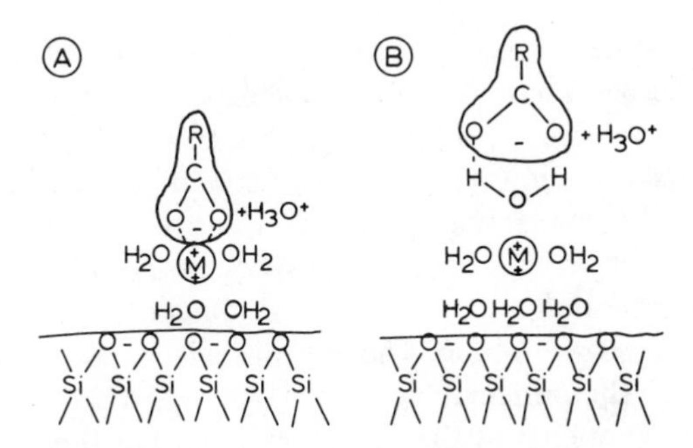

Interaction of anions with mica-type clay surfaces through polyvalent cation bridges.
(A) anion associated directly with cation
(B) anion associated with cation via a water bridge

Fig. 4.1.
Hypothesized interactions of humic anions with mica-type surfaces through polyvalent cation bridges (Greenland, 1971). Reprinted with permission by Williams and Wilkins Co., Baltimore.

modification of the microsite pH by the accumulation of divalent ions around the negatively charged colloidal particle, a mechanism which will be discussed in the following text as contributing to binding of proteins to colloidal particles.

Proteins may be associated with colloidal particles by a variety of mechanisms, including ionic associations, salt bridges, covalent and hydrogen bonding. Generally proteins are associated with clay particles through ionic bonds, where appropriate charges exist, and salt bridges. Associations between humic acids and proteins involve predominantly covalent and hydrogen bonding mechanisms.

At first glance, considering the fact that both proteins and clays generally carry a predominantly negative charge in most soils, any binding or physical attraction would seem to be unlikely. In studies where clay particles are positively charged due to the prevalent soil pH and with clays that are inherently positively charged as a result of their elemental composition, association of the negatively charge proteins is understood relatively easily. In the majority of the situations, the rather strong association between proteins and clays is explained by the existence of salt bridges between the two negatively charged bodies (Fig. 4.1). In this case, a divalent cation, such as calcium or magnesium, provides the link. A side effect of this binding mechanism is a reduction of the pH in the vicinity of the protein as a result of the interactions of the divalent metals with water.

The interaction of nonbound molecules with these clay associated proteins is clearly demonstrated through studies of their susceptibility to proteases. Although there may be some decline in the hydrolysis of the clay associated protein, such hydrolysis is not prevented (Estermann et al., 1959). Monolayers of lysozyme adsorbed on kaolinite and bentonite were digested readily by chymotrypsin, mixed soil microbial cultures,and pure cultures of *Bacillus subtilis*, *B. mycoides*, and *Pseudomonas* sp. The rate of hydrolysis was retarded in a protein–montmorillonite association. Marshman and Marshall (1981) found that the growth of microorganisms on gelatin, bovine serum albumin, and lysozyme adsorbed to montmorillonite or kaolinite was diauxic. Protein hydrolysis took place in the second growth phase. The effect of adsorption of the proteins onto the clay varied with the protein : clay ratio. High ratios had no effect on growth whereas at intermediate ratios the growth rate but not the final cell yield was reduced. At low protein:clay ratios, the protein was not available for hydrolysis.

A number of metabolic enzymes are adsorbed by clay minerals. The activity of glucose oxidase and invertase were found to be suppressed 1.5 to 11 fold by clay minerals (Zvyagintsev and Velikanov, 1968). Montmorillonite and kaolinite were found to rapidly adsorb bacteriolytic but not proteolytic enzymes produced by *Myxobacter virescens* (Haska, 1975). The bacteriolytic activity could be eluted from the clay with high pH or high ionic strength solutions. Contradictory results can be obtained when comparing the effect of clay minerals on enzymatic activities with pure clay preparations and native soil samples. Ross (1983) found that invertase and α-amylase activity was inhibited by a number of clay minerals. The order of inhibition was muscovite $<$ allophane $<$ illite $<$ montmorillonite. Also, all of the clay samples inhibited β-amylase activity. But, with top soils containing mica-vermiculite or mica-beidellite as the main clay component, no adsorption or inhibition was observed. The author concluded that it was unlikely that the clay minerals of the well developed top soils had any direct affect on the activity of freshly incorporated invertase or amylase.

The association of proteins with humic acids is more complex than was noted previously for their reaction with clay minerals. A variety of bonding mechanisms can be proposed. Since protein-humic acids linkages are chemically stable, covalent bonding is apparently a major mechanism involved with this association. Both hydrogen and covalent linkages easily are envisioned to occur. Covalent linkages would include predominantly peptide bonding between the N-terminal amino acid of the protein and carboxy groups of the humic acids. Other probable associations include peptide bonds between γ-amino nitrogen as well as, to some degree, peptide bond formation between free protein carboxyl groups and amino groups on the humic acids.

Rowell et al. (1973) constructed a number of model enzyme-humic acid analogues to determine the effect of binding of the protein on enzymatic activity. Not only were the K_m and pH optima affected by the association, without exception the complexed enzymes were less active than the free enzymes (Table 4.1). The proportion of activity varied with the nature of the complex. Interestingly, in all cases reported the K_m was lowered by the association. The shift in pH optima indicated the affect of the humic acid moieties on the ionic character of the microenvironment.

Table 4.1.
Relative Enzyme Activities and Properties of Lyophilized Enzyme-Humic Acid Analogue Complexes (Rowell et al., 1973).[a]

Enzyme or Complex	Relative Enzyme Activities[b]					With BAA Substrate		
			Protease					
	Peptidase (BAA)	Esterase (BAEE)	(Casein)	(Lactalbumin)	(Bovine Serum Albumin)	pH Optimum	Temperature Optimum (°C)	Michaelis Constant (K_m)
Trypsin	100	100	100	100	100	8.0	30	2.0×10^{-3} M
I-BQ trypsin	44	21	3	23	10	11.0	50	0.8×10^{-3} M
S-BQ trypsin	11	4	4	16	2	9.5	—	—
Pronase	100	100	100	100	100	9.0	45	9.0×10^{-4} M
I-BQ pronase	28	4	11	35	6	—	45	—
S-BQ pronase	68	29	68	54	31	9.0	40	2.9×10^{-4} M

[a]Reprinted with permission of Pergamon Press.
[b]Relative activities expressed as a percental ratio of the complexed to the appropriate free enzyme on a basis of equal nitrogen content. Free enzyme expressed as 100. (I-BQ enzyme = soluble benzoquinone polymer complex products; S-BQ enzyme = soluble product) 50,000 m.w.; BAA = benzoyl-L-arginineamide; BAEE = benzoyl-L-arginine ethyl ester).

As with the protein–clay associations, humic acid bound proteins are vulnerable to proteolytic attack (Sowden, 1970; Brisbane et al., 1972). It must be noted though that the sensitivity of the protein to hydrolytic attack varied with the protease examined. Brisbane et al. (1972) found that pronase and thermolysin released about one sixth of the acid-hydrolyzable amino acids contained in the humic acids, papain had slight activity, whereas chymotrypsin, trypsin, and subtilopeptidase were inactive. These data may be explained by a steric hindrance of the protease reaching the vulnerable peptide bonds of the substrate protein and to a certain degree, binding of the protease itself to the humic acids. Data of Ladd and Butler (1971) suggest a cation–exchange mechanism in which protease amino groups are linked to humic acid carboxyl groups as a mechanism of inhibition. Support for their conclusion was derived from data showing that acetyltrypsin and acetylcarboxypeptidase were unaffected by soil humic acids in concentrations that were markedly inhibitory to nonacetylated proteases.

4.1.5. Enzyme Stability

Individual enzyme proteins are stabilized through their association with soil organic matter, whereas overall soil enzymatic activity patterns are reasonably constant or stabilized by the physical, chemical, and biological constancy of a steady-state ecosystem. Proteins may acquire long-term resistance to denaturation conditions and proteolytic activity as a result of the protection of their tertiary structure through association with colloidal organic matter (Skujins and McLaren, 1969; Pettit et al., 1977). Measurable urease activities were detected in 8715- and 9550-year-old radiocarbon dated Alaskan permafrost samples. No activity was detected in a 32,000-year-old sample (Skujins and McLaren, 1969). This stabilization of enzyme molecules is a general characteristic of extracellular enzymes released into soils.

Accompanying this preservation of individual protein molecules is an overall constancy of enzymic parameters which appears to be descriptive of the specific soil sample. Under steady-state conditions, although the enzyme activities may exhibit some fluctuation, their activity pattern exhibited within a specific soil ecosystem remains essentially constant. Modification of the steady-state conditions may cause a decline or increase in a specific enzymatic activity, but if the original soil conditions are restored, the enzyme profile also returns to levels approximating the predisturbed levels. Thornton et al. (1975) suggested this soil enzyme trait as a potential mechanism for forensic identification of soil samples. They examined phosphatase, arylsulfatase, urease, invertase, and trypsin activities in three soil series (Dublin adobe clay loam, Columbia sandy loam, and Hanford sandy loam) and found that soils collected from sites within close proximity could be distinguished by their enzyme patterns. Also, the K_m of the various activities appeared to be useful in this analysis in that the value varied with soil type and was sample-size independent. This use of the enzyme kinetic parameters would reduce the quantities of soil needed to make a positive comparison.

4.1.6. Kinetic Properties of Soil Enzymes

Evaluation of the kinetic properties of a particular enzymatic activity in soils is a valuable tool for determination of the significance of that enzyme activity *in situ* and the properties that must exist for expression of the activity. Before evaluating some of the data that has been collected relating to enzyme kinetic parameters, some limitations of these measurements must be examined. As was indicated previously, it is rare to separate an enzymatic activity from adhering soil organic matter, especially humic acids. These associated substances have a highly significant effect on the ability of the enzyme to combine with substrate and catalyze the anticipated conversion to product. Thus even with partially purified enzyme preparations from soil it must be remembered that the data represent apparent K_m (K_{app}) and V_{max} values. Even minor variations in assay procedure may cause significant increases or decreases in the kinetic constants. This was demonstrated by Tabatabai and Bremner (1971) in their study of arylsulphatase and phosphatase activity in a variety of Iowa surface soils. When the incubation procedure did not involve shaking of the assay mixture the K_{app} value for arylsulfatase activity in the nine soils ranged from 1.37 to 5.69×10^{-3} M and the K_{app} for phosphatase activity ranged from 1.26 to 4.58×10^{-3} M. Shaking the assay mixture during the incubation decreased the K_{app} for both activities as well as the variation between soil types. Shaking of the assay mixture also resulted in an increase in the V_{max}. This effect of shaking would be anticipated considering the particulate nature of the soil being assayed. Settling of the soil sample to the bottom of the assay container would limit the free association of substrate and enzyme. They also found that the K_{app} was not correlated significantly with pH, cation-exchange capacity, percentage organic carbon, percentage clay, or percentage sand. Kinetic parameters have been recorded for a variety of enzymatic activities in a number of soils (e.g., see Frankenberger, 1983; Tabatabai, 1973; Frankenberger and Tabatabai, 1980). Although the K_m and V_{max} values are highly variable between soil types and are sensitive to variations in assay procedures, they do provide a measure of the quantities of enzyme present in various soil samples (V_{max}) and the optimal substrate concentration for enzymatic activity (K_m). The variation between soil types results from the heterogeneity of organic matter and enzyme distribution and those factors directly affecting enzyme and substrate interaction and overall enzyme stability, that is, organic matter and clay effects, microsite pH, ionic strength, and so on.

Nannipieri et al. (1982) used kinetic analysis of urease, phosphatase, and proteolytic activity in two soils to evaluate the occurrence of isozymes within the soil samples. As was indicated previously, the limitations of enzyme purification from soil samples preclude ready determination of the number of forms and thereby the number of potential enzyme sources within the soil ecosystem. Use of Eadie–Scatchard plots provided a graphical technique for detecting deviations from Michaelis-Menten kinetics. This deviation was observed as two or more straight lines in the graphical representation of the data. The number of straight lines represented the number of enzyme or enzyme groups catalyzing the reac-

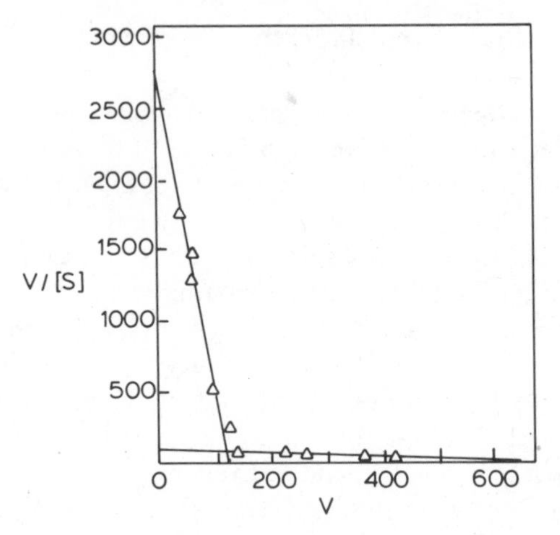

Fig. 4.2.
Eadie–Scatchard plot of benzoylarginineamide hydrolysing protease in Pania Sud soil extract. *V* and *S* are expressed as nmol NH_4^+-N released/hour/ml and as μM, respectively. (Nannipieri et al., 1982). Reprinted with permission of Pergamon Press.

tion (Fig. 4.2). The occurrence of two or more distinct straight lines in the plots is limited to cases where the V_{max} and K_m values of the enzymes are significantly different. The occurrence of enzymes with widely different kinetic constants within an ecosystem would be highly significant. The greater the range between the two K_m values, the wider the range of substrate concentrations that can interact with the enzyme. The lower the K_m the higher the affinity of the enzyme for the substrate. Thus the enzyme with the lower K_m would be active at lower substrate concentrations within the soil; therefore, different enzymes would be active within the soil sample depending upon the substrate availability. *In situ* substrate concentrations would determine which of the enzymes would be active.

4.2. ENZYME ACTIVITIES AND ECOSYSTEM DEVELOPMENT

The ideal situation for interpretation of soil enzymatic activity would be if soil microbial activity and soil enzyme activities were related directly. As will be indicated in the following text, in some situations this assumption is true, but generally the relationship between soil enzymatic activity and microbial growth and activity is complicated by the existence of stabilized extracellular enzymes and the inclusion of enzymes contained within inactive or resting cells in the total enzymatic activity measurements. Under most conditions, enzymatic activity directly associated with intact microbial cells, and not expressed outside of the living cell, can be anticipated to relate in part with overall microbial respiration. This conclusion is based on the observation that these intracellular proteins are highly susceptible to the denaturing conditions existent outside the cell. Attainment of highly significant correlations is limited by the quantities of such enzymes contained within inactive cells and the degree of activation of these enzymes during the enzymatic assay procedure. This situation is observed with dehydrogenase activity (see Casida, 1977; Skujins, 1973). When no exogenously supplied substrate is included in the assay mixture, dehydrogenase provides an estimate of actual microbial respiration under *in situ* conditions (Tate and Terry, 1980a; Tate and Terry, 1980b). Casida noted that increases of dehydrogenase activity in soils amended with limited concentrations of metabolizable carbon sources peaked earlier than did respiration as measured with a Warburg respirometer. He concluded that this discrepancy likely resulted from the excluding of carbon dioxide from the respiration measurements. Ladd and Paul (1973) found that proteolytic activity increased in soils amended with glucose at rates that coincided with bacterial population growth. Other enzymatic activities also relate to microbial activity under some conditions. Frankenberger and Dick evaluated the relationship between the activities of 11 soil enzymes and microbial respiration, biomass, viable plate counts, and soil properties of 10 diverse soils. Alkaline phosphatase, amidase, α-glucosidase, and dehydrogenase activities significantly ($p < 0.01$) correlated to microbial respiration in soils that had been amended with glucose. Phosphodiesterase, arylsulphatase, invertase, γ-galactosidase, and catalase also correlated at the 5 percent level, whereas acid phosphatase and urease were not significantly correlated to microbial respiration. None of the 10 enzymatic activities correlated with carbon dioxide evolution rates. Only phosphodiesterase and α-glucosidase were correlated to microbial numbers ($p < 0.05$). Alkaline phosphatase, amidase and catalase were highly correlated ($p < 0.01$) with microbial biomass. It must be noted that this study involved analyses of soils receiving a readily metabolizable carbon source. Thus the damping effect of enzymes contained within resting cells and stabilized enzymatic activity would have been overcome by the induction of microbial growth accompanying the glucose metabolism. Although it has been the general experience that little correlation is possible between most extracellular enzymatic activities and microbial activity in steady-state ecosystems, evaluation of changes in enzyme activities in disturbed ecosystems has proven to be useful in assessment of the impact of ecosystem disturbance on microbial activity. For example, Klein et al. (1985) have shown the value of such enzymatic activities in estimating toxic effects of mine residues on mineland reclamation.

Since soil enzymes are organic substances produced by a component of soil organic matter, a relationship between soil organic matter content and enzymatic activity is anticipated. This has been demonstrated for a number of enzymatic activity including nucleases (Arutyunyan and Abramyan, 1975) and urease (Myers and McGarity, 1968). Examples of exceptions to this observation are found in situations where the extracellular enzymatic activity and/or soil microbes are leached within the soil profile to depth containing little organic matter.

Types of enzymatic activities present within a given soil ecosystem and their levels of activity are controlled in part by the management techniques in anthropogenically disturbed ecosystems and by the successional stage in undisturbed sites. This is demonstrated dramatically by comparing enzyme changes in developing organic soils, by examining the impact of lycopodium fairy rings on soil enzyme activities, and by comparing the biochemical activities in the soil profile of forest soils.

A major factor controlling enzyme activity expression in organic soils is the degree of humification of the organic matter. These soils can contain as much as 100 percent partially decomposed plant and animal remains. When the soils are first drained for development, a major portion of the organic material has retained at least a portion of its original cellular structure and chemical composition. Thus soil enzyme synthesis reflects this high concentration of easily decomposable substrates. This was shown by Ross and Spier (1979) in their study of organic soils from subantarctic tussock grasslands on Campbell Island where they found high levels of enzymes associated with cellular polysaccharide decomposition. Invertase, amylase, cellulase, hemicellulase, urease, phosphatase, and sulphatase activities were all high in the top horizons of the organic soils collected from six sites on the island. Activities generally declined with soil depth at a rate more rapid than the soil organic carbon contents which would be expected in areas where deep deposits of Histosols exist because the depth of the organic matter layer reflects past accumulation of organic matter not products of soil formation factors such as those generally observed in mineral soils. The enzyme activities correlated negatively with the degree of peat decomposition. The correlations with soil carbon and total nitrogen contents were not significant. Although the soils contained a high degree of activity resulting from the content of decomposible organic matter, litter inputs from tussock grass (*Poa litorosa*) stimulated the activities of all enzymes studied except sulphatase activity which declined under the grass and urease activity which varied from site to site. A similar stimulation of acid phosphatase, invertase, xylanase, cellulase and amylase activities in the well-humified Pahokee muck of the Everglades Agricultural Area by plant inputs was observed (Duxbury and Tate, 1981) (Table 4.2). For each of these activities, cropping to sugarcane (*Saccharum* sp.), St. Augustinegrass [*Stenotaphrum secundatum* (Walt) Kuntz], or paragrass [*Brachiaria mutica* (Forsk.) Stapf.] resulted in significant increases in activities. Note that the greatest activity was found in the soil surface horizons where organic inputs from aboveground biomass and belowground root production would be the greatest. Peroxidase and polyphenol oxidase activities, two enzymatic activities implicated to participate directly in the humification process, were also stimulated by inputs of organic matter (Fig. 4.3, Table 4.3). Because of the low activities of the peroxidase in uncropped soil, the possibility was raised that plant roots may be the predominant source of this activity in Pahokee muck (Duxbury

Table 4.2.
Variation in Hydrolytic Enzymatic Activity in Pahokee Muck with Crop and Position in Soil Profile (Duxbury and Tate, 1981)[a]

Depth, cm	Treatment							
	Fallow		Sugarcane		St. Augustinegrass		Paragrass	
Acid Phosphatase, μmol p-NO_2 Phenol/hour/cm³ Soil								
0–5	1.24ab[b]	(a)	2.34a	(b)	5.39a	(c)	4.31	(cb)
30–35	1.58a	(a)	1.71ab	(ab)	2.83b	(b)	2.36ab	(ab)
60–65	0.84b	(a)	0.49bc	(a)	0.93c	(ab)	1.59b	(b)
80–85	0.12c	(a)	0.14c	(ab)	0.30c	(ab)	0.34c	(b)
Invertase, μmoles Reducing Sugar/hour/cm³ Soil								
0–5	1.08a	(a)	1.44a	(a)	4.04a	(b)	3.94a	(b)
30–35	0.73b	(a)	0.51b	(a)	0.47b	(a)	0.71b	(a)
60–65	0.07c	(a)	0.05c	(a)	0.14c	(a)	0.25bc	(b)
80–85	0.03c	(a)	0.03c	(a)	0.09c	(a)	0.07c	(a)
Xylanase, μmoles Reducing Sugar/hour/cm³ Soil								
0–5	0.086a	(a)	0.225a	(ab)	0.360a	(b)	0.443a	(b)
30–35	0.138a		0.179a		0.184b		0.315a	
60–65	0.027b		0.022b		0.041c		0.032b	
80–85	0.020b		0.030b		0.039c		0.023b	
Cellulase, μmoles Reducing Sugar/hour/cm³ Soil								
0–5	0.028		0.070		0.050		0.030	
30–35	0.060		0.071		0.050		0.030	
Amylase, μmoles Reducing Sugar/hour/cm³ Soil								
0–5	0.007		0.023		0.027		0.025	
30–35	0.022		0.028		0.020		0.047	
60–65	0.006		0.003		0.012		0.005	
80–85	0.005		0.004		N.D.[c]		0.002	

[a]Reproduced from SOIL SCIENCE SOCIETY OF AMERICA JOURNAL, Volume 45, 1981, pp. 322–328 by permission of the Soil Science Society of America.
[b]NOTE: Values are means of measurements over a one-year period and those without letters or followed by the same letter are not significantly different at the 5 percent level; letters without parentheses apply to the effect of soil depth for each crop separately and letters in parentheses apply to the effect of crop for each solid depth separately.
[c]N.D. = not detectable

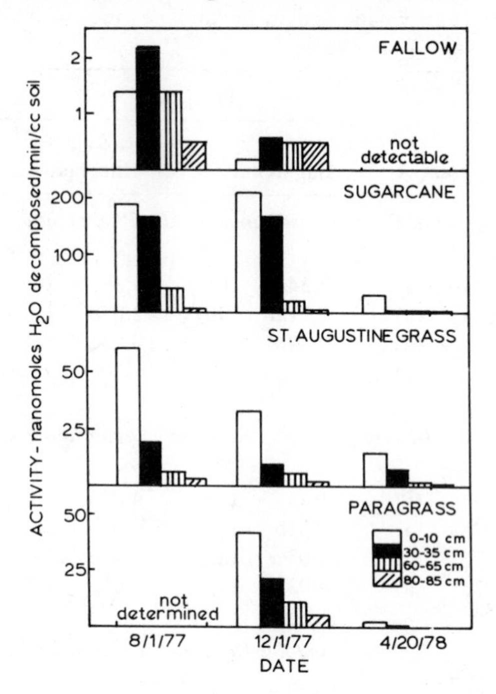

Fig. 4.3.
Effect of crop and position in the soil profile on peroxidase activity in Pahokee muck (Duxbury and Tate, 1981). Reproduced from SOIL SCIENCE SOCIETY OF AMERICA JOURNAL, Volume 45, 1981, pp. 322–328, by permission of the Soil Science Society of America.

and Tate, 1981), but further studies of the microorganisms of this soil and the kinetics of peroxidase activity induction in the soil indicated a microbial source for the activity (Mangler and Tate, 1982). A similar relationship of enzyme activity with degree of humification of the soil organic matter was noted in peat and muck soils of *Alnus glutinosa* communities in northwest Germany (Moller, 1979). In this case, the urease activity correlated with the degree of decomposition, soil pH, and carbon/nitrogen ratio.

Fungi of the *Lycopsium* species grow in soil as fairy rings. Spalding et al. (1975) examined the effect of this fungal growth on soil organic matter through assays of soil respiration and enzymatic activities. The active metabolism of the ring at its point of growth was shown by the augmented soil respiration under the ring as compared with soil either inside or outside of the ring. Phosphatase, protease, dehydrogenase, catalase, polyphenol oxidase, amylase, and dextrinase activities were independent of soil location. But, cellulase, invertase, polygalacturonase, and peroxidase activities were elevated under the fairy rings. These data demonstrate the differential stimulation of the soil enzymes by the variable nutrient inputs resulting from growth of this fungal structure.

Table 4.3.
Polyphenoloxidase in Pahokee Muck as Affected by Soil Depth and Crop (Duxbury and Tate, 1981)[a]

Soil Depth, (cm)	v_{max} Soil Source: Fallow	Sugar-cane	St. Augus-tinegrass	Para-grass
		μliters O_2/hour/cm^3 Soil		
0–5	59a (ab)[b]	44a (a)	82a (bc)	112a (c)
30–35	50a (ab)	49a (a)	64b (b)	91a (c)
60–65	21b (a)	61a (b)	35c (a)	20b (a)
80–85	32ab (a)	60a (b)	38c (ab)	28b (a)

[a]Reproduced from SOIL SCIENCE SOCIETY OF AMERICA JOURNAL, Volume 45, 1981, pages 322–328 by permission of the Soil Science Society of America.
[b]Values are means of measurements over a one-year period. Different letters indicate that values are significantly different at the 5 percent level; letters without parentheses refer to the effect of depth for each crop separately; and letters with parentheses refer to the effect of crop for each soil depth separately.

Invertase, amylase (Ross and Roberts, 1973), and polyphenol oxidase (Ross and McNeilly, 1973) activities were also shown to vary with factors limiting microbial activity in hard beech (*Nothofagus truncata*) forests of New Zealand. Enzyme activities declined throughout the soil profile. Maximal activities were detected in those months when soil conditions were optimal for microbial growth and activity. Invertase activities correlated positively with amylase activities in all soil horizons. Activities also correlated significantly with moisture and organic carbon in some of the horizons. An appreciable portion of the variation in activity of these two enzymatic activities was explained by variations in pH, moisture, organic carbon, and total polyphenols. Polyphenol oxidase activities increased in the soils with increased pH. Generally, this enzyme activity did not correlate with the other enzyme activities or could a significant portion of the variation in its activity be explained by pH, moisture, organic carbon, or total polyphenols. The polyphenol oxidase activity did correlate negatively and significantly with contents of catechol phenolics.

Analysis of these three examples of studies of soil enzymatic activity demonstrate the utility of these activities as indicators of microbial activity and thereby changes in soil organic matter fractions. The variability of the data between sample types, and the failure of various enzyme activities which *a priori* may be associated with microbial activity indicate the complications of data analysis resulting from stabilized enzymatic activities and activities residing in resting cells. It must be concluded that soil enzymatic analysis is a valuable tool for evaluation of soil organic matter transformations, but as a result of the complexity of the system, care and common sense must be exercised in interpretation and extrapolation of the results.

As an extension of these more basic studies, a common desire of those attempting to predict cropping success, reclamation success, or overall ecosystem development capabilities is to find an enzymatic activity predictive of aboveground plant productivity. Klein (1985) has indicated successful use of soil enzymes as predictors of reclamation success. Data relating a variety of enzymatic activities to soil fertility or crop yield is generally less favorable. Moore and Russell (1972) evaluated the potential for use of dehydrogenase activity as a general predictor of soil fertility. The dehydrogenase activity did not correlate with soil properties known to influence crop growth, did not change when soil nutrient status was adjusted from that known to be low to a more favorable condition for plant growth, and was highly variable between the various samples; that is, precision of the assays was low. Since dehydrogenase is a measure of general microbe respiration and many microbial activities which do not contribute to soil fertility directly occur in soil, it is reasonable to accept the failure of this enzymatic activity to relate to soil fertility. Verstraete and Voets (1977) evaluated a number of soil microbial and biochemical characteristics in two fields over a five-year period and compared the data with soil management practices and crop yields. Of the various activities measured, phosphatase, saccharase, β-glucosidase, and urease enzymatic activities, as well as N-mineralization and soil respiration rates were found to be useful in soil characterization. These traits increased with increases in soil organic matter, clay, and $CaCO_3$ contents. They also responded to organic fertilization. Multiple regression analysis showed that the variation was explained in large part by the soil alkalinity and humus content. Regression analysis also indicated that winter wheat yields correlated positively with phosphatase activity, whereas sugarbeet yields and urease activity correlated negatively.

4.3. CONCLUSIONS

To understand fully the biochemical transformations of organic matter occurring within the soil ecosystem and the properties of the soil which control the synthesis and expression of these reactions, an accurate assessment of a variety of enzymatic activities is necessary. Interpretation of the data is difficult because of both the high potential for development of artifacts in the assay procedure and the extreme variability of these activities *in situ*. Problems associated with the assay procedure relate to the collection of representative soil samples, sample storage, and extrapolation of the data. Assay procedures may limit severely the general applicability of the data. Questions arise relating to the validity of estimating enzyme activities in highly buffered solutions with optimal substrate concentrations. In many cases, where sensitive methods for product analysis exist, quantification of the activity under conditions approximating those *in situ* is desirable. The compartmentilization of enzymatic activities within active and inactive soil fractions makes it difficult to relate measurable activities with levels expressed in the native soil site. But in spite of all of these difficulties, soil enzymatic analyses have proven valuable in delimiting microbial processes occurring in soil and relating these processes to a number of overall properties of the ecosystem and to a variety of management techniques.

REFERENCES

Arutyunyan, E. A., and S. A. Abramyan, 1975. Activity of soil nucleases. Izv. S-Kh. Nauk. 18: 49-53.

Bartha, R., and L. Bordeleau, 1969. Cell-free peroxidases in soil. Soil Biol. Biochem. 1: 139-143.

Bordeleau, L. M., and R. Bartha, 1972. Biochemical transformations of herbicide-derived anilines in culture medium and in soil. Can. J. Microbiol. 18: 1857-1864.

Brisbane, P. G., M. Amato, and J. N. Ladd, 1972. Gas chromatographic analysis of amino acids from the action of proteolytic enzymes on soil humic acids. Soil Biol. Biochem. 4: 51-61.

Burns, R. G. (ed.) 1978. Soil Enzymes. Academic Press, London, 380 pp.

Burns, R. G. 1982. Enzyme activity in soil: Location and possible role in microbial ecology. Soil Biol. Biochem. 14: 423-427.

Burns, R. G., and J. A. Edwards, 1980. Pesticide breakdown by soil enzymes. Pestic. Sci. 11: 506-512.

Cacco, G., and A. Maggioni, 1976. Multiple forms of acetyl-naphthyl-esterase activity in soil organic matter. Soil Biol. Biochem. 8: 321-325.

Cantarella, H., and M. A. Tabatabai, 1983. Amides as sources of nitrogen for plants. Soil Sci. Soc. Am. J. 47: 599-603.

Casida, L. E., Jr. 1977. Microbial metabolic activity in soil as measured by dehydrogenase determinations. Appl. Environ. Microbiol. 34: 630-636.

Ceccanti, B., P. Nannipieri, S. Cervelli, and P. Sequi. 1978. Fractionation of humus-urease complexes. Soil Biol. Biochem. 10: 39-45.

Chunderova, A. N. 1970. Enzyme activity and pH of soil. Sov. Soil Sci. 1970: 308-314.

Dick, W. A., and M. A. Tabatabai, 1983. Activation of soil phosphatase by metal ions. Soil Biol. Biochem. 15: 359-363.

Duxbury, J. M., and R. L. Tate III, 1981. The effect of soil depth and crop cover on enzymatic activities in Pahokee muck. Soil Sci. Soc. Am. J. 45: 322-328.

Estermann, E. F., G. H. Peterson, and A. D. McLaren, 1959. Digestion of clay-protein, lignin-protein, and silica-protein complexes by enzymes and bacteria. Soil Sci. Soc. Am. Proc. 23: 31-36.

Frankenberger, W. T., Jr. 1983. Kinetic properties of L-histidine ammonia-lyase activity in soils. Soil Sci. Soc. Am. J. 47: 71-74.

Frankenberger, W. T., Jr., and F. T. Bingham, 1982. Influence of salinity on soil enzyme activities. Soil Sci. Soc. Am. J. 46: 1173-1177.

Frankenberger, W. T., Jr., and W. A. Dick, 1983. Relationship between enzyme activities and microbial growth and activity indices in soil. Soil Sci. Soc. Am. J. 47: 945-951.

Frankenberger, W. T., Jr., and J. B. Johanson, 1982. Effect of pH on enzyme stability in soils. Soil Biol. Biochem. 14: 433-437.

Frankenberger, W. T., Jr., and M. A. Tabatabai, 1980. Amidase activity in soils: II. Kinetic parameters. Soil Sci. Soc. Am. J. 44: 532-536.

Galstian, A. S. 1974. Enzymatic activity in soils. Geoderma 12: 43-48.

Greenland, D. J. 1971. Interactions between humic and fulvic acids and clays. Soil Sci. 111: 34-41.

Hankin, L., D. C. Sands, and D. E. Hill, 1974. Relation of land use to some degradative enzyme activities in soil bacteria. Soil Sci. 118: 38-44.

Harrison, A. F. 1983. Relationship between intensity of phosphatase activity and physico-chemical properties of woodland soils. Soil Biol. Biochem. 15: 93-99.

Hartenstein, R. 1982. Soil macroinvertebrates, aldehyde oxidase, catalase, cellulase and peroxidase. Soil Biol. Biochem. 14: 387-391.

Haska, G. 1975. Influence of clay minerals on sorption of bacteriolytic enzymes. Microbial Ecol. 1: 234-245.

Juma, N. G., and M. A. Tabatabai, 1977. Effects of trace elements on phosphatase activity in soils. Soil Sci. Soc. Am. J. 41: 343-346.

Kaplan, D. L., and R. Hartenstein, 1979. Problems with toluene and the determination of extracellular enzyme activities in soil. Soil Biol. Biochem. 11: 335-338.

Kiss, S., M. Dragan-Bularda, and D. Radulescu, 1975. Biological significance of enzymes accumulated in soil. In N. C. Brady (ed.) Adv. Agron. 27: 25-87. Academic Press, New York.

Klein, D. A., D. L. Sorensen, and E. F. Redente, 1985. Soil enzymes: A predictor of reclamation potential and progress. In R. L. Tate III and D. A. Klein (eds.), Soil Reclamation Processes: Microbiological Analyses and Applications. pp. 141-172. Marcel Dekker, New York.

Ladd, J. N. 1972. Properties of proteolytic enzymes extracted from soil. Soil Biol. Biochem. 4: 227-237.

Ladd, J. N., and J. H. A. Butler, 1971. Inhibition by soil humic acids of native and acetylated proteolytic enzymes. Soil Biol. Biocehm. 3: 157-160.

Ladd, J. N., and E. A. Paul, 1973. Changes in enzymic activity and distribution of acid-soluble, amino acid-nitrogen in soil during nitrogen immobilization and mineralization. Soil Biol. Biochem. 5: 825-840.

Lloyd, A. B. 1975. Extraction of urease from soil. Soil Biol. Biochem. 7: 357-358.

Malcolm, R. E. 1983. Assessment of phosphatase activity in soils. Soil Biol. Biochem. 15: 403-408.

Mangler, J. E., and R. L. Tate III, 1982. Source and role of peroxidase in soil organic matter oxidation in Pahokee muck. Soil Sci. 134: 226-232.

Marshman, N. A., and K. C. Marshall, 1981. Bacterial growth on proteins in the presence of clay minerals. Soil Biol. Biochem. 13: 127-134.

Martin, J. P. and K. Haider, 1980. A comparison of the use of phenolase and peroxidase for the synthesis of model humic acid-type polymers. Soil Sci. Soc. Am. J. 44: 983-988.

Mathur, S. P. 1971. Characterization of soil humus through enzymatic degradation. Soil Sci. 111: 147-157.

Mathur, S. P., and R. B. Sanderson, 1980. The partial inactivation of degradative soil enzymes by residual fertilizer copper in Histosols. Soil Sci. Soc. Am. J. 44: 750-755.

McLaren, A. D., A. H. Pukite, and I. Barshad, 1975. Isolation of humus with enzymic activity from soil. Soil Sci. 119: 178-180.

Moller, H. 1979. Untersuchungen zum saccharase- und ureasegehalt von bruchwaldtorfen und anmoorhumus nordwestdeutscher Erlenwalder. Telma 9: 175-192.

Moore, A. W., and J. S. Russell, 1972. Factors affecting dehydrogenase activity as an index of soil fertility. Plant Soil. 37: 675-682.

Myers, M. G., and J. W. McGarity, 1968. The urease activity in profiles of five great soil groups from northern New South Wales. Plant Soil 28: 25-37.

Nannipieri, P., B. Ceccanti, S. Cervelli, and C. Conti, 1982. Hydrolases extracted from soil: Kinetic parameters of several enzymes catalysing the same reaction. Soil Biol. Biochem. 14: 429-432.

Nannipieri, P., B. Ceccanti, S. Cervelli, and E. Matarese, 1980. Extraction of phosphatase, urease, proteases, organic carbon, and nitrogen from soil. Soil Sci. Soc. Am. J. 44: 1011-1016.

Nannipieri, P., S. Cervelli, and F. Pedrazzini, 1975. Concerning the extraction of enzymatically active organic matter from soil. Experientia 31: 513-515.

Paulson, K. N., and L. T. Kurtz, 1969. Locus of urease activity in soil. Soil Sci. Soc. Am. Proc. 33: 897-901.

Paulson, K. N., and L. T. Kurtz, 1970. Michaelis constant of soil urease. Soil Sci. Soc. Am. Proc. 34: 70-72.

Pettit, N. M., L. J. Gregory, R. B. Freedman, and R. G. Burnss 1977. Differential stability of soil enzymes: Assay and properties of phosphatase and arylsulphatase. Biochim. Biophys. Acta 485: 357-366.

Ross, D. J. 1970. Effect of storage on dehydrogenase activity of soils. Soil Biol. Biochem. 2: 55-61.

Ross, D. J. 1983. Invertase and amylase activities as influenced by clay minerals, soil-clay fractions and topsoils under grassland. Soil Biol. Biochem. 15: 287-293.

Ross, D. J., and A. Cairns, 1982. Effects of earthworms and ryegrass on respiratory and enzyme activities of soil. Soil Biol. Biochem. 14: 583-587.

Ross, D. J., and B. A. McNeilly, 1973. Biochemical activities in a soil profile under hard beech forest. 3. Some factors influencing the activities of polyphenol-oxidising enzymes. N. Z. J. Sci. 16: 241-257.

Ross, D. J., and H. S. Roberts, 1973. Biochemical activities in a soil profile under hard beech forest. 1. Invertase and amylase activities and relationships with other properties. N. Z. J. Sci. 16: 209-224.

Ross, D. J., and T. W. Spier, 1979. Biochemical activities of organic soils from subantarctic tussock grasslands on Campbell Island. 2. Enzyme activities. N. Z. J. Sci. 22: 173-182.

Rowell, M. J., J. N. Ladd, and E. A. Paul, 1973. Enzymically active complexes of proteases and humic acid analogues. Soil Biol. Biochem. 5: 699-703.

Schinner, F. A., R. Niederbacher, and I. Neuwinger, 1980. Influence of compound fertilizer and cupric sulfate on soil enzymes and carbon dioxide-evolution. Plant Soil 57: 85-93.

Sjoblad, R. D., and J. M. Bollag, 1981. Oxidative coupling of aromatic compounds by enzymes from soil microorganisms. In E. A. Paul and J. N. Ladd (eds.), Soil Biochem. 5: 113-152. Marcel Dekker, New York.

Skujins, J. J. 1967. Enzymes in soil. In A. D. McLaren and G. A. Peterson (eds.), Soil Biochem. 1: 371-414. Marcel Dekker, New York.

Skujins, J. 1973. Dehydrogenase: an indicator of biological activities in soil. Bull. Ecol. Res. Comm. NFR Statens Naturvetensk. Forskningsrad 17: 235-241.

Skujins, J. 1976. Extracellular enzymes in soil. Crit. Rev. Microbiol. 4: 383-421.

Skujins, J. J., and A. D. McLaren, 1969. Assay of urease activity using ^{14}C-urea in stored, geologically preserved, and in irradiated soils. Soil Biol. Biochem. 1: 89-99.

Sowden, F. J. 1970. Action of proteolytic enzymes on soil organic matter. Can. J. Soil Sci. 50: 233-241.

Spalding, B., J. M. Duxbury, and E. L. Stone, 1975. Lycoipodium fairy rings: Effect on soil respiration and enzymatic activities. Soil Sci. Soc. Am. Proc. 39: 65-70.

Stevenson, F. J. 1982. Humus chemistry. John Wiley & Sons. New York. 443 pp.

Sulfita, J. M., and J. M. Bollag, 1980. Oxidative coupling activity in soil extracts. Soil Biol. Biochem. 12: 177-183.

Tabatabai, M. A. 1973. Michaelis constants of urease in soils and soil fractions. Soil Sci. Soc. Am. Proc. 37: 707-710.

Tabatabai, M. A., and J. M. Bremner, 1971. Michaelis constants of soil enzymes. Soil Biol. Biochem. 3: 317-323.

Tate, R. L. III. 1985. Carbon mineralization in acid, xeric forest soils: Induction of new activities. Appl. Environ. Microbiol. 50: 454–459.

Tate, R. L. III, and A. L. Mills, 1983. Cropping and the diversity and function of bacteria in Pahokee muck. Soil Biol. Biochem. 15: 175–179.

Tate, R. L. III, and R. E. Terry, 1980a. Variation in microbial activity in Histosols and its relationship to soil moisture. Appl. Environ. Microbial. 40: 313–317.

Tate, R. L. III, and R. E. Terry, 1980b. Effect of sewage effluent on microbial activities and coliform populations of Pahokee muck. J. Environ. qual. 9: 673–677.

Thornton, J. I., D. Crim, and A. D. McLaren, 1975. Enzymic characterization of soil evidence. J. Forensic Sci. 20: 674–692.

Verstraete, W., and J. P. Voets. 1977. Soil microbial and biochemical characteristics in relation to management and fertility. Soil Biol. Biochem. 9: 253–258.

Zantua, M. I., and J. M. Bremner, 1975a. Preservation of soil samples for assay of urease activity. Soil Biol. Biochem. 7: 297–300.

Zantua, M. I., and J. M. Bremner, 1975b. Comparison of methods of assaying urease activity in soils. Soil Biol. Biochem. 7: 291–296.

Zvagintsev, D. G., and L. L. Velikanov, 1968. Influence of soils and clay minerals on the activity of glucose oxidase and invertase. Soviet Soil Sci. 1968: 789–794.

FIVE

SOURCE AND TRANSFORMATIONS OF READILY METABOLIZED ORGANIC MATTER

Soil organic matter can be divided into two major pools based on relative susceptibility to biological decomposition: (1) easily (or readily) metabolized pool, and (2) relatively biodegradation resistant pool. Although both organic matter pools contribute to ecosystem function and stability, the reactions involving the readily decomposed pool have the greatest impact on nutrient cycling within the ecosystem; that is, the turnover of this carbon pool accounts for the vast majority of those plant nutrients originating in the soil organic matter component. Reaching a definition of what constitutes a "readily or easily metabolized organic component" is difficult. Most of the problem arises from the assignment of some of the more resistant organic matter components to one of the two organic matter pools. **Readily metabolized soil organic matter** is that soil organic component which provides a carbon and energy source for the soil microbial community with a minimal energy expenditure by the microbe. This energy expenditure includes that needed to synthesize catabolic enzymes as well as activate the substrate molecule to enhance its degradation susceptibility. The activation processes include the combination of polysaccharides and hexoses with adenosine triphosphate to form hexose-monophosphate as well as substrate level phophorylations of the Embden-Meyerhoff-Parnas pathway for glucose catabolism. A reaction sequence involving the participation of a large number of complex enzymes for the conversion of a substrate to a state where energy can be derived from its oxidation is more energy intensive than a sequence where the action of a single or only a few enzymes are needed before the microbe derives an energy benefit. Thus a compound, such as glucose, is more readily metabolized than lignin. In reality, a gradient of "desirability" of substrates as energy sources exists. A glucose, or hexose, molecule is metabolized more easily than a simple aromatic. But simple aromatics are more susceptible to biological decomposition than is a complex polysaccharide, such as cellulose, or polyaromatics. Because of the relationship between energy and carbon supply of these compounds, the products of metabolism of readily metabolized organic compounds are predominantly carbon dioxide and soil microbial biomass. For example, Sorensen and Paul (1971) found that in a heavy clay soil all amended acetate had been metabolized after six days incubation at 25°C, but only 70 percent of the carbon had been evolved as carbon dioxide. The bulk of the remaining carbon was detected in carbohydrates and amino acids. In a related study, Kassim et al. (1981) found that after 12 weeks in a Greenfield sandy loam and a Steinbeck loam, 82 and 91 percent of the carbon of 1-^{14}C-acetate had been evolved as carbon dioxide, respectively, with 3.9 and 8.0 percent of the added carbon contained in the microbial biomass, respectively. Thus only 5 to 10 percent of the original substrate carbon was incorporated into other soil organic matter components, such as the humic fraction. The potential did remain for movement of a portion of the acetate carbon initially found in microbial cell carbon into biodegradation resistant soil humic substances following death and decomposition of the cell structure. Similar results were observed for glycine, serine, cysteine, alanine, leucine, tyrosine, protein, uracil, cystine, and uridine. Although cellulose is frequently considered to be among the more biodegradation resistant plant components as a result of its crystalline structure and the complexity of the enzymes involved in its decomposition, Kassim et al. (1981) found little difference in the distribution of carbons from glucose, cellulose, glucosamine, wheat straw polysaccharides, and the polysaccharides from a number of microorganisms between carbon dioxide and biomass carbon. Carbon evolved as carbon dioxide after 12 weeks incubation in Greenfield sandy loam and Steinbeck loam ranged from 58 to 83 percent (with values for glucose and cellulose being 76 and 83, and 68 and 72 percent, respectively). Between 3.5 and 7.2 percent of the carbon was incorporated into microbial biomass. These data can be compared to the results of metabolism of melanin, a relatively biodegradation resistant compound, in Steinbeck loam and Greenfield sandy loam where 5 and 13 percent of the carbon was evolved as carbon dioxide after one year incubation for the two respective soils with 0 and 0.7 percent of the carbon in microbial biomass, respectively (Stott et al., 1983).

A variety of biochemical literature sources and textbooks are available which describe the degradative reactions of these compounds. [A summary of the reactions involving plant substituents and their impact on the soil microbial community is provided by Alexander (1977).] Also a number of literature sources can be found indicating which of these compounds are present in soil organic matter and how to detect them (Gieseking, 1975; Kononova, 1966; Schnitzer and Khan, 1972; Stevenson, 1982). Therefore, the emphasis of the present discussion involves the decomposition kinetics of the easily metabolized compounds in soil and the impact of this decomposition on total ecosystem function. Specific soil organic matter pools that will be examined are (1) fixed carbon inputs from plant debris, (2) native soil organic matter, and (3) soil microbial populations. Along with the nutrient and soil structure effects indicated previously, a commonly discussed and debated impact of metabolism of readily metabolizable substrates is the priming effect. The question of the occurrence of this phenomenon and the explanation of its observance in some cases will be evaluated.

5.1. PLANT DEBRIS

Photosynthetically fixed carbon from higher plants provides the major carbon and energy inputs in most soil ecosystems. A small quantity of carbon is fixed by soil bacteria, blue green algae, and green algae, but these inputs are minimal except perhaps in very limited situations. Frequently aboveground productivity is controlled primarily through tight coupling of litter decomposition processes and plant nutrient uptake. Plant roots are physically located around and within sites of litter decomposition so that essentially all mineralized nutrients (excluding those assimilated by the decomposer community) are incorporated into new plant biomass. Little, if any, of the nutrients are leached from the site. The plant carbon sources of these nutrients enter the soil ecosystem as dead and decaying aboveground biomass, sloughed roots cells, senescent root tissues, or root exudates. Aboveground sources have been most studied, but belowground organic matter sources may be substantial. Griffin et al. (1977), for example, found that between 0.26 and 0.73 mg of organic matter is sloughed weekly per peanut (*Arachis hypogaea* L. cv. Argentine) plant. McClaugherty et al. (1982) estimated fine root (0 to 0.5 mm and 0.5 to 3.0 mm diameter) production in a 53-year-old red pine (*Pinus resinosa* Ait.) plantation and in an adjacent 80-year-old mixed hardwood stand in north-central Massachusetts. Standing crops of live fine roots in the hardwoods and plantations were 6.1 and 5.1 Mg/ha, respectively. Annual production estimates ranged from 4.1 to 11.4 Mg/ha/yr in the hardwoods and from 3.2 to 10.9 Mg/ha/yr in the plantation. (Variation in production estimates related to differences in assumptions made in the calculations.) In a related study, these workers noted that fine roots lost between 20 and 60 percent of their initial mass over a four-year period. These data indicate that fine roots in the forest ecosystem are major contributors to the soil organic matter pools.

The diversity in sources of plant derived organic matter inputs as well as the complexity of these substances result in extreme heterogeneity in the microbial reactions involved in its decomposition (Alexander, 1977). These organic carbon components range from the reasonably readily decomposable components of the cytoplasm to the more biodegradation resistant components of the cell wall. Cell wall components also vary in biodegradation susceptibility from the readily metabolized hemicelluloses and celluloses through the more biodegradation resistant lignins. The general kinetics of biomass decomposition suggest a biphasic process; that is, a rapid catabolic period followed by a period of slow carbon dioxide evolution usually is observed. The initial rapid carbon dioxide evolution rate results from decomposition of easily metabolized substrates, such as amino acids, proteins, simple sugars and polysaccharides. During the latter slow metabolic period, the more biodegradation resistant components are metabolized. Some carbon dioxide evolved in the slow decomposition period results from catabolism of the microbial polymers synthesized during the initial decomposition period. Also, some readily metabolized substances may not be catabolized during the initial rapid decomposition phase because of their physical protection

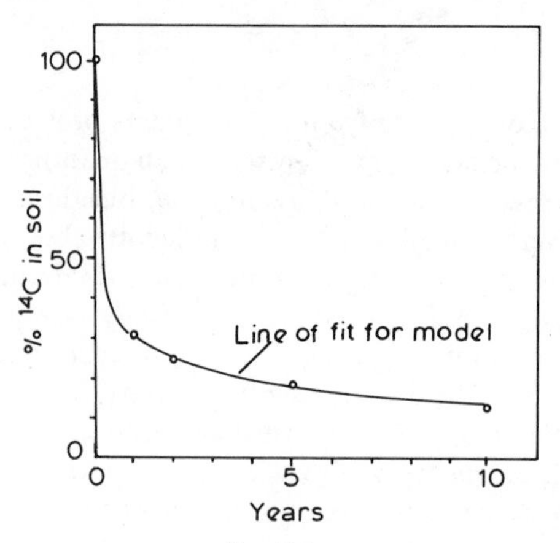

Fig. 5.1.
Decomposition of uniformly labeled ryegrass incubated in soil in the field (Jenkinson and Rayner, 1977). Reprinted by permission of The Williams and Wilkins Co., Baltimore.

from contact with the decomposer community. For example, these compounds may be occluded by a more biodegradation resistant cell wall. As the physical structure of the protective barrier is breached, the readily metabolized substrates enter the nutrient pool.

A typical decomposition pattern for succulent green plant material in soil is presented in Fig. 5.1. For this study, uniformly labeled ryegrass was incubated with a variety of Rothamsted soils (Jenkinson and Rayner, 1977). Of note is the fact that approximately two-thirds of the carbon was lost as carbon dioxide during the initial year of incubation. The carbon remaining in the soil would be detected as nonmetabolized plant substituents and newly synthesized microbial biomass and products. Similar decomposition curves are found for leaf litter, except the percent decomposed in the initial year is greatly reduced (Shanks and Olson, 1961). Significant variation in first-year weight loss of leaves in nylon net bags resulted from species variation (*Fagus grandifolia*, 21 percent; *Acer saccharum*, 32 percent; *Quercus shumardii*, 34 percent; *Quercus alba*, 39 percent; and *Morus rubra*, 64 percent). Slightly less initial decomposition (54 to 57 percent) of blue grama [*Bouteloua gracilis* (H.B.K.) Lag. × Streud.] was detected over a 412 day incubation period than was noted with the ryegrass (Nyhan, 1975). A greater portion of the readily metabolized substrates is protected from attack by the decomposer community during the early decomposition phase as a result of physical occlusion in these leaf tissues than is observed in succulent green tissues of ryegrass.

Murayama (1984) found that rice and barley straw saccharide decomposition rates under field conditions were represented by a first-order kinetics model:

$$Y_t = C_1 e^{-K_1 t} + C_2 e^{-K_2 t}$$

where Y_t is the residue remaining in soil at time t, k_1 and k_2 are the decomposition rate constants for the labile fraction (C_1) and the nonlabile fraction (C_2), respectively. About 82 percent of the total saccharide content of the rice straw was in the labile fraction with a rate constant of 0.64 to 0.81. The values for barley straw were 70 to 92 percent and 0.50 to 0.61, respectively. The nonlabile fraction had a half-life of 9 to 59 months.

Similar decomposition kinetics are observed for ryegrass (*Lolium multiflorum*) and maize (*Zea mays*) tissues incubated in forest soils of Nigeria (Jenkinson and Ayanaba, 1977). With a range of contrasting soils, 20 percent of the original plant material remained after one year of incubation. This was reduced to 14 percent after the second year of incubation. Interestingly, although significant differences in soil temperatures were detected, little difference in decomposition rates was found between shaded soils and soils in open sun. Comparison of the ryegrass decomposition rates with those observed by Jenkinson in England, as previously discussed, revealed a similar decomposition pattern except the decomposition rate was four times faster in Nigeria. With the Rothamsted data, 70 percent of the material decomposed with a half-life of 0.25 years. The remaining plant material had a half-life of eight years. The Rothamsted data fit the following model:

$$\% \text{ retention} = 70.9\ e^{-2.83t} + 29.1\ e^{-0.087t}$$

The Nigerian data could be described by this model except that each decomposition coefficient was multiplied by four.

Decomposition kinetics of highly lignified wood substrates are similar to those observed with the succulent tissues described previously, except the rates and extent of metabolism are reduced (Fig. 5.2) (Allison and Murphy, 1962). During the first 60 days incubation of finely ground wood and bark samples from a variety of hardwood species, 30.3 percent of the wood carbon was released as carbon dioxide in the absence of fertilizer nitrogen and 45.1 percent in its presence. Similarly, with bark tissue, 22.4 percent and 24.5 percent of the carbon was evolved as carbon dioxide, respectively. Rates for the individual tree species are presented in Table 5.1. Although differences were noted in the carbon dioxide production rates between the various species, variation was minimal. This suggests overall properties of wood and bark in general as limitations to microbial attack rather than species differences. Hardwoods were decomposed more rapidly than were the softwoods. Hence, because of the greater nitrogen demand resulting from the increased metabolic activity, nitrogen amendment was required for maximal activity of the microbial populations decomposing the hardwoods, whereas this was not the case with the softwoods. These data indicate that with these hardwoods nitrogen limitations were more significant than lignin effects.

As suggested by the differences in decomposition rates between the ryegrass and the woody products, susceptibility of plant components to biodegradation varies with a number of physical and chemical properties of the plant itself.

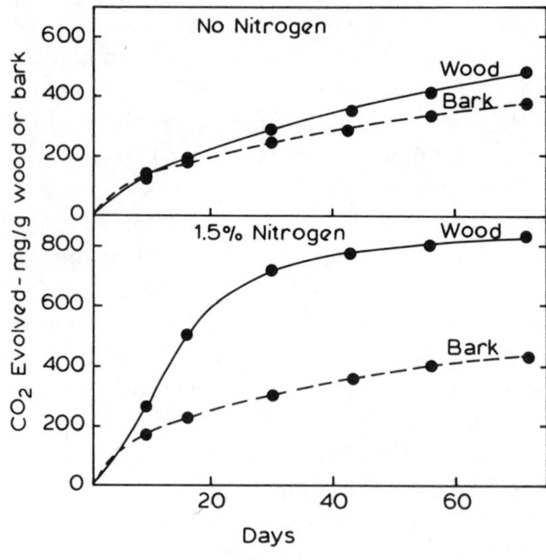

Fig. 5.2.
Decomposition of blackoak wood and bark in Branchville sandy loam. Reproduced from SOIL SCIENCE SOCIETY OF AMERICA PROCEEDINGS, Volume 26, 1962, pp. 463–466 by permission of the Soil Science Society of America.

Plant associated properties that contribute to the control of decomposition rates include the carbon: nitrogen ratio, lignin content, and to a certain degree, the surface area of the substrate, which is observed most frequently in the study of decomposition of woody materials. The high lignin content of these materials results in a slow disruption of woody tissue structure. Thus internal readily metabolized compounds are protected by the external cellular structure. Disruption of the physical structure necessarily enhances the availability of the internal substrates. There is a surface area effect associated with decomposition of more succulent plant tissues, but it is not as readily observed because of the more rapid disruption of the cell wall structure and subsequent liberation of internal substituents. Thus any process increasing the surface area of the debris before incorporation into the soil increases the decomposition rate. This impact of surface area was noted by Cheshire et al. (1974) in a study of hemicellulose decomposition in soil. Incorporation of hemicellulose into soil resulted in approximately 70 percent of the carbon being evolved as carbon dioxide after 448 days incubation, whereas fine grinding of the hemicellulose prior to incorporation increased the decomposition to 80 percent.

The effect of the association of readily metabolized plant substituents with lignin on biodecomposition has long been appreciated (e.g., see Reevey and Norman, 1948). The mechanism of biodecomposition inhibition by lignin primarily involves steric hindrance of microbial and enzymatic interaction with the more metabolizable molecules. The magnitude of the effect of this association on de-

Table 5.1.
Carbon Dioxide Evolved from Hardwoods Incubated in Branchville Sandy Loam (Allison and Murphy, 1962).[a]

Tree Species	Wood (percent)	Bark (percent)
Black oak	24.9	21.3
White Oak	38.1	27.4
Red Oak	31.9	22.9
Post Oak	28.1	23.2
Hickory	33.7	11.7
Red Gum	31.3	21.2
Yellow Poplar	30.6	37.5
Chestnut	26.6	23.1
Black Walnut	27.1	13.7

[a]Reprinted by permission of the Soil Science Society of America.

composition was suggested previously when comparing decomposition rates of readily decomposable substrates free and associated with polyphenolic compounds. Variation in decomposition of various plant parts is related to both lignin content and carbon:nitrogen ratio (Fogel and Cromack, 1977). In the Fogel and Cromack study, highest correlation was found with the carbon:nitrogen ratio. They used the exponential decay model to determine annual decomposition constants for various plant parts. Rate constants for needles, female cones, branches, and bark decomposition in seven stands representing four mature vegetation types in western Oregon were 0.22 to 0.31, 0.047 to 0.083, 0.059 to 0.089, and 0.005 to 0.083 per year, respectively.

The effect of tissue nitrogen content results from the microbial requirement of fixed nitrogen for cell synthesis. Plant materials with a high carbon:nitrogen ratio do not provide sufficient nitrogen for metabolism of the decomposer populations under conditions of rapid microbial activity. As the readily metabolized substrates are exhausted, the nutrient limitation shifts from nitrogen to carbon. This type of nutrient limitation was observed by Knapp et al. (1983a; 1983b) in their study of microbial decomposition of wheat straw (*Triticum aestovim* L. var Nugaines). They found that decomposition was nitrogen limited during the initial decomposition period and carbon limited after prolonged incubation. During the initial decomposition period when the microbes are metabolizing carbon-rich/nitrogen-poor substrates, such as carbohydrates, straw metabolism was stimulated by amendment with mineral nitrogen. As the readily decomposed carbon pool was exhausted and the microbial metabolic rate declined, thereby reducing the nitrogen demand, sufficient nitrogen was contained in the substrate for continued microbial catabolism of the more resistant plant components. Because of the greater energy expenditures necessary to metabolize these substrates, the microbes at that point became carbon-limited. Hence, amendment of the soil with a carbon source stimulated microbial respiration.

This differential need for nitrogen by the microbial community during decomposition of carbon rich plant substituents is also demonstrable through evaluation of changes in carbon, hydrogen, and nitrogen content of litter following incubation in soil. Incubation of sweet chestnut (*Castanea sativa* Mill.) and beech (*Fagus sylvatica* L.) leaves in litter for one year resulted in an increase in the percent nitrogen content (Anderson, 1973). Only the beech leaves exhibited a net increase in the total nitrogen content. The percentage increases in nitrogen content resulted primarily from the more rapid losses of non-nitrogenous leaf constituents, whereas the quantities of nitrogen originally existing in the leaves were preserved. The percentages of carbon and hydrogen of both leaf types changed less than 1 percent throughout the one-year incubation period, suggesting that carbohydrate losses were directly proportional to the weight loss.

Following the period of net immobilization of nutrients, decomposition of plant litter becomes a source of plant nutrients (for examples of the magnitude, see Dalal, 1979; Gosz et al., 1973). Gosz et al. (1973) found that the rate of nutrient release from decomposing branch and leaf litter correlated with nutrient concentration in the current litter fall. As with the previous study, the concentration of nitrogen, sulfur and phosphorus in the litter increased with time. Maximum nutrient release from current litter occurred in the autumn and summer seasons.

5.2. ABIONTIC SOIL ORGANIC FRACTION

A pool of readily metabolized soil organic compounds that exists free of living cells, but is derived either from plant products or exudates or from microbial cellular material or products, has a significant impact on steady-state microbial respiration, especially in soils receiving minimal carbonaceous inputs. The substances are synthesized by living microbial and plant cells, but as a result of the death of the cells in soil and/or active excretion of cellular synthesized products through the cell wall, they become stabilized into native soil organic matter. Since they cannot be associated with any particular plant or microbial fraction, they are termed "abiontic"; that is, they are produced biologically but exist free of the living cell. With the state of the art of biochemical analysis procedures (Stevenson, 1982), documentation of the occurrence and quantification of these readily decomposable molecules has been relatively common for a number of decades. Small quantities of most any monomer of cell polymers have been isolated from soil. These include amino acids (Schmidt et al., 1960; Piper and Posner, 1968), and purines and pyrimidines (Anderson, 1961; Cortez and Schnitzer, 1979). Because of the continued synthesis of these molecules during microbial growth and the common capability for their decomposition by the soil microbial community, significant levels are generally anticipated to exist free in the soil, but under normal conditions, large concentrations do not accumulate.

An exception to this rule of minimal accumulation of abiontic compounds in soil is the occurrence of extracellular polysaccharides. A variety of polysaccharides are found in soil organic matter. These may be unchanged or slightly modified plant polysacharides occurring free in soil or stabilized by incorporation into humic acids (Morita and Levesque, 1980; Cheshire et al., 1973), remnants of microbial cells (Casagrande and Park, 1978), or polymers synthesized and excreted by microorganisms. The composition of these polymers can vary from the homopolymers of cellulose to the rather complex heteropolymers common to microbial exopolysaccharides. Gupta et al. (1963) in their analysis of carbohydrate substituents within a variety of soil profiles found glucose to be the dominant monomer (42 to 54 percent). Other monomers in decreasing order of occurrence were galactose, mannose, arabinose, xylose, rhamnose, and fucose-ribose. These polymers also exhibit a broad distribution within the soil matrix. Foster (1981) treated ultrathin sections of natural soil fabrics with heavy metal stains specific for carbohydrates to examine their distribution among soil particulate fractions. Carbohydrates were found associated with living and dead cells of plants and microbes, and clay particles, and located in submicron size mineral aggregates. This association with aggregates and clay particles provided a partial explanation of the degadation stability of these polymers.

The differential stability of various soil organic matter fractions to biodegradation was shown by Anderson and Paul (1984) in their radiocarbon dating analysis of organo-mineral complexes. They measured the radiocarbon date of organic fractions from a clayey Haploboroll soil (Indian Head) collected in 1963 and 1978. Changes in the ^{14}C radiocarbon dates of the fractions between the two sample times were taken as the turnover rate. Enrichment of ^{14}C in the atmosphere due to nuclear explosions in the 1960s and its subsequent incorporation into organic matter resulted in a general 20 percent decline in the total soil organic matter radiocarbon age. Oldest and least active fractions were nonhydrolyzable carbon and the aromatic humic acid-A fraction. The reduced radiocarbon age of humin and the clay associated humic acid-B fraction indicated their participation in short-term carbon cycling. The organic fraction associated with the coarse clay fraction includes the more recently synthesized microbial cell wall polymers and related carbohydrates as well as unmetabolized plant polysaccharides. For example, Cheshire et al. (1973) found rye straw carbohydrates associated with the soil humin fractions after 224 days of incubation.

Soil manipulation procedures that would tend to disrupt this mineral–organic matter association increases the rate of catabolism of these soil organic matter fractions. For example, repeated wetting and air-drying cycles disrupt colloid–organic matter associations, thereby increasing biodegradation of the organic fraction. Sorenson (1974) found that air-drying/rewetting cycles of 30 days over 260 to 500 days increased carbon dioxide evolution from 16 to 121 percent over soils maintained at a constant moisture.

Other mineral association that influence decomposition rates of anionic organic matter substituents include reactions with cationic metals. Association of microbial and plant polysaccharides with a variety of metal ions increases their biodegradation stability (Martin et al., 1966). Effects of zinc, copper, iron, and aluminum salts or complexes of a variety of bacterial and plant polysaccharides containing from 6 to 83 percent uronic acids on biodecomposition of the polysaccharide in Greenfield sandy loam was evaluated. Significant inhibition of decomposition rates by some of the metals was noted for most of the polysaccharides. For example, zinc and aluminum had little influence on decomposition of *Azotobacter chroococcum* polysaccharide, whereas decomposition of copper and iron complexes was reduced about 50 percent over noncomplexed carbohydrates. In contrast, zinc had a major impact on decomposition of karaya gum. A much smaller concentration than what was shown to be noninhibitory with the above polysaccharide reduced karaya gum decomposition to between 0 and 3 percent of the control rates.

These studies demonstrate the existence of a pool of reasonably available organic matter within what could be termed native soil organic matter. Because of the interactions with soil colloidal particles and metals, these materials are less available for biodegradation than exogenously supplied fixed carbon or microbial biomass. These native soil organic matter pools do provide a source of energy and nutrients for the microbial community in soils receiving little or no organic inputs.

5.3. MICROBIAL BIOMASS

The fact that soil microorganisms may be a controlling factor in plant nitrogen availability, especially those microbes whose activity is stimulated by amendment of low nutrient soils with succulent green plant remains, or other rapidly metabolized carbon sources, has been appreciated for decades. Amendment of soils containing little available mineral nitrogen with a carbon-rich/nitrogen-poor substrate results in immobilization of the previously available mineral nitrogen into the soil microbial biomass tissue. Thus the aboveground biomass productivity declines until either an input of fixed nitrogen into the ecosystem occurs or that nitrogen immobilized into the microbial cells becomes available through death and decomposition of the microbial cells.

Quantities of plant nutrients retained within the microbial community may be substantial. Anderson and Domsch (1980) estimated nutrient values for the soil microbial population in 29 soils by determining the microbial cell carbon content of the soils and calculating the associated nitrogen, phosphorus, potassium and calcium values by measuring (1) the average carbon to mineral content of 24 pure cultures of soil microbes, and (2) the relative contribution of bacterial and fungal populations to microbial activity in 17 of the soils. Their data indicate that the microbial biomass of the soils contained between 0.27 and 4.8 percent of the total soil carbon with a mean value of approximately 2.5 percent. These carbon values corresponded to 0.5 to 15.3 percent of the total soil nitrogen (mean was approximately 5.0 percent). The average quantities of nitrogen, phosphorus, potassium, and calcium in the microflora of the top 12.5 cm of the soils were approximately 108, 83, 70, and 11 kg/ha, respectively. Small quantities of nutrients were found in three forest soils. Between 0.52 and 0.91 percent of the total carbon and 0.16 and 0.22 percent of the total nitrogen was located in the microbial biomass in these soils.

The availability of the nutrients secluded in microbial biomass to plant populations is dependent upon the stability of the microbial cell. As is presented in the following text, a number of studies have involved the evaluation of the decomposition rate of whole microbial cells as well as various cell substituents. In most of these reports, axenically grown microbial cells were added to discrete soil samples. An anticipated result from the amendment of soil with populations that may or may not be native to the soil community and are certainly added at levels far beyond what would be normally present is that the microbes die and decay. Of greater interest from the view of plant nutrition is the release of nutrients as the result of a die-off of native soil microbial populations. Under steady-state conditions, the population densities of soil microorganisms are relatively constant. The microbes are reproducing at rates essentially equivalent to their death rates. The size of this population is characteristic of the soil ecosystem and the physical, chemical, and biological conditions of the soil; that is, the nutrients available within a given site allow for the development of a specific number of microbial cells. Maintenance of this steady-state population density is also dependent upon the physical and chemical parameters delimiting the site and the rate of predation. Predators limit maximum population density but do not deplete their food sources (Alexander, 1981). van Veen et al. (1984) developed a model for microbial carbon and nitrogen turnover in soils based on laboratory and field studies of a number of different Australian soils. Significant effects of wet/dry cycles, as discussed previously, on soil organic matter decomposition and microbial biomass turnover were observed. Their data indicated that the soils had a capacity to preserve quantities of organic matter and microbial biomass. The mechanisms for preservation of portion of the microbial population include protection against predation as well as amelioration of adverse soil conditions. A number of biotic and abiotic interactions including preservation in refuge sites unavailable to predator populations are involved in microbial biomass preservation (Alexander, 1981). This conservation of a basic microbial population density indicates that for undisturbed systems much of the nutrients contained in the microbial biomass is retained by that population. Of greater interest in evaluating plant nutrient supplying capacity of the soil microbial cells is that biomass is synthesized above this steady-state level due to a temporary influx of carbonaceous nutrients. Once the nutrients are exhausted, the interactions of predators and population death due to nutrient deprivation results in a return of the population density to the predisturbance level. The question then becomes, what are the kinetics of this decomposition and mineralization of cellular nitrogen contained within the excess microbial populations?

The release of nutrients from primary decomposer biomass depends in part on the nature of the secondary decomposer population (Woods et al., (1982); that is, the secondary decomposers may convert the microbial cell nutrients into

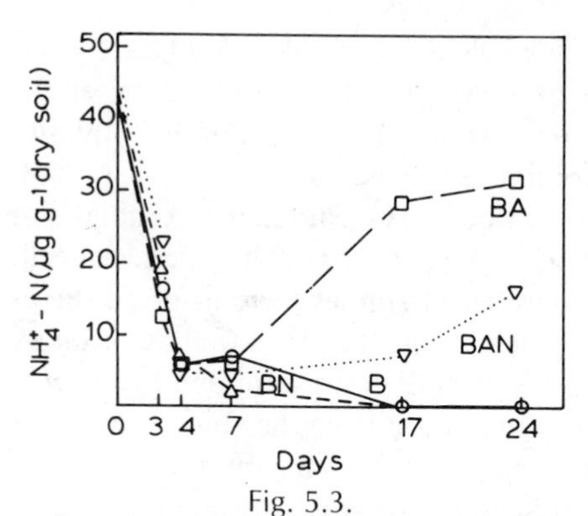

Fig. 5.3.
Effect of trophic interactions on net N-mineralization in polyprolene oxide sterilized Olney sandy loam. (B = bacteria alone; BA = bacteria + amoebae; BN = bacteria + nematodes; BAN = bacteria + amoebae + nematodes).

forms which are still not available to the plant community. Growth of *Pseudomonas cepacia* in soil resulted in net nitrogen immobilization into the microbial cell substituents (Fig. 5.3). Grazing by amoebae (*Acanthamoeba polyphaga*) always reduced bacterial biomass, increased soil respiration, and increased nitrogen mineralization. Grazing by nematodes (*Mesodiplogaster lheritieri*) resulted in reduced bacterial population densities and increased respiration, but nitrogen mineralization did not increase until the nematode population declined. A detailed nitrogen budget indicated that the nematode biomass was not sufficient to account for the nitrogen mineralized from the microbial cells but not appearing as mineral nitrogen in the incubation chamber. The authors postulated a change in the excretory pathways of the nematodes as the bacterial nutrient source became limiting. The nitrogen budget also indicated that the bacterial and amoebal biomass accounted for the organic nitrogen when only those populations were present. These data show the complexity of modeling nutrient regeneration from microbial populations once the nutrients are immobilized into soil biomass.

A number of studies have involved evaluation of the decomposition of various microbial cellular products and the mineralization of plant nutrients (McGill et al., 1975; Nakas and Klein, 1979; Nelson et al., 1979; Shields et al., 1973). Mineralization kinetics resemble those discussed previously for plant material decomposition. A rapid initial mineralization rate resulting from catabolism of readily available cell components is followed by the slower mineralization of the more resistant substituents. An example of the decomposition rate of exogenously added microbial substituents is provided by Marumoto et al. (1982). They evaluated the decomposition of ^{14}C- and ^{15}N-labeled fungal and bacterial cells in two soils. Mean decomposition for the carbon substituents after 10 days incubation was 43 and 34 percent for parabrown earth and the chernozem soils, respectively. After 28 days incubation, approximately half of the carbon had been mineralized. Cell wall structure does have a controlling effect on this decomposition rate. Mineralization rates for melanoid fungal biomass are significantly reduced in comparison with hyaline fungal tissues (Hurst and Wagner, 1969; Kapoor and Haider, 1982; Malik and Haider, 1982). Malek and Haider (1982) noted that the decomposition of fungal melanins was considerably less than that of the cell wall or cytoplasm and that carbon contributions of these cell components to soil humic acids was low. Most of the carbon was recovered in the soil humin fraction.

These studies indicate that, in general, the biochemical mechanisms involved in the biological decomposition of microbial cell substituents do not in reality differ from those observed with succulent plant tissue. Exceptions to this observation involve the effects of differences between the plant and microbial carbon : nitrogen ratios and the greater protection of the microbial cells by the soil structure. The mechanism of protection of the microbial population in soils was presented in the preceding text. Characteristically bacterial cells have a carbon : nitrogen ratio of approximately 4 or 5 : 1 and the ratio for fungal cells lies between 10 and 15 : 1. These ratios are well within the range necessary to observe net nitrogen mineralization rather than immobilization (Alexander, 1977). Hence, in contrast to the situation with green plant material where a net decrease in available soil nitrogen occurs during the early phase of biomass mineralization, mineral nitrogen pools increase during decomposition of microbial biomass in systems where microbial populations decline.

5.4. EFFECT OF EXOGENOUS ORGANIC MATTER ON DECOMPOSITION OF NATIVE SOIL ORGANIC MATTER

With the development of isotopic labeling procedures for plant residues, it became possible to evaluate the source of mineral nitrogen and carbon dioxide evolved from plant amended soil samples. For the first time, the differentiation of carbon dioxide from amended plant material from that produced from native soil organic matter oxidation became feasible. In an early study, Broadbent and Norman (1946) found that more nonlabeled carbon dioxide (that is, soil organic matter derived carbon dioxide) was evolved from soil receiving labeled plant material than would have been anticipated from total carbon dioxide evolution rates from soil samples not receiving plant amendments. The prevailing assumption was that total carbon dioxide from plant residue amended soil should consist of that carbon dioxide normally produced as a result of native soil organic matter oxidation plus that resulting from catabolism of the organic amendments. This apparently did not occur because Broadbent and Norman detected greater oxidation of soil organic matter carbon than was anticipated. This stimulation of the catabolism of native soil organic matter by exogenously supplied substrates was termed the **priming effect**. Since Broadbent and Norman's report, a number of supportive studies have been published (e.g., Broadbent, 1947; Broadbent and Nakashima, 1974; Hallam and Bartholomew, 1953; Terry,

1980). Accompanying these publications have been several studies where this priming effect has not been observed (e.g., see Bingman et al., 1953; Martin and Haider, 1979). Typical results supportive of the priming effect are depicted in Fig. 5.4 (Broadbent and Nakashima, 1974). In this study, doubly labeled barley (*Hordeum vulgare* L.) grown in a closed chamber containing ^{13}C-enriched carbon dioxide and ^{15}N-enriched nutrient solutions were added to Columbia fine sandy loam. Catabolism of native soil organic matter, as was indicated previously by the increase in total carbon dioxide that contributed by decomposition of the labeled plant substances, was augmented by an increment approximately equal to that contributed by the oxidation of the amended plant materials. Evaluation of the kinetics of this stimulation suggested that the duration of the priming effect would be approximately one year. Other workers have recorded much smaller but significant priming effects of shorter duration. Contrasting data was presented by Martin and Haider (1979) when they examined decomposition of cornstalk lignins, phenols, model phenolase humic polymers, and fungal melanins and the influence of readily available carbon sources on this decomposition in a number of mineral soils. They concluded that the decomposition of phenols, model humic acid polymers, fungal melanins, and lignins under favorable conditions was not significantly influenced by availability of readily metabolized carbon sources.

The repeated observation of the priming effect in a variety of situations by different investigators indicates that it merits serious consideration as a real soil biological phenomenon. The variability in its occurrence makes assessment of the true environmental impact difficult. The inconsistency in detecting the priming effect suggest that it is limited to either special environmental conditions or perhaps even to certain types of laboratory studies. The potential explanations include (1) the phenomenon occurs in laboratory incubated soils only as an artifact of the incubation conditions (natural cycling of partially decomposed substrates between soil organic fractions may be inhibited. This solubilized organic carbon, rather than being returned to the soil organic matter fraction, is oxidized to carbon dioxide); (2) the phenomenon results from an imbalance between microbial growth, substrate amendment, and increased availability of native soil organic matter due to soil manipulation. (The soil microbes would encounter less substrate than they need for synthesis of cell mass or soil organic components. The excess carbon dioxide would result from death and decay of these excess microbial cells as well as the carbon dioxide produced from native soil organic matter oxidation by these cells prior to their death); (3) the excess carbon dioxide may result from temporary changes in the composition of the microbial community due to induction of growth of inactive microbes by the substrate amendment (Organisms which are not competitive under the new conditions die and are decomposed at rates higher than occurred prior to amendment. The cell carbon of these poor competitors becomes the source of the excess carbon dioxide associated with the priming effect); and (4) changes in microbial growth resulting from the limitations imposed by growth on two limiting sub-

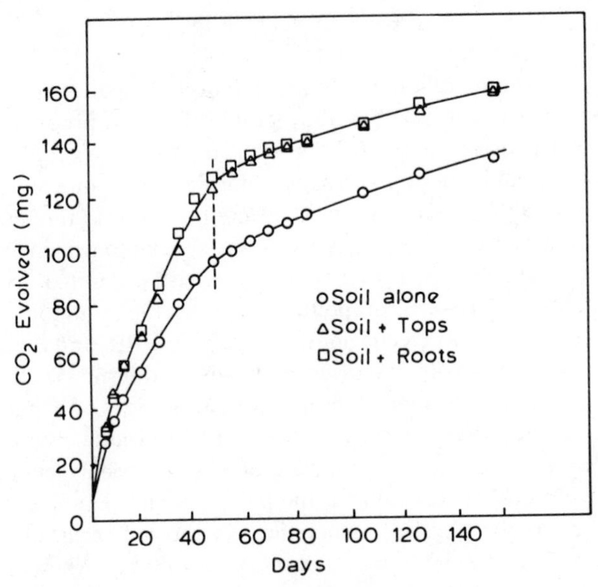

Fig. 5.4.
Carbon dioxide evolution from Colombia fine sand loam, unamended or amended with barley tops and roots. Reproduced from SOIL SCIENCE SOCIETY OF AMERICA PROCEEDINGS, Volume 38, 1974, pp. 313–315 by permission of the Soil Science Society of America.

strates. The latter explanation was presented by Parnas (1976) in his evaluation of effects of variation in the carbon : nitrogen ratio on population biomass.

5.5. OTHER CARBON INPUTS

With modern agricultural procedures, ecosystem management, waste disposal plans, and accidental industrial spills, a number of other biodegradable, carbonaceous compounds enter the soil ecosystem. Although from the view of the total ecosystem, these materials are of minor or perhaps even insignificant consequence, at their point of entry they may have a major localized impact on soil organic matter processes. This probability is limited generally to some waste disposal procedures, such as land farming of oily wastes and accidental industrial spills. Except when included in the category of accidental spills, few pesticides or related soil amendments would be used at concentrations sufficient to impact overall organic matter levels.

5.6. CONCLUSIONS

Mineralization of the readily metabolized soil organic matter fraction provides most of the nutrients necessary for aboveground biomass production. This soil organic matter fraction is derived from plant debris and exudates, animal remains, the soil microbial population, and abiontic soil organic matter fractions. Readily decomposable organic matter is differentiated from the more biodegradation resistant organic matter fractions by its capacity to provide a carbon and energy source for the soil microbial community with a minimal expenditure of energy. This class of organic compounds includes the common cellular polymers, such as proteins, polysaccharides, and nucleic acids, and their monomers. Many compounds are easily classified as being readily metabolized, whereas some materials, such as the more complex polysaccharides, are metabolized at a rate that makes their designation in either class of organic matter debatable. Rate of biodegradation and thus the rate of plant nutrient generation is controlled by the physical associations of the polymers and monomers in the plant material as well as the physical associations in the soil. Degradation of plant debris is complicated by the nitrogen and lignin contents, whereas soil components, including microbial cells, may be preserved by association with colloidal mineral fractions or within soil aggregates. Therefore, control of the rate of aboveground biomass production is limited by the mineralization of the plant debris by the microbial community which is in turn controlled by the quality of the litter produced by the plant itself as well as the physical and chemical properties of the soil.

REFERENCES

Alexander, M. 1977. Introduction to Soil Microbiology. John Wiley & Sons, New York, 544 pp.

Alexander, M. 1981. Why microbial predators and parasites do not eliminate their prey and hosts. Ann. Rev. Microbiol. 35: 113–153.

Allison, F. E., and R. M. Murphy, 1962. Comparative rates of decomposition in soil of wood and bark particles of several hardwood species. Soil Sci. Soc. Am. Proc. 26: 463–466.

Anderson, D. W., and E. A. Paul, 1984. Organo-mineral complexes and their study by radiocarbon dating. Soil Sci. Soc. Am. J. 48: 298–301.

Anderson, G. 1961. Estimation of purines and pyrimidines in soil humic acid. Soil Sci. 91: 156–161.

Anderson, J. M. 1973. The breakdown and decomposition of sweet chestnut (*Castanea sativa* Mill.) and beech (*Fagus sylvatica* L.) leaf litter in two deciduous woodland soils. II. Changes in the carbon, hydrogen, nitrogen, and polyphenol content. Oecologia (Berlin) 12: 275–288.

Anderson, J. P. E., and K. H. Domsch, 1980. Quantities of plant nutrients in the microbial biomass of selected soils. Soil Sci. 130: 211–216.

Bingeman, C. W., J. E. Varner, and W. P. Martin, 1953. The effect of the addition of organic materials on the decomposition of an organic soil. Soil Sci. Soc. Am. Proc. 17: 34–38.

Broadbent, F. E. 1947. Nitrogen release and carbon loss from soil organic matter during decomposition of added plant residues. Soil Sci. Soc. Am. Proc. 12: 246–249.

Broadbent, F. E., and T. Nakashima, 1974. Mineralization of carbon and nitrogen in soil amended with carbon-13 and nitrogen-15 labeled plant material. Soil Sci. Soc. Am. J. 38: 313–315.

Broadbent, F. E., and A. G. Norman, 1946. Some factors affecting the availability of the organic nitrogen in soil—a preliminary report. Soil Sci. Soc. Am. Proc. 11: 246–267.

Casagrande, D. J., and K. Park, 1978. Muramic acid levels in bog soils from the Okefenokee swamp (Georgia). Soil Sci. 12: 181–183.

Cheshire, M. V., C. M. Mundie, and H. Shepherd, 1973. The origin of soil polysaccharide: Transformation of sugars during the decomposition in soil of plant material labelled with ^{14}C. J. Soil Sci. 24: 54–68.

Cheshire, M. V., C.M. Mundie, and H. Shepherd, 1974. Transformation of sugars when rye hemicellulose labelled with C decomposes in soil. J. Soil Sci. 25: 90–98.

Cortez, J., and M. Schnitzer. 1979. Purines and pyrimidines in soils and humic substances. Soil Sci. Soc. Am. J. 43: 958–961.

Dalal, R. C. 1979. Mineralization of carbon and phosphorus from carbon-14 and phosphorus-32 labelled plant material added to soil. Soil Sci. Soc. Am. J. 43: 913–916.

Fogel, R., and K. Cromack, Jr. 1977. Effect of habitat and substrate quality on Douglas fir litter decomposition in western Oregon. Can. J. Bot. 55: 1632–1640.

Foster, R. C. 1981. Polysaccharides in soil fabrics. Science (Washington, D.C.). 214: 665–667.

Gieseking, J. E. (ed.) 1975. Soil Components, Volume 1. Organic Components. Springer Verlag. New York, 534 pp.

Gosz, J. R., G. E. Likens, and F. H. Bormann, 1973. Nutrient release from decomposing leaf and branch litter in the Hubbard Brook Forest, New Hampshire. Ecol. Monogr. 43: 173–191.

Griffin, G. J., M. G. Hale, and F. J. Shay, 1977. Nature and quantity of sloughed organic matter produced by roots of axenic peanut plants. Soil Biol. Biochem. 8: 29–32.

Gupta, U. C., F. J. Sowden, and P. C. Stobbe, 1963. The characterization of carbohydrate constituents from different soil profiles. Soil Sci. Soc. Am. Proc. 27: 380–382.

Hallam, M. J., and W. V. Bartholomew, 1953. Influence of rate of plant residue addition in accelerating the decomposition of soil organic matter. Soil Sci. Soc. Am. Proc. 17: 365–368.

Hurst, H. M., and G. H. Wagner, 1969. Decomposition of ^{14}C-labeled cell wall and cytoplasmic fractions from hyaline and melanic fungi. Soil Sci. Soc. Am. Proc. 33: 707–711.

Jenkinson, D. S., and A. Ayanaba, 1977. Decomposition of carbon-14 labeled plant material under tropical conditions. Soil Sci. Soc. Am. J. 41: 912–915.

Jenkinson, D. W., and J. H. Rayner, 1977. The turnover of soil organic matter in some of the Rothamsted classical experiments. Soil Sci. 123: 298–305.

Kapoor, K. K., and K. Haider, 1982. Mineralization and plant availability of phosphorus from biomass of hyaline and melanic fungi. Soil Sci. Soc. Am. J. 46: 953–957.

Kassim, G., J. P. Martin, and K. Haider, 1981. Incorporation of a wide variety of organic substrate carbons into soil biomass as estimated by the fumigation procedure. Soil Sci. Soc. Am. J. 45: 1106–1112.

Kononova, M. M. 1966. Soil organic matter. Its role in soil formation and in soil fertility. Pergamon Press, New York, 544 pp.

Knapp, E. B., L. F. Elliott, and G. J. Campbell, 1983a. Microbial respiration and growth during the decomposition of wheat straw. Soil Biol. Biochem. 15: 319–323.

Knapp, E. B., L. F. Elliott, and G. S. Campbell, 1983b. Carbon, nitrogen, and microbial biomass interrelationships during the decomposition of wheat straw: A mechanistic simulation model. Soil Biol. Biochem. 15: 455–461.

Malik, K. A., and K. Haider, 1982. Decomposition of ^{14}C-labeled melanoid fungal residues in a marginally sodic soil. Soil Biol. Biochem 14: 457–460.

Martin, J. P., J. O. Ervin, and R. A. Shepherd, 1966. Decomposition of the iron, aluminum, zinc, and copper salts or complexes of some microbial and plant polysacharides in soil. Soil Sci. Soc. Am. Proc. 30: 196–200

Martin, J. P., and K. Haider, 1979. Biodegradation of ^{14}C-labeled model and cornstalk lignins, phenols, model phenolase humic polymers, and fungal melanins as influenced by a readily available carbon source and soil. Appl. Environ. Microbiol. 38: 283–289.

Marumoto, R., J. P. E. Anderson, and K. H. Domsch, 1982a. Decomposition of ^{14}C- and ^{15}N-labeled microbial cells in soil. Soil Biol. Biochem. 14: 461–467.

Marumoto, T., J. P. E. Anderson, and K. H. Domsch, 1982b. Mineralization of nutrients from soil microbial biomass. Soil Biol. Biochem. 14: 469–475.

McGill, W. B., J. A. Shields, and E. A. Paul, 1975. Relation between carbon and nitrogen turnover in soil organic fractions of microbial origin. Soil Biol. Biochem. 7: 57–63.

McClaugherty, C. A., J. D. Aber, and J. M. Melillo, 1982. The role of fine roots in the organic matter and nitrogen budgets of two forested ecosystems. Ecology 63: 1481–1490.

McClaugherty, C. A., J. D. Aber, and J. M. Melillo, 1984. Decomposition dynamics of fine roots in forested ecosystems. Oikos 42: 378–386.

Morita, H., and M. Levesque, 1980. Monosaccharide composition of peat fractions based on particle size. Can. J. Soil Sci. 60: 285–289.

Murayama, S. 1984. Decomposition kinetics of straw saccharides and synthesis of microbial saccharides under field conditions. J. Soil. Sci. 35: 231–242.

Nakas, J. P., and D. A. Klein, 1979. Decomposition of microbial cell components in a semi-arid grassland soil. Appl. Environ. Microbiol. 38: 454–460.

Nelson, D. W., J. P. Martin, and J. O. Ervin, 1979. Decomposition of microbial cells and components in soil and their stabilization through complexing with model humic-acid type phenolic polymers. Soil Sci. Soc. Am. J. 43: 84–88.

Nyhan, J. W. 1975. Decomposition of carbon-14 labeled plant materials in a grassland soil under field conditions. Soil Sci. Soc. Am. J. 39: 643–648.

Parnas, H. 1976. A theoretical explanation of the priming effect based on microbial growth with two limiting substrates. Soil Biol. Biochem. 8: 139–144.

Piper, T. J., and A. M. Posner, 1968. On the amino acids found in humic acid. Soil Sci. 106: 188–192.

Reevey, W. S., and A. G. Norman, 1948. Influence of plant materials on properties of the decomposed residue. Soil Sci. 65: 209–226.

Schmidt, E. L., H. D. Putnam, and E. A. Paul, 1960. Behavior of free amino acids in soil. Soil Sci. Soc. Am. Proc. 24: 107–109.

Schnitzer, M., and S. U. Khan, 1972. Humic substances in the environment. Marcel Dekker, New York, 327 pp.

Shanks, R. E., and J. S. Olson, 1961. First year breakdown of leaf litter in southern Appalachian forests. Science (Washington, D.C.) 134: 194–195.

Shields, J. A., E. A. Paul, and W. E. Lowe, 1973. Turnover of microbial tissue in soil under field conditions. Soil Biol. Biochem. 5: 753–764.

Sorensen, L. H. 1964. Rate of decomposition of organic matter in soil as influenced by repeated air drying—rewetting and repeated additions of organic matter. Soil Biol. Biochem. 6: 287–292.

Sorensen, L. H., and E. A. Paul, 1971. Transformation of acetate carbon into carbohydrate and amino acid metabolites during decomposition in soil. Soil Biol. Biochem. 3: 173–180.

Stevenson, F. J. 1982. Humus chemistry: Genesis, composition, reaction. John Wiley & Sons, New York, 443 pp.

Stott, D. E., J. P. Martin, D. D. Focht, and K. Haider, 1983. Biodegradation, stabilization in humus, and incorporation into soil biomass of 2,4-D and catechol carbons. Soil Sci. Soc. Am. J. 47: 66–70.

Terry, R. E. 1980. Nitrogen mineralization in Florida Histosols. Soil Sci. Soc. Am. J. 44: 747–750.

van Veen, J. A., J. N. Ladd, and M. J. Frissel, 1984. Modelling C and N turnover through the microbial biomass in soil. Plant Soil 76: 257–274.

Verma, L., and J. P. Martin, 1976. Decomposition of algae cells and components and their stabilization through coupling with model humic acid type phenolic polymers. Soil Biol. Biochem. 8: 85–90.

Woods, L. E., C. V. Cole, E. T. Elliott, R. V. Anderson, and D. C. Coleman, 1982. Nitrogen transformations in soil as affected by bacterial-microfaunal interactions. Soil Biol. Biochem. 14: 93–98.

SIX

HUMIFICATION AND ORGANIC MATTER STABILITY

The discussions of soil organic matter components to this point have involved an evaluation of the mineralization of readily metabolizable organic matter in soils and the basic chemical properties of this organic matter fraction which control its susceptibility to biological decomposition. These limitations to biodegradation included traits, such as chemical structure of the biological polymer, its carbon : nitrogen ratio, and its lignin contents. Changes in the chemical complexity of organic matter which increase its biodegradation resistance may also be catalyzed in soil. These reactions are generally referred to as humification. **Humification** is the biological, microbial, or chemical conversion of organic residues to humus. Humus is the relatively biodegradation resistant, predominantly dark brown to black, fraction of soil organic matter. The major sources of organic compounds which are humified are nascent plant and animal components incorporated into soil, the organic intermediates formed during the decomposition of these biological remains, and microbially synthesized products and biomass. Other less common, but perhaps environmental or economically significant materials which are humified are components of industrial and petroleum spills, soil amended pesticides, and other industrially synthesized soil amendments. The primary impact of humification on ecosystem development is that, in many cases, relatively simple organic compounds are removed from the easily metabolized soil organic matter pool and added to the relatively stable soil organic matter fractions.

In soils receiving minimal exogenous nutrient inputs, because of their intermediate biodegradation susceptibility between nonhumified organic carbon substrates and humic substances, humified cell polymers become a significant nutrient source. The size of this pool was suggested by recent analysis of sulfur distribution in soil organic matter fractions. The impact of humification on sulfur availability was investigated by Bettany et al. (1979, 1980). Carbon : nitrogen : sulfur ratios ranged from 113 : 10.6 : 1 to 270 : 23.3 : 1 in conventional humic acid as compared to a range of 61 : 6.5 : 1 to 112 : 9.7 : 1 for total soil organic matter. The conventional humic acid fraction was considered to be the terminal

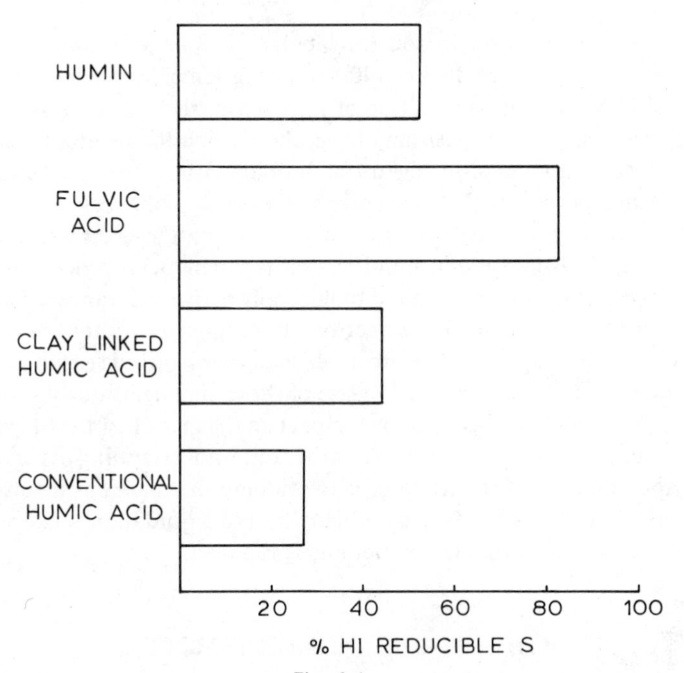

Fig. 6.1.
Average labile sulfur contents of organic fractions from an Ap-horizon of well-drained, cultivated soils of Saskatchawan, Canada (Bettany et al., 1979).

product of humification. Based on the average hydrogen iodide-reducible portion of the sulfur fraction (Fig. 6.1), it was proposed that the majority of the potentially labile sulfur was contained in the fulvic acid, clay associated humic acid, and humin fractions (Bettany et al., 1979). Cultivation of a Udic Haploboroll soil increased the resistance of the soil organic sulfur fraction to biological decomposition (Bettany et al., 1980). Cultivation reduced organic sulfur concentrations in all organic matter fractions analyzed, but 80 percent of the loss was accounted for in the conventional humic acid, clay associated humic acid, and humin fractions. In contrast, the fulvic acid fractions that contained more hydrogen iodide-reducible sulfur, had a narrower carbon : nitrogen : sulfur ratio, and contained what was thought to be the most biodegradation sensitive sulfur, only contributed 14 percent of the loss. It was postulated that the apparent stability of the fulvic acid fraction resulted from decomposition of the humic acids and humin fraction into smaller molecular weight substances which entered the fulvic acid fraction.

Plant chemical composition may affect the humification rate of the residues. For example, high polyphenolic content of the residues may increase the potential for humification. The effect of polyphenol content of plant residues on the quantity of nitrogen retained in humic substances was shown in a study of the decomposition of tea litter in soil (Silvapalan, 1982). Mineralization of high phenolic containing tea residues resulted in formation of large amounts of humic matter and increased the proportion of nitrogen in the humic substances. The data suggested that decomposition of the phenol-rich tea residues leads to formation of large amounts of nitrogen-rich humic matter. It would be interesting to evaluate this process with other polyphenolic-rich plant residues.

From the view of increasing native soil organic matter levels and the positive soil properties derived therein, humification is a beneficial process and should be encouraged. But, humification of biologically active substances, such as herbicides, can have a detrimental effect on subsequent ecosystem development. Herbicides properly applied to control weeds in one growing season may be stabilized into soil humic materials. Release of these chemicals during subsequent growing seasons could have a negative impact on the growth of the plant community developing at that time. Thus to improve our understanding of this basic soil process, the objective of this chapter is to evaluate the reactions involved in humification and the mechanism by which the biodegradation susceptibility of readily metabolizable organic compounds is reduced.

6.1. HUMIFICATION: THE PROCESS

Reactions of interest are those relating to incorporation of preexisting organic compounds into an established pool of humus. Those reactions involving *de novo* synthesis of humic and fulvic acids and humin are discussed in Chapter 8. Because of the greater study to date of humification of the more common cellular components, this discussion will emphasize those reactions relating to humification of proteins, carbohydrates, and major microbial cellular components. An example of the effect of humification on one class of man-made compounds, pesticides, will be presented in Section 6.4. Reactions evaluated herein pertain to most terrestrial ecosystems. Exceptions relate to highly disturbed sites, such as mine wastes, where essentially all native pools of organic matter have been destroyed, and newly created ecosystems, such as volcanic ash or lava flows. In the latter ecosystems, pools of humic substances are established through *de novo* synthesis and humification reactions, but due to the low levels of the reactants, the reactions are difficult to detect.

The probability that a given substrate will be humified depends upon its basic chemical properties and its longevity in soil. Since many humification reactions are controlled by the rate of collision of the substrate molecules, which is in many cases limited by diffusion, the more biologically stable compounds are most likely to be incorporated directly into humic substances than those which are rapidly metabolized. For example, a greater portion of the slowly degraded lignin and its biodegradation intermediates are humified than are readily metabolized plant components (Martin et al., 1980). Distribution of coniferyl alcohol units of cornstalk lignins, free coniferyl alcohols, wheat straw, and the polysaccharide fraction of wheat straw into carbon dioxide, humic acids, fulvic acids, extracted soil, and microbial biomass were examined in Greenfield silt loam and Steinbeck loam (Table 6.1). The readily decomposable fraction of the wheat straw was converted primarily to carbon dioxide (80 and 84 percent of added carbon in Greenfield silt loam and Steinbeck loam, respectively) over a two-year incubation period, where lignin carbon conversion to carbon dioxide ranged from 37 to 69 percent depending upon the location of the ^{14}C-label in the lignin molecule. The greatest proportion of lignin carbon oxidized to carbon dioxide originated from the more labile lignin moieties. Degradation of free coniferyl alcohol to carbon dioxide was intermediate between the two polymeric substrates. With the increased residence time of lignin in soil, that proportion incorporated into humic acids also increased as much as 10-fold over that observed with the more labile plant components. It should also be noted that of the carbon contained in the readily decomposable substrates retained in the soil a larger proportion was incorporated into microbial biomass. As much as 7.8 percent of the soil residual readily decomposable substrates were incorporated into microbial biomass. But, as a result of the greater conversion of these substrates to carbon dioxide, actual quantities of substrate incorporated into microbial biomass carbon, albeit readily decomposible or lignin, were quite similar; that is, less than 0.6 verses 0.8 to 1.2 percent. Most of the wheat straw and wheat straw polysaccharides remained in a form susceptible to 6 N HCl hydrolysis. This suggests that the majority of the residual polysaccharides either remained in the nascent chemical form or was incorporated into newly synthesized soil polysaccharides. The redistribution of the ^{14}C-label of glucose in polysaccharides into other soil polysaccharides was suggested by data of Cheshire et al. (1969). They found that between 60 and 80 percent of the carbon of ^{14}C-labeled glucose and starch incubated in soil was evolved as carbon dioxide in 14 days with glucose and 84 days with the starch. Analysis of the ^{14}C-labeled residues in the soil showed rapid redistribution of the glucose carbons into galactose and mannose.

In a related study (Stott et al., 1983a), the distribution of various plant polymers between carbon dioxide, humic acids, and microbial biomass was examined in Steinbeck loam, Fallbrook sandy loam, Coachella sandy loam, and Holtville sandy loam. Near neutral soil pH levels (pH 7.0 to 7.8) stimulated carbon dioxide evolution, especially early in the incubation. Carbon dioxide evolution was reduced in acid soils (pH 5.0 to 5.5). After one year incubation, the majority (36 to 54 percent) of the residual lignin was retained in the humic acid fraction, whereas only 17 to 20 and 16 to 27 percent of the polysaccharide and protein carbons were found in this fraction. Relatively larger proportions of the readily decomposable compounds were detected in the soil biomass fractions: 4.9 to 7.8, 4.6 to 13.4, 0.1 to 0.7 percent from polysaccharide, protein, and lignin, respectively. The exact quantity depended on the soil type examined.

Residence times of biodegradable and relatively resistant plant carbons in soil vary over a large range. Differences between the long residence times of lignin and those of more biodegradation susceptible substrates are shown in a related study using the Greenfield and Steinbeck soils (Kassim et al., 1981) (Table 6.2).

Table 6.1.
Distribution of Carbons Lignin, Wheat Straw, and Wheat Straw Polysaccharide Fraction as Percent of Labeled Carbon in Carbon Dioxide, Humic Acids, and Microbial Biomass After a Two-Year Incubation Period[a]

Substrate (Label Position)	Greenfield Silt Loam			Steinbeck Loam		
	CO_2	Humic Acid	Biomass	CO_2	Humic Acid	Biomass
	Percent[b]					
	Coniferyl Alcohol Units of Cornstalk Lignins					
1-^{14}C	63	17	—	57	22	—
2-^{14}C	41	28	<0.6	41	31	<0.6
Ring-^{14}C	41	28	<0.6	37	35	<0.6
$O^{14}CH_3$	69	15	—	55	27	—
	Wheat Straw					
Unlabeled	74	9.4	1.2	71	15	0.9
	Polysaccharide Fraction of Wheat Straw					
Unlabeled	84	3.5	1.2	80	5.8	0.8

[a]Source: Martin et al., 1980.
[b]Percentage of applied ^{14}C evolved as carbon dioxide over a two-year incubation period.

For these compounds, 61 to 99 percent of added carbon was oxidized to carbon dioxide over a 12-week incubation period as opposed to the one- and two-year incubation periods used in the previously cited work. Most of the carbon retained in the soil was incorporated into humic materials in that only 6.1 to 34.2 percent of the residual carbon (0.1 to 7.2 percent of the total labeled carbon added) was detected in microbial biomass. In comparison, over the 12-week incubation period 72, 72, and 62 percent of cellulose, glucosamine, and wheat straw carbon, respectively, evolved as carbon dioxide. Proportions of the carbon incorporated into microbial biomass were similar to the more easily metabolized substrates. Phenolic compounds were more stable in the Greenfield and Steinbeck soils over a 12 week incubation (Table 6.3) (Kassim et al., 1982). The percent carbon converted to carbon dioxide ranged from 26 to 88 percent. The proportion retained as microbial biomass was similar to that previously noted with the other carbon sources. The similarity in actual quantities of carbon incorporated into the microbial biomass from the more degradable substrates suggests the relationship of the carbon in this soil organic matter fraction to total soil organic carbon and the transient nature of the microbial biomass. Microbial biomass comprises a small portion of total organic carbon. More importantly, the carbon in microbial biomass is turning over constantly resulting in a redistribution of the carbon atoms into new microbial biomass, carbon dioxide, and soil humic materials. Thus the labeled carbon from the amended substrates en-

Table 6.2.
Biodegradation and Incorporation into Biomass of Easily Decomposable Carbon Compounds Following Incubation for 12 Weeks in Greenfield Silt Loam and Steinbeck Loam[a]

Substrate	Soil	C Loss as CO_2[b]	Biomass[b]
Glycine	Steinbeck	83	1.6
Serine	Steinbeck	88	2.8
Cysteine	Steinbeck	68	4.6
Alanine	Steinbeck	78	2.6
Leucine	Greenfield	94	1.1
	Steinbeck	96	1.6
Tyrosine	Steinbeck	92	0.6
	Greenfield	91	0.2
Uracil	Steinbeck	97	0.1
Cytosine	Steinbeck	99	0.1
Uridine	Steinbeck	85	3.1
Cellulose	Greenfield	72	7.2
	Steinbeck	68	6.1
Glucosamine	Greenfield	72	3.5
	Steinbeck	67	6.4
Wheat straw	Greenfield	62	4.9
	Steinbeck	43	4.0

[a]Reproduced from SOIL SCIENCE SOCIETY OF AMERICA JOURNAL, Volume 46, 1982, pp. 305–309 by permission of the Soil Science Society of America.
[b]As percent of applied carbon.

ters the microbial cellular carbon only to be eventually dislodged by unlabeled carbon from subsequent substrates metabolized by the organisms.

The fate of the soil amended carbon is also concentration dependent (Table 6.4) (Kassim et al., 1982). In all cases as the carbon concentrations increased, that proportion yielded as carbon dioxide increased. A similar effect on incorporation of the carbon into the microbial population was not observed. More significantly, a time effect was observed with the biomass. With long incubation times, the proportion of the carbon in microbial biomass decreased. This supports the hypothesis of an active conversion of microbial biomass to humic substances following death of the cell. Variously labeled monomers of lignin also behaved in a manner similar to that of lignin when amended to soil. After one year incubation in the Greenfield silt loam, 93 to 96 percent of the residual 2-^{14}C ferulic acid, ring ^{14}C ferulic acid, and $O^{14}CH_3$ ferulic acid, respectively, were found in the humus fraction, whereas 76 to 86 percent of the glucose residual carbon was found in the same fraction. The majority of the ferulic acid was retained in the humic acid fraction, whereas the glucose was retained in the ex-

Table 6.3.
Incorporation into Microbial Biomass and Biodegradation of Aromatic Ring Containing Compounds Following 12 Weeks Incubation in Steinbeck Loam and Greenfield Silt Loam[a]

Compound and Label	Soil	C Loss as CO_2[b]	Biomass[b]
2-^{14}C ferulic acid	Greenfield	54	5.1
	Steinbeck	54	5.5
3-^{14}C ferulic acid	Greenfield	70	0.3
	Steinbeck	78	0.3
Ring-^{14}C ferulic acid	Greenfield	56	1.9
	Steinbeck	61	4.2
$O^{14}CCH_3$ ferulic acid	Greenfield	69	1.1
	Steinbeck	77	0.5
Ring-^{14}C catechol	Greenfield	26	1.2
Ring-^{14}C anisic acid	Greenfield	88	2.2
	Steinbeck	82	3.4

[a]Reproduced from SOIL SCIENCE SOCIETY OF AMERICA JOURNAL, Volume 46, 1982, pp. 305–309 by permission of the Soil Science Society of America.
[b]Percentage of ^{14}C carbon evolved as carbon dioxide.

tracted soil (humin fraction) suggesting a polysaccharide nature of the product resulting from glucose metabolism.

Necessarily, organic compounds containing more reactive groupings are also more likely to be humified than less reactive compounds. Although many substances become associated with humic substances through noncovalent bonding mechanisms, such as hydrogen bonding, ionic bonding, and van der Waals forces, greatest stabilization results from formation of covalent bonding associations. Primary covalent linkage of polysaccharides to humic acids occurs via ester linkages (R_1COOR_2), whereas proteins become associated with humic acids through a number of bonding mechanisms. These include formation of peptide bonds, direct binding to aromatic rings, and linkage to quinone rings. For a more complete discussion the chemical properties of these associations, see Stevenson (1982). Development of modern techniques for analysis of humic components has resulted in contrasting conclusions relating to the importance of various humified materials in humic and fulvic acid fractions. For example, using $^{18}O/^{16}O$ ratios, Dunbar and Wilson (1983) suggest that the principle source of oxygen in humic and fulvic acids is cellulose or other plant carbohydrates and not lignin. In contrast, Wilson et al. (1983) using ^{13}C-cross polarization nuclear magnetic resonance spectroscopy with magic angle spinning suggest that "humification pathways in which carbohydrates are incorporated into humic substances via nonhydrolyzable linkages are not important for soils investigated in this work." These data demonstrate the problems of generalization of data col-

Table 6.4.
Effect of Substrate Concentration and Length of Incubation Period of Decomposition of Glucose and Ferulic Acid Carbons and Incorporation of these Carbons into Soil Microbial Biomass[a]

		Weeks of Incubation					
		2		12		52	
Substrate	Concentration	CO_2	Biomass	CO_2	Biomass	CO_2	Biomass
	μg/g Soil	Percent[b]					
2-^{14}C ferulic acid	10	34	9.3	50	3.3	63	1.9
	1,000	40	7.8	58	5.0	72	2.0
	10,000	44	—	66	2.5	78	0.9
Ring-^{14}C ferulic acid	1	26	4.9	39	2.5	54	1.8
	1,000	51	3.7	62	2.2	74	1.8
	10,000	55	—	70	0.8	82	0.7
$O^{14}CH_3$ ferulic acid	1,000	75	0.8	82	0.6	88	0.6
UL Glucose	1	41	24.0	60	10.8	68	7.7
	1,000	63	10.4	76	14.4	81	3.0
	10,000	68	—	83	3.2	89	1.5

[a]Kassim et al., 1982.
[b]Percent of total ^{14}C added to soil.

lected from a few soil types and ecosystems to overall soil organic matter properties. As modern techniques are applied to a greater number of soil samples, the chemical pathway of humification processes will be clarified. Of greater interest for this study of humification is the fact that readily decomposable substrates do become covalently bonded to humic and fulvic acids and the impact of this association on biodegradation resistance.

6.2. BIODEGRADATION EFFECTS OF HUMIFICATION

At one time the infallibility of the microbial community in decomposition of organic compounds was essentially a maxim; that is, microorganisms were considered capable of developing mechanisms to decompose all organic compounds. With the advent of the modern organic chemical industry and the resultant widespread distribution of synthetic organics in the soil environment, it was soon realized that this maxim was, at best, of limited validity. A better statement regarding microbial biodegradation capabilities would be that microorganisms are capable of decomposing biologically synthesized organic compounds plus a variety of synthetic organics providing the environmental conditions are right for growth and development of the microbe. Preservation of human remains in the extreme cold of northern climatic zones or the extremely xeric conditions of desert regions is well known. This mineralization capacity of the microbial community does not imply rapid disappearance of the compounds from the ecosystem. Degradation rates can vary from the extremely rapid catabolism commonly detected with simple sugars amended to soil to the nearly imperceptible catabolism of the more complex polymers, such as lignin and humic acids. The rate at which an organic compound is degraded to carbon dioxide, water, ammonia, and so on in soil depends both on its basic structure and the nature of its interaction with other soil components. The most stabilizing interaction results from humification of the compound or its derivatives, but minor modifications of the chemical structure may also retard mineralization.

Most of the studies of the impact of minor structural variation are autecological in that they are conducted with axenic cultures of the microorganism of interest. This provides a basic understanding of the biochemical mechanisms involved in the catabolic reactions but gives a limited view of the actual situation in native soils. Frequently, a chemical structure that provides a barrier or limitation in the oxidation of a given substrate by one microbial isolate is readily breached by another microbe. Thus it must be realized that most types of metabolic limitations are not universally operative within the soil microbial community. Generally, the actual degradation rate of any specific compound, if an accurate measure of the half life of the compound is needed, must be assessed in the soil of interest.

Minor changes in the molecular structure of rather simple organic compounds can reduce the mineralization rate drastically. This has been exploited in the soil additive industry where it is desirable to produce compounds with a life in soil sufficient to achieve the primary objective but with biodegradation susceptibility such that the compound does not accumulate in the ecosystem. Examples of structural changes affecting stability are provided by studies of methyl substitution of linear styrene dimers (Higashimura et al., 1983), benezene sulfonate (Ripin et al., 1975), and ethyl esters of chlorinated carboxylic acids (Paris et al., 1984). Degradation of linear unsaturated dimers prepared from styrene and *o*-, *m*-, and *p*-methyl styrene by *Alcaligenes* sp. strain 559 and *Pseudomonas* sp. strain 419 was examined by Higashimura et al. Both isolates decomposed all styrene dimers, and methylstyrene-styrene dimers, but were unable to catabolize methylstyrene-methylstyrene dimers. Degradation rates depended upon the molecular structure of the dimer and the microbial strain tested. Also, the metabolism of methylstyrene-styrene dimers by *Alcaligenes* sp. was stimulated by the presence of styrene dimers, whereas styrene dimers had no effect on this catabolic reaction by the *Pseudomonas* sp.

Alkyl benzene oxidation by *Pseudomonas testosteroni* H-8 is inversely proportional to the length of the alkyl group (Ripin et al., 1975). Significant respiration of benzene sulfonate, toluene sulfonate, and ethylbenzene sulfonate by this bacterium was observed, but propylbenzene sulfonate and higher homologs were not metabolized.

Metabolism of carboxylic acid esters with a fixed aromatic moiety and an increasing chain length of the alkyl group and ethyl esters of chlorine-substituted carboxylic acids by *Pseudomonas putida* and a number of isolates from waters has been examined (Paris et al., 1984). The microbial rate constants for these compounds varied over a 50-fold range. Rate constants for decomposition of carboxylic acid esters with a fixed aromatic moiety increased with increasing length of the alkyl substituents.

Irrespective of the basic catabolic rate for the simple organics in soil, complexing with humic acid polymers greatly reduces their decomposition susceptibility. This has been demonstrated for a variety of simple cell substituents, such as amino sugars and proteins, as well as components of algal and fungal cells.

Proteins, peptides, and amino acids are rapidly metabolized in soil. Verma et al. (1975) found that after 4 to 12 weeks incubation, between 71 and 95 percent of the ^{14}C-labeled proteins, peptides, and amino acids amended to soil were oxidized to ^{14}C-labeled carbon dioxide. Conversion rates of greater than 90 percent were only detected with ^{14}C-labeled carboxy groups of the amino acids. Metabolism of the amino acids could be reduced by a variety of associations with soil humic acids, but greatest reduction was found when the substrates were covalently combined with the humic acids. Simply mixing the freeze-dried amino acid compounds with freeze-dried humic acid reduced protein decomposition by approximately 35 percent but had little effect on amino acid oxidation. A more intimate association of the humic acid-amino acid mixture achieved by freeze-drying solutions of amino acids, peptides, and proteins plus the humic acids, reduced protein catabolism by 60 percent. But, amino acid oxidation was only slightly reduced. Covalent linkage of the proteins, peptides, and individual amino acids into model phenolase polymers (humic acid-like compounds) resulted in a reduction of the degradation by 80 to 90 percent. The amino acid portions of the polymer molecules were a little more susceptible to biological degradation than were the aromatic fractions. Position in the polypeptide chain in relationship to the covalent bonding to the aromatic polymer also affected biodegradation susceptibility. Although all residues of di- and tripeptides linked to the humic acid-like polymer were metabolized, the N-terminal amino acid, which was linked to the phenolic polymer, was more stable than the carboxy terminal amino acid. This suggests greater steric hindrance to the enzyme approaching the N-terminal amino acid.

Amino sugars are stabilized into humic acid type polymers through nucleophilic addition of the amino group to the aromatic nuclei of quinones (Bondietti et al., 1972). Over a 12-week incubation in Greenfield sandy loam, 15 to 23 percent of the ^{14}C-label of glucosamine covalently linked to humic acid-like polymers was evolved as ^{14}C-labeled carbon dioxide. Over the same incubation period over 70 percent of the labeled glucosamine was mineralized to carbon dioxide in the absence of humic acid-like polymers or simply mixed with the humic polymer. Polyglucosamine (chitosan) mineralization was also reduced by linkage to the humic acid-like polymer. In this instance, 31 percent of the covalently bonded chitosan was evolved as carbon dioxide after eight weeks incubation in the soil as compared to 53 percent in the absence of the humic acid-like polymer.

Decomposition of microbial cellular components is also affected by humification reactions. This was demonstrated in an examination of the decomposition of algal cellular components in Greenfield sandy loam (Verma and Martin, 1976). Relative decomposition rates of cytoplasmic and cell wall fractions, both free and complexed with humic acid-like compounds are presented in Table 6.5. The complexity of the cellular components and the range in susceptibility to biological oxidation is shown by approximate three-fold difference in decomposition rates of *Anabaena flos-aquae* cell material during the early phase of the incubation. Fourteen percent of the cell wall carbon and 42 percent of the cytoplasmic carbon was oxidized to carbon dioxide during the first few days of incubation. The difference in the oxidation rate of these two cell fractions was less obvious with the *Nostoc muscorum*, although a nearly two-fold difference was still observed. Complexing of either cell wall or cytoplasmic components with humic acid-like polymers reduced this initial decomposition by approximately 50 percent to as much as 85 percent, depending upon the component and the algal species. This diminished decomposition rate was maintained throughout the incubation period. After 22 weeks the percent mineralization of the complexed cell components was still 24 to 57 percent of that observed with the noncomplexed materials.

6.3. HUMIFICATION OF PESTICIDES

Short term organic matter effects on pesticide efficacy have long been known. These interactions may result in the need to use larger dosages than are commonly used in low organic matter soils to achieve the desired effect. Also, the longevity of the pesticide may be increased as a result of limitations to biodegradation resulting from the physical association of the pesticide with humus. For example, the levels of 2-chloro-6-(trichloromethyl)pyridine, a nitrification inhibitor, necessary to inhibit nitrification in Pahokee muck, a well humified organic soil, are 50- to 100-fold those recommended as effective in mineral soils (Tate, 1977). This interference presumably results from the association of the inhibitor with the soil organic matter thereby reducing the quantity of active ingredient that the microorganisms encounter. Similarly, interactions with the organic matter of organic soils may retard pesticide decomposition. Lindane degradation was found to be slower in organic soils compared to mineral soils (Mathur and Saha, 1977). Both of these phenomena result by protecting the pesticide from encounter with the microbial community by physically blocking the approach to the substrate by the microbial cells and enzymes.

Of greater concern is the potential for long-term delayed actions of a pesticide within an ecosystem, wherein the pesticide could have an affect upon subsequent crops or on the ecological succession. Because of the potential for crop loss, due

Table 6.5.
Decomposition of Cell Wall and Cytoplasm of *Anabaena flos-aquae* and *Nostoc muscorum*, Free and Complexed with Model Phenolase Polymers, in Greenfield Sandy Loam[a]

	Percent Mineralization						
	Weeks						
Amendment	0.36	1	2	4	6	12	22
	Anabaena flos-aquae						
Cell walls	14	25	36	45	46	55	61
Cell walls-C[b]	5	9	15	17	23	33	35
Cytoplasm	42	53	61	67	70	73	76
Cytoplasm-C	6	8	10	12	14	17	18
	Nostoc muscorum						
Cell walls	23	36	44	52	57	67	70
Cell walls-C	13	17	23	26	30	39	40
Cytoplasm	35	55	58	65	69	74	75
Cytoplasm-C	8	11	15	18	22	26	27

[a]Verma and Martin, 1976.
[b]Complexed with model phenolase polymers.

to immediate death of the plant in the field—in the case of herbicides—as well as enforced destruction of the harvested product due to pesticide contamination, retention of pesticides in soil over one or more growing seasons could be viewed as both a public health and economic concern. The proposed scenario involves (1) the proper utilization of a soil applied pesticides followed by (2) stabilization of that pesticide in the soil organic matter. This pesticide thereby is retained in the soil in an unchanged form for one or more growing seasons. Later, as the organic component to which the pesticide is attached is decomposed or chemically modified in a manner to reduce the attraction of the pesticide, this pesticide is again released into the soil. At that time, it may be incorporated into the growing crop. If the crop is one that is expected to contain residues of that pesticide, there is no problem. Alternatively, should the pesticide not be appropriate for the subsequent crop, the product will likely not be marketable. This has been demonstrated to occur (Lichtenstein, 1980; Bartha, 1980). A number of pesticides have been shown to become unextractable, that is, bound to soil humus and subsequently released and incorporated into soil flora and fauna. For example, Lichtenstein (1980) reported the incorporation of bound parathion into oats and earthworms.

The pesticides may become associated with soil organic matter via a number of temporary interactions, such as hydrogen bonding and van der Waals forces, but the most significant association from the view of long-term stability involves covalent bonding of the pesticide to the humic acid. Wolf and Martin (1976) found that chloropropham and 2, 4-dichlorophenoxyacetic acid (2, 4-D) is bound into melanins of fungi which may be subsequently incorporated into soil humic acids. Subsequent work by Stott et al. (1983) demonstrated that 2, 4-D was rapidly decomposed in soil and not polymerized into model humic acid polymers, therefore direct bonding of this herbicide to soil humic acids appeared unlikely. But, 2, 4-dichlorophenol, a decomposition product of 2,4-dichlorophenoxyacetic is incorporated into humic acid-like polymers by phenolase of *Rhizoctonia praticola* (Sjoblad and Bollag, 1977).

Soil bound 3, 4-dicloroaniline, a biodegradation intermediate of a number of herbicides, has been indicated as a source of contamination of rice grain (Still et al., 1980). With the positive evidence of crop contamination by humus-bound pesticide residues, the duration of this contaminant source in the soil becomes of issue. Land management practices that lead to reducing levels of this bound pesticide product may become necessary. Stimulation of microbial mineralization of the residues appears to be a potential solution in that this humus bound pesticide may, at least in part, be susceptible to microbial mineralization. You and Bartha (1982) found that mineralization of both free and humus-bound 3, 4-dicloroaniline by *Pseudomonas putida* was enhanced by aniline. The aniline was postulated to enrich for microbial populations and to induce pathways responsible for the cometabolic catabolism of 3, 4-dicloroaniline. This herbicide decomposition product is apparently associated with a more readily decomposable soil humic acid fraction. Saxena and Bartha (1983) found that intact humic acid-3, 4-dichloroaniline complexes were decomposed considerably faster than the average turnover rate of native soil organic matter. Washing and treating the pesticide-humic acid complexes to remove adsorbed pesticide or that bound with susceptible chemical bonds yielded a product that was more stable than native soil humic acid, but the decomposition rates were identical to those of identically treated humic acid. The authors therefore concluded that extensive accumulation of the pesticide or similarly derived halogenated anilines in soil organic matter is unlikely. The assumption is that decomposition of the initial more biodegradable product is sufficient to preclude including the pesticide containing polymer into more stable humic acid fractions.

These studies demonstrate a practical, economically important aspect of the humification process. For a more detailed examination of humification of pesticides, see the Lichtenstein (1980) and Bartha (1980) reviews. With a better understanding of the bonding mechanism of the pesticide with the humic acid and the subsequent potential for biodegradation, it will be possible to clarify the economic impact of this soil organic matter formation reaction.

6.4. CONCLUSIONS

Humification is a generally favorable soil process because it augments resources of soil organic matter and the plant nutrients contained therein. Although a number of weak chemical associations of plant, animal, and microbially synthesized organics with soil humus occur, the primary linkage involved in long-term stabilization of the organic matter is the covalent bond. This bonding may be similar to biologically synthesized linkages—for example, peptide bond formation between amino acids, peptides, and proteins and humic acid—or more unique to the abiontic and abiotic system, such as the bonding of proteins to quinone rings. Properties controlling the probability that any individual organic compound becomes linked to soil humus include the primary structure of the cellularly synthesized product as well as its longevity within the soil ecosystem. These associations, be they similar or dissimilar to enzymatically formed linkages, result in biological stabilization of the previously readily decomposable product. Such stabilization could decrease the mineralization susceptibility as much as 10-fold or more.

REFERENCES

Bartha, R. 1980. Pesticide residues in humus. ASM News. 46: 356-360.

Bettany, J. R., J. W. B. Stewart, and S. Saggar, 1979. The nature and forms of sulfur in organic matter fractions of soils selected along an environmental gradient. Soil Sci. Soc. Am. J. 43: 981-985.

Bettany, J. R., S. Saggar, and J. W. B. Stewart, 1980. Comparison of the amounts of sulfur in soil organic matter fractions after 65 years of cultivation. Soil Sci. Soc. Am. J. 44: 70-75.

Bondietti, E., J. P. Martin, and K. Haider, 1972. Stabilization of amino sugar units in humic-type polymers. Soil Sci. Soc. Am. J. 36: 597-602.

Cheshire, M. V., C. M. Mundie, and H. Shepherd, 1969. Transformation of ^{14}C glucose and starch in soil. Soil Biol. Biochem. 1: 117-130.

Dunbar, J., and A. T. Wilson, 1983. The origin of oxygen in soil humic substances. J. Soil Sci. 34: 99-103.

Higashimura, T., M. Sawamoto, T. Hiza, M. Karariwa, A. Tsuchii, and T. Suzuki, 1973. Effect of methyl substitution on microbial degradation of linear styrene dimers by two soil bacteria. Appl. Environ. Microbiol. 46: 386-391.

Kassim, G., J. P. Martin, and K. Haider, 1981. Incorporation of a wide variety of organic substrate carbons into soil biomass as estimated by the fumigation procedure. Soil Sci. Soc. Am. J. 45: 1106-1102.

Kassim, G., D. E. Stott, J. P. Martin, and K. Haider, 1982. Stabilization and incorporation into biomass of phenolic and benzenoid carbons during biodegradation in soil. Soil Sci. Soc. Am. J. 46: 305-309.

Lichtenstein, E. P. 1980. "Bound" residues in soils and transfer of soil residues in crops. Residue Rev. 76: 147-153.

Martin, J. P., K. Haider, and G. Kassim, 1980. Biodegradation and stabilization after 2 years of specific crop lignin, and polysaccharide carbons in soils. Soil Sci. Soc. Am. J. 44: 1250-1255.

Mathur, S. P., and J. G. Saha, 1977. Degradation of lindane-^{14}C in a mineral soil and in an organic soil. Bull. Environ. Contamn. Toxicol. 17: 424-430.

Paris, D. F., N. L. Wolfe, and W. C. Steen, 1984. Microbial transformation of esters of chlorinated carboxylic acids. Appl. Environ. Microbiol. 47: 7-11.

Ripin, M. J., T. M. Cook, K. F. Noon, and L. E. Stark, 1975. Bacterial metabolism of arylsulfonates: Role of meta cleavage in benzene sulfonate oxidation by *Pseudomonas testosteroni*. Appl. Microbiol. 29: 382–387.

Saxena, A., and R. Bartha, 1983. Microbial mineralization of humic acid-3,4-dichloroaniline complexes. Soil Biol. Biochem. 15: 59–62.

Sivapalan, K. 1982. Humification of polyphenol-rich plant residues. Soil Biol. Biochem. 14: 309–310.

Sjoblad, R. D., and J.-M. Bolag, 1977. Oxidative coupling of aromatic pesticide intermediates by a fungal phenol oxidase. Appl. Environ. Microbiol. 33: 906–910.

Stevenson, F. J. 1982. Humus Chemistry. John Wiley & Sons. New York, 443 pp.

Still, C. C., T.-S. Hsu. and R. Bartha, 1980. Soil-bound 3,4-Dichloroaniline: Source of contamination in rice grain. Bull. Environ. Contam. Toxicol. 24: 550–554.

Stott, D. E., G. Kassim, W. M. Jarrell, J. P. Martin, and K. Haider, 1983a. Stabilization and incorporation into biomass of specific plant carbons during biodegradation in soil. Plant Soil 70: 15–26.

Stott, D. E., J. P. Martin, D. D. Focht, and K. Haider, 1983b. Biodegradation, stabilization in humus, and incorporation into soil biomass of 2, 4-D and chlorocatechol carbons. Soil Sci. Soc. Am. J. 47: 66–70.

Tate, R. L. III 1977. Nitrification in Histosols: A potential role for the heterotrophic nitrifier. Appl. Environ. Microbiol. 33: 911–914.

Verma, L., J. P. Martin, and K. Haider, 1975. Decomposition of carbon-14-labeled proteins, peptides, and amino acids; free and complexed with humic polymers. Soil Sci. Soc. Am. Proc. 39: 279–284.

Verma, L., and J. P. Martin, 1976. Decomposition of algal cells and components and their stabilization through complexing with model humic acid-type phenolic polymers. Soil Biol. Biochem. 8: 85–90.

Wilson, M. A., S. Heng, K. M. Goh, R. J. Pugmire, and D. M. Grant, 1983. Studies of litter and acid insoluble soil organic matter fractions using ^{13}C-cross polarization nuclear magnetic resonance spectroscopy with magic angle spinning. J. Soil Sci. 34: 83–97.

Wolf, D. C., and J. P. Martin, 1976. Decomposition of fungal mycelia and humic-type polymers containing carbon-14 from ring and side-chain labeled 2, 4-D and chloropropham. Soil Sci. Soc. Am. J. 40: 700–704.

You, I.-S., and R. Bartha, 1982. Stimulation of 3, 4-dichloroaniline mineralization by aniline. Appl. Environ. Microbiol. 44: 678–681.

SEVEN

LIGNIN: DECOMPOSITION AND HUMIFICATION

Because of structural and chemical similarities, lignin has long been considered to be a humic acid precursor (Stevenson, 1982). This is in contrast to the fate in soil of most common plant and animal biochemicals which are predominantly oxidized to carbon dioxide and water during microbial cellular energy production with a small portion incorporated into microbial biomass. Thus the basic molecular structure of lignin as well as its preponderance for incorporation into humic acids are unique among biological polymers. Lignin, which is synthesized from sinapyl, coniferyl, and coumaryl alcohols, contains a variety of complex organic linkages that are considerably less common in other major plant or animal biochemicals. For example, spruce (*Picea abies*) lignin has been indicated to be comprised of 48 percent arylglycerol-β-aryl ethers, 6 to 8 percent noncyclic benzyl aryl ethers, 9.5 to 11 percent biphenyl, 7 percent 1, 2-diarylpropane structures, 9 to 12 percent phenylcoumaran structures, and 3.5 to 4 percent diphenyl ethers (Crawford, 1981). This complexity of components is accentuated by the high degree of randomness of the polymer as well as the extreme variation in molecular composition of lignins isolated from various plant sources. This randomness and molecular complexity greatly retard biological degradation of this polymer. Biodecomposition susceptibility and rates are also affected by the covalent attachment of a variety of other plant polymers to the lignin molecule, such as the association of cellulose with lignin in lignocellulose. The diversity in molecular structure and randomness of the polymer increase the biological stability through steric hindrance of the enzymes approaching the point of expression of their activity on the lignin molecule and by increasing the number of enzymatic activities required for lignin mineralization. For decomposition of a simple polymer containing uniform linkages, such as the peptide bond of protein, a few enzymes may be all that are absolutely necessary for cleavage of the polymer. But with lignin, the complexity of the molecule and the diversity of the subunits and linkages between the subunits predict requirements for a number of different enzymatic activities and mechanisms for depolymerization. This is further complicated by the energy expenditures necessary for cleavage of aro-

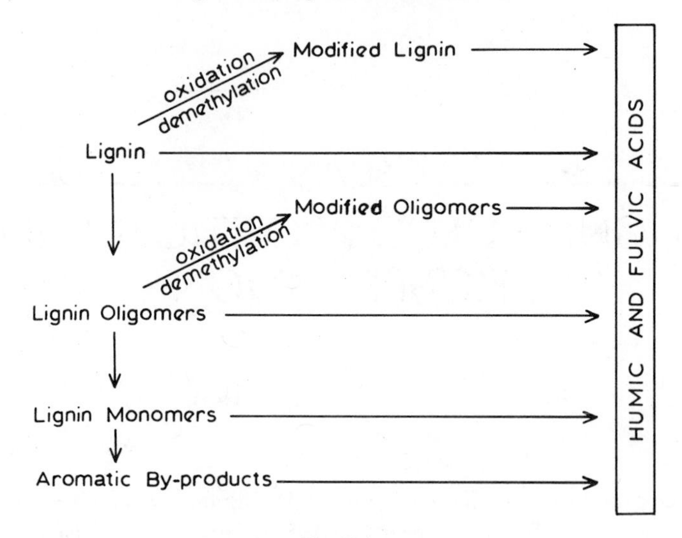

Fig. 7.1.
Major pathways for the conversion of lignin to humic substances.

matic ring structures before the microbe derives any energy return. These enzymatic activities include a variety of etherases and polyphenol oxidases. Also, for lignocellulose, the cellulase complex is necessary for total biodecomposition of the molecule. The complexity of the substrate, the diversity of the enzymes necessary for its decomposition, and the high energy investment required by the microbe before it gains energy for cellular growth and development, all contribute to the long half-life of lignin in soil. As was indicated in Chapter 6, increased longevity of biopolymers or their substituents in soil augments the probability of their becoming associated with soil humic substances through humification reactions. Thus lignin has been implicated as a primary precursor of soil humic substances.

Lignin may be linked into humic acids in its nascent form, or lignin carbons may be humified through the bonding of a variety of degradation intermediates or monomer units Fig. 7.1. For example, Ertel et al. (1984) found that the relative distribution of phenols in humic substances extracted from river and lake waters suggested a lignin origin. Recognizable lignin was detected primarily in humic acids as opposed to the fulvic acid fractions. Examples of incorporation of various lignin intermediates as well as the nascent lignin structure into soil organic fractions are provided in Section 7.3 where the fate of lignin in the soil ecosystem is presented.

Understanding of the biochemical mechanisms of lignin decomposition and the environmental fate of this compound has been greatly enhanced in recent years through the development of techniques for synthesis of ^{14}C-labeled native lignin and lignocelluloses from a variety of plant sources and of methods to separate soil organic fractions into biotic and abiontic fractions; that is, the soil fumigation procedure (see Section 3.2.2). Natural ^{14}C-labeled lignocelluloses, with the isotopic label in the lignin moiety, are prepared by feeding the plant labeled lignin precursors, such as phenylalanine (Crawford and Crawford, 1976), phenylalanine, tyrosine, or cinnamic acid (Benner et al., 1984a), and coumaric acid (Haider et al., 1977). A measure of the decomposition of specific molecular components of the lignin is provided by using precursors with individual carbons labeled—for example, methoxy groups, side chains, or ring labels. The labeled precursor apparently is incorporated solely into the lignin moiety of the lignocellulose although some have questioned the absolute specificity of the labeling. The specificity of lignocellulose substituent labeling was tested by Crawford and Crawford (1976) through the use of *Thermonospora fusca* ATCC 27730, an organism capable of catabolism of the cellulose but not the lignin moiety of lignin cellulose. They found that none of the ^{14}C-label from the lignin precursors was evolved as carbon dioxide during the catabolism of the labeled lignocellulose by this organism. But, decomposition of the lignin portion of this labeled lignocellulose by *Polyporus versicolor* resulted in the production of ^{14}C-labeled carbon dioxide. Thus this procedure allows production of a lignin or lignocellulose substrate whose metabolism can be traced in complex ecosystems. For such studies, optimization of the substrate concentration in the reaction mixture is apparently necessary in that Baker (1983) found that varying the amount of labeled substrate or leaf material containing the labeled substrate resulted in nonlinear changes in the lignocellulose mineralization rate. Questions relating to specificity of the label were raised by Benner et al. (1984b) when they evaluated the distribution of ^{14}C-labeled phenylalanine, tyrosine, and cinnamic acids in nonhumic components of labeled lignocelluloses prepared from a variety of aquatic macrophytes. They found 8 to 24 percent of the label from phenylalanine and tyrosine in protein, whereas only 3 percent of the labeled cinnamic acid was found in protein. They proposed that microbial catabolism of the protein associated label could result in an over estimation of lignolytic activity.

A variety of other assay procedures, including use of alternate substrates and decolorization of dyes, have also been proposed as measures of lignin mineralization decomposition. Crawford et al. (1981) suggested the use of polyguaiacol as a model polymer for lignin decomposition studies. They prepared a polymer of ring-labeled (^{14}C) *o*-methoxyphenol (^{14}C) guaiacol through peroxidase-hydrogen peroxide catalyzed oxidation of the labeled monomers. The polymer contained 67.71 percent carbon, 5.09 percent hydrogen, 27.49 percent oxygen, 25.44 percent methoxyl, and 8.60 percent phenolic hydroxyls. The average molecular weight ranged from 5000 to 15,000 daltons, as estimated by gel chromatography. The polymer was comprised primarily of *o*-*o*, and *p*-*p* linked guaiacol units, but occasional *o*-*p*-biphenyl and some *p*-diphenoquinone structures were detected. The authors concluded that the polymer had many of the characteristics of a synthetic lignin. The polymer was degraded by the lignolytic system of *Phanerochaete chrysosporium*. Thus a lignin-like polymer, which is apparently metabolized in a similar manner as native lignin preparations and can be prepared easier with less expense than preparations of labeled native plant lignins, is available for studies of lignin decomposition. Similarly, Ohta (1979) developed a dehydrogenated polymer of coniferyl alcohol as a synthetic lignin which they used to evaluate lignin decomposition capacity of *Fusarium solani* M-13-1.

Recently, Glen and Gold (1983) suggested that polymeric dyes Poly B-411, Poly R-481, and Poly Y-606 may be used as alternatives to radiolabeled lignin in biodegradation assays. Decolorization of the dyes, like lignin degradation, occurred during the secondary metabolic phase of *Phanerochaete chrysosporium*, was suppressed by high fixed nitrogen concentrations in the growth medium, and was strongly dependent upon oxygen concentrations of the cultures. Dye decolorization was also inhibited by a number of lignin catabolic inhibitors. A mutant of *P. chrysosporium* lacking phenol oxidase, thereby being incapable of metabolizing lignin, was also unable to decolorize the dyes. Based on these similarities in expression of the decolorization and lignolytic activity, the authors suggested that these dyes may serve as indicators of lignolytic activity.

These introductory comments document the complexity of the lignin molecular structure and suggest that this complexity has a major impact on the fate of the molecule in soil. Several excellent reviews of the biochemistry and structural aspects of lignin chemistry and biochemistry have been prepared (Hurst and Burges, 1967; Zeikus, 1978; Crawford, 1981; Stevenson, 1982). Thus the discussions of this chapter are limited to an evaluation of the microorganisms involved in lignin catabolism and the biochemical aspects of this catabolism as they impinge upon the fate of lignin in the soil ecosystem. Although some literature is descriptive of microbial catabolism of lignin in soil ecosystems, much of the biochemical and microbiological studies of lignin relate to activities of various wood root organisms. The overall objective of this analysis is to evaluate the impact of lignin on soil organic matter pools as well as the basic biochemical mechanisms involved in these transformations.

7.1. MICROBIAL SPECIES ASSOCIATED WITH LIGNIN DECOMPOSITION

The properties of the enzymatic pathways associated with lignin decomposition as well as the energetics of the process precluded isolation and characterization of the microbes involved for a number of decades. With the development of procedures to synthesize radiolabeled lignin, it has been possible to detect the slow, in many cases cometabolic, catabolism of this plant polymer. Much of the research to date involves evaluation of the properties of wood rot fungi, in that these organisms would *a priori* be anticipated to be capable of decomposing lignin to some degree. Attempts to correlate lignin decomposition capability with the metabolic capacity to depolymerize, demethylate, or cleave aromatic rings, thereby reducing the number of microorganisms that would have to be screened in search of lignin catabolizers, has also proven to be negative (Kaplan and Hartenstein, 1980). These workers evaluated a variety of soil bacteria, fungi imperfecti, and basidiomycetes in an attempt to correlate growth on aromatic carbon sources and activities of *p*-methoxyphenol demethylase, polyphenoloxidases, monooxygenases, and dioxygenases with their capability to decompose specifically labeled dehydrogenated polymerizates (synthetic lignins). Although a number of white rot basidiomycete fungi and bacteria were capable of decomposing the polymer, none of the enyzmatic activities correlated with this capability. In spite of this difficulty, a variety of fungi, actinomycetes, and bacteria have been isolated with lignolytic capability. The organisms involved include plant pathogenic fungi, a number of common soil fungal genera, yeasts, and a variety of common soil actinomycetes and bacteria. The only common trait apparent between the organisms, is that in most cases, they are classed in microbial genera known for their ability to decompose a variety of complex organic substrates.

7.1.1. Fungi and Yeasts

As is true of lignin decomposition, in general, the bulk of the research involving fungal species relates to those fungal species involved with wood rot. For example, Eslyn et al. (1975) evaluated the decay of blocks of alder, poplar, and pine woods by *Graphium* sp., *Monodictys* sp., *Paecilomyces* sp., *Papulospora* sp., *Thielavia terrestris*, and *Allescheria* sp. Decayed wood blocks were analyzed for weight loss and lignin cellulose, xylan and mannan contents. As would be anticipated, carbohydrates were decomposed faster than lignin. Generally, among the carbohydrates, cellulose was depleted faster than the major hemicellulose, xylan. All of the fungi metabolized the lignin. Pine wood was more resistant to decomposition by all the fungi. Less than 15 percent decomposition of this wood was detected with *Paecilomyces* sp., *Papulospora* sp., and *Thielavia terrestris*, and no metabolism of pine wood was detected for the remaining three fungi.

Blanchette (1984) while screening wood attacked by 26 white rot fungi, found that these fungi selectively removed lignin from a variety of coniferous and hardwood tree species. The micromorphological characteristics of the delignified wood was distinct thereby allowing for easy differentiation of this delignified wood from other types of wood decay.

The basidiomycetes, NRRL 6464, *Pleurotus ostreatus*, and *Phanerochaete chrysosporium*, are capable of metabolizing both the lignin and the cellulose moiety of lignocellulose (Freer and Detroy, 1982). After 20 days incubation, *Pluerotus ostreatus* and *Phanerochaete chrysosporium* mineralized 20 percent of the lignin, whereas the NRRL 6464 isolate converted 40 percent of the lignin component to carbon dioxide. Lignin metabolism of the *Pleurotus ostreatus* and the *Phanerochaete chrysosporium* isolates was inhibited by 20 mM nitrogen amendment to the medium, whereas as the NRRL 6464 isolate was relatively unaffected by nitrogen concentration variation. Cellulose decomposition from the lignocellulose by all three fungi was enhanced by fixed nitrogen amendment.

Cyathus stercoreus (Schw.) de Toni NRRL 6473, isolated from cattle dung decomposed 45 percent of the lignin from wheat straw during a 62-day incubation period at 25°C (Wicklow et al., 1980). Decomposition of ^{14}C-labeled lignin from kenaf by a variety of *Cyanthus* species correlated with the fungal capability of metabolizing low molecular weight lignin decomposition products (Abbott and Wicklow, 1984). One species, *Cyanthus canna* preferred lignin degradation over the other plant products.

A number of common soil fungi are capable of mineralizing soluble lignocarbohydrate complexes from wheat straw (Milstein et al., 1983). *Aspergillus japonicum* efficiently degraded phenolics and carbohydrates in the mixture, whereas *Trichoderma* sp. was only capable of decomposing a portion of the carbohydrate fraction and none of the aromatics. *Polyporus versicolor* metabolized the aromatic and carboyhydrate polymers along with polymerizing the low molecular weight aromatics. Depolymerization of the lignin by these fungi was enhanced by amending of the growth medium with xylose. Aromatic polymer formation by *Polyporus versicolor* was also stimulated by xylose amendment and was accompanied by laccase production by the fungus. The xylose catabolism apparently stimulated aromatic intermediate formation by the fungus, whereas laccase catalyzed the polymerization reaction.

Fusarium solani isolate AF-W1 was shown to metabolize the side chain and ring structure of synthetic lignins, lignins, and aromatic acids (Norris, 1980). Lignin decomposition was inhibited by D-glucose. Only small portions of the lignin carbon was incorporated into the fungal cell mass. Data demonstrated that the isolate was capable of using lignin as a sole carbon source.

Yeast species have also been shown to mineralize lignin. A *Candida* sp., isolated from decaying leaves, when growth in a medium containing lignin and glucose, decomposes the lignin to 2-and 3-ring aromatic structures (Clayton and Srinivasan, 1981). These polyaromatics are metabolized into smaller structures that enter the yeast cells for mineralization.

These data demonstrate that although a variety of wood "diseases" are attributable to lignolytic fungi, a variety of common soil fungi and yeasts are also capable of mineralizing this complex polymer. In several cases, mineralization was limited by the presence of nitrogen or required a readily metabolizable carbon source, such as glucose or xylose, whereas with other isolates lignin catabolism provided the energy the microbe needed for growth. Generally, the complex lignin structure was mineralized totally to carbon dioxide and water.

7.1.2. Actinomycetes

A number of streptomycete strains can decompose the lignin and lignocellulose. This was first demonstrated by Crawford (1978). He selected three streptomycete strains out of 30 actinomycete cultures capable of decomposing lignocellulose while growing on an agar medium with newsprint as the primary carbon and energy source for study of lignocellulose catabolic properties. Although the organisms primarily oxidized cellulose to carbon dioxide, lignin was also mineralized. Depending on the culture studied, approximately 3.5 percent of the lignin was oxidized to carbon dioxide during a 42.7 day incubation period. Subsequently, Antai and Crawford (1981) found that *Streptomyces viridosporus* T7A and *Streptomyces setonii* 75V.2 decomposed lignin and cellulose when grown on grass, soft-, or hardwood lignocellulose. Grass lignocelluloses were preferred by the isolates. Genetic manipulation of these two isolates resulted in production of 4 of 19 recombinants tested with enhanced production of an acid precipitable polymeric lignin from corn stover lignocellulose over that produced by the wild type *Streptomyces viridosporus* strain (Pettey and Crawford, 1984). The acid precipitable polymeric lignin is a lignin degradation intermediate with potential commercial value. Thus the production of variants through protoplast fusion appears to be a valuable technique for selection of genetic variants with maximum capacity to accumulate this lignin product. Accumulation of the acid precipitable polymer can be substantial. Borgmeyer and Crawford (1985) noted that 30 percent of the lignin initially present in corn stover lignocellulose was converted to this product by *Streptomyces viridosporus*. The kinetics of the acid precipitable polymeric lignin varied between streptomycete culture examined. *S. viridosporus* produced the acid precipitable product in solid-state fermentation over a 6 to 8 week incubation, whereas *S. badius* produced as much or more of the product, but only in liquid culture over a 7- to 8-day incubation period. The *S. viridosporus* product was more lignin-like than that of *S. badius*. The data suggest that differing mechanisms of lignin degradation may exist between the two streptomycete strains.

As opposed to the white rot fungi which do not degrade lignin in the absence of readily metabolizable carbon substrates, *Streptomyces badius* strain 252 is capable of decomposing milled-wood lignin in a minimal medium (Barder and Crawford, 1981). Also, lignin degradation was greatest in the presence of high organic nitrogen levels. Further stimulation of the mineralization rates occurred in media where the organic nitrogen was supplemented with low levels of nitrate.

7.1.3. Bacteria

As with the other microbial taxonomic groups previously discussed, a variety of common soil bacteria are capable of decomposing lignin and lignocelluloses. A number of Gram-negative, aerobic, motile or nonmotile, nonendospore forming bacteria of the genera *Pseudomonas* and *Flavobacterium* capable of decomposing native lignin were isolated by Sorensen (1962). Isolation of bacteria was favored by use of a neutral growth medium containing a suspension of fine particles of lignin solidified with silica gel. A *Nocardia* sp. from Finnish soils is capable of decomposing lignin and assimilating the degradation products as carbon sources (Trojanoswki et al., 1977). This organism produced ^{14}C-labeled carbon dioxide from ^{14}C-labeled lignin methoxyl groups, side chains and ring carbons of coniferyl alcohol dehydropolymers and from specifically labeled lignin of plant material. The lignin degradation products, phenolcarboxylic and cinnamic acids and alcohols were also metabolized to carbon dioxide. Similarly, a variety of *Nocardia* and *Pseudomonas* spp. capable of mineralizing corn stalk lignins and dehydropolymer coniferyl alcohol have been isolated from lignin enriched lake water (Haider et al., 1978) and soils (Odier et al., 1981). Kerr et al. (1983) isolated an *Arthrobacter* sp. capable of degrading lignin as a sole carbon source. Both the lignin and the cellulose components of lignocellulose of *Spartina alterniflora* and kraft lignin from slash pine were mineralized by this isolate.

7.1.4. Other Lignin Decomposers

An analysis of lignin decomposition would be incomplete without indicating the role of organisms other than microbes in its degradation. The termite *Nasutitermes exitiosus* (Hill) when fed natural and synthetic ^{14}C-labeled lignins produced labeled carbon dioxide (Butler and Buckerfield, 1979). Conversion of the labeled lignin to labeled intermediates varied with the position of the ^{14}C label in the parent compound. An average of 7 percent of the ring-labeled phenate, 63 percent of the methoxyl-labeled maize lignin and 64 percent of ring-labeled ferulic acid were metabolized to carbon dioxide and intermediates over of 6 to 69 days. Termite bodies contained only a small portion of the added radioactivity at the end of the study. Analysis of termites separate from the voided feces indicated that most of the lignin decomposition occurred in the termites and not externally in the feces. Measurement of the termite respiration rates indicates that a significant portion of the carbon in litter lignin is returned directly to the atmosphere. Further study of the biochemical reactions and the source of the enzymes involved in the process would be interesting. As with other termite mediated processes, the role of the intestinal flora and actual enzymatic activities contributed by the animal cells needs to be elucidated.

The variety of microorganisms isolated from plant materials, soils, and waters demonstrates the versatility of mechanisms developed to derive energy or carbon from the complex lignocellulose molecule. With some organisms, the lignin serves as an energy source, whereas others require a readily decomposable substrate for energy. Some organisms apparently only decompose the lignin cometabolically, whereas others incorporate lignin carbon into their cellular structures. Fixed nitrogen requirements vary as well as the impact of this nitrogen on the catabolic rates. This diversity of lignolytic processes with axenic cultures provides evidence that lignin is also metabolized in soil under a variety of conditions and that a variety of intermediates that may be polymerized in humic substances are produced.

7.2. BIOCHEMISTRY OF LIGNIN DECOMPOSITION

As with most biochemical processes, a number of lignin associated enzymatic capabilities are common to most of the microorganisms involved in lignin decomposition. For example, aerobic pathways for aromatic ring degradation generally involve participation of mono- and dioxygenases and/or peroxidases. This situation is predicted by the laws of comparative biochemistry. With lignin decomposition, although many similar enzymatic reactions may be commonly found among the microorganisms involved, the complexity of the lignin substrate predicts that a large amount of variation may exist between the exact pathways of lignin decomposition expressed by each individual microbial species. This diversity depends on the point of attack of the enzyme on the lignin molecule as well as the capability of the microbe to modify the lignin components (i.e., demethylate, or catabolize side chains, etc.) Hence, an analysis of the biochemical reactions involved in lignin decomposition, albeit a plant pathogenic fungus or a soil bacterium, will help elucidate the transformations of lignin leading to its incorporation into soil humic substances. Greater limitation is found in predicting the specific environmental conditions required for induction or control of the rates of lignolytic activity in soils. This results from the wide variety of metabolic requirements of microbes for induction of lignolytic enzymes. Some plant pathogens require a supply of readily metabolizable organic matter and others do not, some are inhibited by fixed mineral nitrogen sources and others are not, and some metabolize the lignin as a source of carbon and energy, and others catabolize lignin cometabolically. Implications of such diversity in control of this metabolic pathway on soil lignin reactions are generally unknown. Thus at this time, the properties of the various lignolytic organisms and the control of the activity will be examined in light of the potential interrelatedness to processes occurring in soil. Exact correlation must await more detailed study of lignin and lignin product behavior in soil.

7.2.1. Cultural Parameters for Induction of Lignin Degradation

Most of the work with axenically grown microbial cultures to date involves evaluation of the parameters controlling growth of wood rot fungi. This work has been reviewed adequately (Zeikus, 1978; Crawford, 1981; Stevenson, 1982). Therefore, instead of presenting a detailed evaluation of the data relating to microbial cultural parameters, the range of parameters controlling microbial decomposition of lignin and the impact of these parameters on the catabolism of lignin in the soil ecosystem will be discussed.

The most studied lignolytic organism is the white rot fungus, *Phanerochaete chrysosporium*. Early work revealed that this organism requires a readily decomposable substrate, such as glucose or cellulose as a carbon and energy source, to catabolize lignin to carbon dioxide (Kirk et al., 1976). Under optimal growth conditions, the oxidation of 5 mg of synthetic lignin required the metabolism of approximately 100 mg glucose (Kirk et al., 1978). Furthermore, there was an obligatory requirement for molecular oxygen. With 5 percent O_2 in N_2 in the gas phase above a nonagitated culture there was essentially no lignin catabolism (Kirk et al., 1978). Lignin metabolism was enhanced two- to three-fold over that in air (21 percent O_2) in 100 percent O_2. But, agitation of the culture to enhance mixing of air with the growth medium suppressed lignin decomposition. This effect of agitation most probably results from interference with the association of the microorganism with the lignin substrate. The pH optimum for lignin degradation is 4 to 4.5 with marked reduction in lignolytic activity above pH 5.5 and below 3.5. The source of fixed nitrogen is of little importance, but its concentration greatly affected rates of lignin decomposition. Increasing the nitrogen concentration from 2.4 mM N to 24 mM N reduces lignolytic activity 65 to 75 percent.

Induction of the lignolytic enzyme system in *Phanerochaete chrysosporium* occurs following nitrogen starvation of the fungus. Induction of the enzyme system in the fungus grown under optimal conditions was examined by Keyser et al. (1978). They observed a linear growth phase lasting 24 hours during which the fixed nitrogen of the growth medium was depleted. During the subsequent 24 hours of incubation, linear growth ceased and ammonium permease was derepressed (an indicator of nitrogen starvation). Following this nitrogen starvation period, the lignolytic enzymes were induced as demonstrated by the development of capability to oxidize synthetic lignin to carbon dioxide. Amendment of the starving culture with ammonium delayed synthesis of the lignolytic enzymes. Ammonium amendment following the onset of enzyme synthesis resulted in a temporary decline in activity. Starvation for carbohydrates or sulfur also resulted in induction of lignolytic activity (Jeffries et al., 1981). Phosphorus deprivation had no effect on the enzymatic activity. Amendment of a carbohydrate limited culture with a carbohydrate source resulted in a transient repression of the lignolytic activity. The proper "starvation" conditions are apparently found in rotting wood because Reid (1983a) found that wood degradation by *Phanerochaete chrysosporium* was maximal in unsupplemented wood. Reid (1983b) also noted that the behavior of the fungus in response to nutrient amendment in native wood samples differed from that observed with synthetic lignin in defined medium suggesting an *in situ* adaptation to the properties of the wood which is not possible in culture.

Because of the participation of oxygenase enzymes in lignin decomposition, *Phanerochaete chrysosporium* lignolytic activity is sensitive to the partial pressure of oxygen. Increasing the O_2 pressure from 1 to 2 atmospheres had little effect on lignin decomposition but 3 atmospheres of O_2 was inhibitory and greater than 4 atmospheres was fatal (Reid and Seifert, 1980). The oxygen effect related to its concentration and not the increased pressure in that lignin degradation in 5 atmospheres of air was similar to that in 1 atmosphere of O_2.

In contrast to these rather fastidious enzyme induction requirements for the soft rot fungus, lignin degradation appears to be somewhat easily inducible in *Streptomyces badius* strain 252 (Bader and Crawford, 1981) and *Fusarium solani* (Norris, 1980). The streptomycete degraded milled-wood lignin in a minimal medium. Lignin degradation by this organism was enhanced by amendment of the medium with organic nitrogen and an organic carbon cosubstrate. In fact, lignin degradation was maximal in the presence of high levels of organic nitrogen. Interestingly, greatest lignin degradation was found in a medium containing organic nitrogen supplemented with low levels of nitrate. With the *Fusarium solani* strain, lignin decomposition did not occur in cultures where glucose was present as a growth substrate. Small amounts of the carbon from the aromatic acids catabolized by the *Fusarium solani* isolate were incorporated into cell mass. The data suggested that *Fusarium solani* AF-W1 used lignin as its sole carbon source.

7.2.2. Anaerobic Lignin Decomposition

Lignin monomers are converted to carbon dioxide and methane under strict anaerobic conditions (Healy and Young, 1979; Healy et al., 1980; Kaiser and Hanselmann, 1983; Taylor, 1983). Whereas the work of Healy and associates noted catabolism of lignin monomers, such as ferulic acid under methanogenic conditions, Taylor observed vanillic acid, a lignin decomposition product, catabolism to carbon dioxide by a *Pseudomonas* sp. growing with nitrate as the terminal electron acceptor (nitrate respiration). Although a failure to observe lignin decomposition under anaerobic conditions is commonly reported (Hackett et al., 1977; Odier and Monties, 1983), anaerobic decomposition of cornstalk lignin to carbon dioxide and methane in an anaerobic digestor was reported by Boruff and Buswell (1934). Lignin oligomers solubolized by alkaline heat treatment and separated into molecular size fractions are fermented under methanogenic conditions to carbon dioxide and methane (Colberg and Young, 1985a). The smaller the molecular size fraction, the more extensive the catabolism to gaseous end products. Up to 30 percent of the solubilized lignin-derived carbon products were anaerobically mineralized to carbon dioxide and methane. Acclimation of anaerobic cultures for two years resulted in development of microbial populations able to catabolize lignin-derived substrates with a molecular weight of 600 as a sole carbon source (Colberg and Young, 1985b). Their data suggest that the enriched microbial consortium had the capability to cleave the β-aryl-ether bond, the most common intermonomeric linkage in lignin. Benner et al. (1984c) found that after 294 days in anaerobic sediments, 16.9 percent and 30.0 percent of the lignin and cellulose components, respectively, of *Spartina alterniflora* lignocellulose were degraded to carbon dioxide and methane. Lignocellulose prepared from a hardwood, *Rhizophora mangle* was more resistant to biodegradation. These data suggest slow lignin decomposition under anaerobic conditions in sediments.

These observations of lignin product catabolism under anaerobic conditions are of greater practical significance for industrial considerations than from the soil organic matter view point. The industrial significance relates to the possible of utilization of lignified plant products as an economically feasible substrate for biological methane production. Methanogenesis from lignin or even lignin monomers in soil ecosystems is limited, at best, as a result of the extreme conditions necessary for conversion of plant lignins to carbon dioxide and methane. A highly reducing environment is necessary for methanogenesis to occur. This is most commonly encountered in swamps, bogs, and some man-made ecosystems, such as landfills. Such reducing conditions are unusual in the more common soil ecosystems. Benner et al. (1984c) have shown that methanogenesis from lignin or lignin products may even be of limited occurrence in swampy ecosystems. In such sites, the soil organic component contains a variety of readily decomposable substrates which are contained within the cellular matrix of the partially decomposed plant debris, or in the case of land fills, within the buried waste materials. Decomposition of the more complex lignin compounds would be delayed by the preferential decomposition of the more easily metabolized substrates.

7.2.3. Enzymology of Lignin Degradation

From the viewpoint of soil organic matter synthesis, the most significant lignolytic enzymatic activities are those associated with hydroxylation of the aromatic rings and with depolymerization of the structure itself, albeit oligomer production or total depolymerization to constituent monomers. Importance of these enzymatic activities is derived from (1) incorporation of hydroxyl groups into the aromatic ring structure which may subsequently enhance polymerization of the aromatics into humic and fulvic acids, and (2) the production of aromatic products which may be incorporated directly into humic and/or fulvic acids. The general pathways involved with the incorporation of lignin carbons into soil humic and fulvic acids are depicted in Fig. 7.1. Lignin or lignin decomposition products may be incorporated into soil humic acids through the actions of a variety of soil phenol oxidase activities (Liu et al., 1981; Sulfita and Bollag, 1981). Other lignolytic enzymatic activities contribute to humic and fulvic acid synthesis by increasing the diversity of products which may be humified.

Both laccase (polyphenol oxidase) and peroxidase activities are associated with lignin decomposition. For example, phenol oxidase activity is instrumental in lignin decomposition by *Sporotrichum pulverulentum*. This was demonstrated through the selection of a phenol oxidase-less mutant of the fungus (Ander and Ericksson, 1976). The mutant was incapable of decomposing kraft lignin or wood, whereas the oxidase positive revertant regained these degradative capabilities. Amendment of agar media containing kraft lignin inoculated with the phenol oxidase-less mutant with laccase resulted in lignin decomposition by the mutant strain. Peroxidase also participates in hydroxylation of the aromatic rings. An extracellular peroxidase is excreted into the growth medium of *Phanaerochaete chrysosporium* growing on wood. This peroxidase has a molecular weight of 42,000 (Tien and Kirk, 1983). In the presence of hydrogen peroxide the enzyme catalyzes the following reaction:

1,2-bis (3-methoxy-4 (^{14}C) methoxyphenyl-propane-1,
3-diol $\rightarrow$ vanillin methyl ether

Forney et al. (1982) demonstrated that the peroxidase activity was located in the periplasmic space of cells from lignolytic cultures grown for 14 days in nitrogen limited medium. No such activity was detected in cells lacking lignolytic capabilities. Faison and Kirk (1985) found that this hydrogen peroxide dependent lignolytic activity was synthesized in cultures grown under conditions conducive for induction of lignin decomposition; that is, this activity was expressed in nitrogen starved cultures and suppressed in the presence of excess nutrients, cycloheximide, or culture agitation.

Depolymerization of the lignin polymer to oligomers or monomers also produced humic substance precursors. A number of microorganisms are capable of depolymerizing lignin to various degrees. *Fusarium solani* M-13-1, a soil fungal isolate capable of using dehydrogenation polymer of coniferyl alcohol (a model lignin polymer) as a sole carbon source, metabolized dehydrodiconiferyl alcohol to six aromatic intermediates (Ohta et al., 1979). Also, a *Candida* yeast isolate from decaying leaves degraded lignin to 2- or 3-ring aromatic compounds which were further metabolized intracellularly (Clayton and Srinivasan, 1981). Similar activities have been demonstrated for *Phanerochaete chrysosporium* (Gold et al., 1984). Oligomeric intermediates are also produced during lignin decomposition by *Streptomyces viridosporus* (Crawford et al., 1983). In each case, an aromatic compound with the potential of being polymerized into soil humic substances was produced.

7.3. FATE OF LIGNIN PRODUCTS IN SOIL

Decomposition and/or modification of plant components commences in many cases before the plant debris becomes soil incorporated. Under those circumstances, biological decomposition is controlled by the physical properties of the plant tissue and is catalyzed by phylosphere organisms and by those microbial propagules entering the ecosystem defined by the plant tissue which are capable of functioning therein. Such an ecosystem involving lignocellulose degradative activity on a log of douglas fir (*Pseudotsuga menziesii*) was described by Aumen et al., (1983). Evaluation of the microbial activity with scanning microscopy, plate counts, and degradation of ^{14}C-lignocelluloses revealed that most of the microbial colonization and lignocellulose-decomposing activity occurred on the log surface. Incubation of surface wood samples in defined media demonstrated that mineralization was enhanced by amendment of the medium with either ammonium sulfate or organic nitrogen sources as compared to rates measured in a mineral salts solution augmented with trace minerals only. Nitrate was the most favorable fixed nitrogen source for lignin decomposition. The source of fixed nitrogen did not affect cellulose degradation rates. Both lignin and cellulose decomposition was repressed by glucose amendment of the growth media. Their data indicated that wood decomposition in streams of the Pacific Northwest (U.S.A.) was nitrogen-limited.

Incorporation of plant materials into soil results in selection of microbial populations capable of decomposing lignocelluloses. Crawford et al. (1977) through the use of ^{14}C-labeled lignin in lignocellulose or similarly labeled cellulose in the lignocellulose found that the cellulose portion of the compound was approximately 4 to 10 times more susceptible to microbial decomposition than the lignin moiety. A differential susceptibility to biological attack of the molecule is also found within the lignin molecule itself. Haider et al. (1977) demonstrated the differential sensitivity of various lignin structural groups to microbial oxidation to carbon dioxide. They found that in a neutral sandy loam soil 33, 18, and 20 percent of the CH_2OH, 2-side chain, and ring carbons of coumaryl alcohols, respectively, linked into model lignins were oxidized to carbon dioxide over a 28-day incubation period. Comparable oxidation rates for similarly labeled corn stalk lignins were on the average 7 percent higher than those measured with the synthetic lignins. Between 40 and 60 percent of the residual activity from both the lignins and model polymers was recovered from the humic acid fractions of the soil. Martin et al. (1980) noted that the major portions of lignin carbons incubated in two agricultural soils for a two-year period were incorporated into the more resistant aromatic portion of soil humus. This suggests that few of the lignolytic microbes in his soils were capable of using lignin as a carbon source. This contrasts with the fate of polysaccharides which are primarily oxidized for carbon and energy. Microbial cell incorporated polysaccharides are synthesized predominantly into cell proteins and polysaccharides. Some of the latter compounds are humified upon death and lysis of the microbial cell.

Lignin carbons entering physically stressed ecosystems, such as osmotically or moisture stressed sediments, may also be incorporated directly or partially decomposed then incorporated into humic substances. Maccubbin and Hodson (1980) found that slash pine (*Pinus elliottii*) lignocellulose was mineralized slowly in salt marsh sediment. After approximately 35 days, 1.4 and 3.9 percent of the lignin and cellulose components, respectively, of lignocellulose were converted to carbon dioxide. The lignocellulose of cordgrass (*Spartina alterniflora*) was oxidized approximately three-fold faster under comparable conditions. A bacterial assemblage has been isolated from salt-marsh sediments capable of metabolizing cordgrass lignocellulose (Benner et al., 1984b). The polysacharide component of the lignocellulose was mineralized approximately twice as fast as the lignin moiety. The data suggested that bacteria are the primary mediators of lignocellulose decomposition in salt marshes. With the differential decomposition rate between the lignin and the other plant components demonstrated in these studies, an enrichment of lignin in the salt marsh sediments is anticipated. This was shown by Wilson (1985). This enrichment favors incorporation of the lignin into sediment humic acids fractions.

Nutrient limitations of lignocellulose mineralization was shown in sediments of Toolik Lake, Alaska (USA) (Federle and Vestal, 1980). Nitrogen and phosphorus amendment increased the mineralization rate of white pine (*Pinus strobus*) cellulose, but nitrogen amendment had not effect on lignin catabolism. Phosphorus amendment inhibited lignin decomposition. The differential affect of nutrient limitations in these sediments may also in part explain the enrichment of lignin in the sediments.

7.4. CONCLUSIONS

The complex aromatic plant component, lignin, may be incorporated into soil humic substances through a number of reactions. These processes include humification of the nascent lignin directly as well as incorporation of a variety of lignin oligomers and monomers, both modified and unmodified. Humification of lignin is favored by its relative biodegradation resistance. As was indicated for the more readily metabolized plant components, the longer the soil residence time of a compound, the greater the probably it will be incorporated into humic or fulvic acids.

The data presented herein has shown that lignin carbons contribute to the soil organic matter pool. Although the lignin theory of humic acid synthesis is an attractive idea, care must be taken to avoid the misconception that lignin is the only precursor to native soil humic substances. As will be discussed in the next chapter, lignin is only one of the potential precursors of the aromatic components of soil humic substances. As shown in Chapter 6, and documented in subsequent chapters, a variety of plant products as well as microbial components and products are incorporated into soil humic materials.

REFERENCES

Abbott, T. P., and D. T. Wicklow, 1984. Degradation of lignin by *Cyanthus* species. Appl. Environ. Microbiol. 47: 585–587.

Ander, P., and K.-Eriksson, 1976. The importance of phenol oxidase activity in lignin degradation by the white-rot fungus *Sporotrichum pulverulentum*. Appl. Environ. Microbiol. 109: 1–8.

Antai, S. P., and D. L. Crawford, 1981. Degradation of softwood, hardwood, and grass lignocelluloses by two *Streptomyces* strains. Appl. Environ. Microbiol. 42: 378–380.

Aumen, N. G., P. J. Bottomley, G. M. Ward, and S. V. Gregory, 1983. Microbial decomposition of wood in streams: Distribution of microflora and factors affecting (^{14}C)lignocellulose mineralization. Appl. Environ. Microbiol. 46: 1409–1416.

Baker, K. H. 1983. Effect of selected assay parameters on measurement of lignocellulose mineralization with a radiolabeled substrate. Appl. Environ. Microbiol. 45: 1129–1131.

Barder, M. J., and D. L. Crawford, 1981. Effects of carbon and nitrogen supplementation on lignin and cellulose decomposition by a *Streptomyces*. Can J. Microbiol. 27: 859–863.

Benner, R., A. E. Maccubbin, and R. E. Hodson, 1984a. Preparation, characterization, and microbial degradation of specifically radiolabeled (^{14}C) lignocelluloses from marine and freshwater macrophytes. Appl. Environ. Microbiol. 47: 381–389.

Benner, R., S. Y. Newell, A. E. Maccubbin, and R. E. Hodson, 1984b. Relative contributions of bacteria and fungi to rates of degradation of lignocellulosic detritus in salt-marsh sediments. Appl. Environ. Microbiol. 48: 36–40.

Benner, R., A. E. Maccubbin, and R. E. Hodson, 1984c. Anaerobic biodegradation of the lignin and polysaccharide components of lignocellulose and synthetic lignin by sediment microflora. Appl. Environ. Microbiol. 47: 998–1004.

Blanchette, R. A. 1984. Screening wood decayed by white rot fungi for preferential lignin degradation. Appl. Environ. Microbiol. 48: 647–653.

Borgmeyer, J. R., and D. L. Crawford, 1985. Production and characterization of polymeric lignin degradation intermediates from two different *Streptomyces* spp. Appl. Environ. Microbiol. 49: 273–278

Boruff, C. S., and A. M. Buswell, 1934. The anaerobic fermentation of lignin. J. Am. Chem. Soc. 56: 886–888.

Butler, J. H. A., and J. C. Buckerfield, 1979. Digestion of lignin by termites. Soil Biol. Biochem. 11: 507–513.

Clayton, N. E., and V. R. Srinivasan, 1981. Biodegradation of lignin by *Candida* spp. Naturwissenschaften 68: 97–98.

Colberg, P. J., and L. Y. Young, 1985a. Anaerobic degradation of soluble fractions of (^{14}C-lignin)lignocellulose. Appl. Environ. Microbiol. 49: 345–349.

Colberg, P. J., and L. Y. Young, 1985b. Aromatic and volatile acid intermediates observed during anaerobic metabolism of lignin-derived oligomers. Appl. Environ. Microbiol. 49: 350–358.

Crawford, D. L. 1978. Lignocellulose decomposition by selected *Streptomyces* strains. Appl. Environ. Microbiol. 35: 1041–1045.

Crawford, R. L. 1981. Lignin biodegradation and transformation. John Wiley & Sons, New York, 154 pp.

Crawford, D. L., and R. L. Crawford, 1976. Microbial degradation of lignocellulose: The lignin component. Appl. Environ. Microbiol. 31: 714–717.

Crawford, D. L., R. L. Crawford, and A. L. Pometto III, 1977. Preparation of specifically labeled ^{14}C-(lignin)- and ^{14}C-(cellulose)-lignocelluloses and their decomposition by the microflora of soil. Appl. Environ. Microbiol. 33: 1247–1251.

Crawford, D. L., A. L. Pometto III, and R. L. Crawford, 1983. Lignin degradation by *Streptomyces viridosporus*: Isolation and characterization of a new polymeric lignin degradation intermediate. Appl. Environ. Microbiol. 45: 898–904.

Crawford, R. L., L. E. Robinson, and R. D. Foster, 1981. Polyguaiacol: A useful model polymer for lignin biodegradation research. Appl. Environ. Microbiol. 41: 1112–1116.

Ertel, J. R., J. I. Hedges, and E. M. Perdue, 1984. Lignin signature of aquatic humic substances. Science (Washington, D.C.) 1223: 485–487.

Eslyn, W. E., T. K. Kirk, and M. J. Effland, 1975. Changes in the chemical composition of wood caused by six soft-rot fungi. Phytopathology 65: 473–476.

Faison, B. D., and T. K. Kirk, 1985. Factors involved in the regulation of a ligninase activity in *Phanerochaete chrysosporium*. Appl. Environ. Microbiol. 49: 299–304.

Federle, T. W., and J. R. Vestal, 1980. Lignocellulose mineralization by Arctic lake sediments in response to nutrient manipulation. Appl. Environ. Microbiol. 40: 32–39.

Forney, L. J., C. A. Reddy, and H. S. Pankratz, 1982. Ultrastructural localization of hydrogen peroxide production in lignolytic *Phanerochaete chrysosporium* cells. Appl. Environ. Microbiol. 44: 732–736.

Freer, S. N., and R. W. Detroy, 1982. Biological delignification of ^{14}C-labeled lignocelluloses by basidiomycetes: Degradation and solubolization of the lignin and cellulose components. Mycologia 74: 943–951.

Glenn, J. K., and M. H. Gold, 1983. Decolorization of several polymeric dyes by the lignin-degrading basidiomycete *Phanerochaete chrysosporium*. Appl. Environ. Microbiol. 45: 1741–1747.

Gold, M. H., A. Enoki, M. A. Morgan, M. B. Mayfield, and H. Tanaka, 1984. Degradation of the γ-carboxyl-containing diarylpropane lignin model compound 3-(4'-ethoxy-3'-methoxyphenyl)-2-(4''-methoxyphenyl) propionic acid by the basidiomycete *Phanerochaete chrysosporium*. Appl. Environ. Microbiol. 47: 597–600.

Hackett, W. F., W. J. Connors, T. K. Kirk, and J. G. Zeikus, 1977. Microbial decomposition of synthetic ^{14}C-labeled lignins in nature: Lignin biodegradation in a variety of natural materials. Appl. Environ. Microbiol. 33: 43–51.

Haider, K., J. P. Martin, and E. Rietz, 1977. Decomposition in soil of ^{14}C-labeled coumaryl alcohols: Free and linked in dehydropolymer and plant lignins and model humic acids. Soil Sci. Soc. Am. J. 41: 556–562.

Haider, K., J. Trojanowski, and V. Sundman, 1978. Screening for lignin degrading bacteria by means of ^{14}C-labeled lignins. Arch. Microbiol. 119: 103–106.

Healy, J. B., Jr., and L. Y. Young, 1979. Anaerobic biodegradation of eleven aromatic compounds to methane. Appl. Environ. Microbiol. 38: 84–89.

Healy, J. B., Jr., L. Y. Young, and M. Reinhard, 1980. Methanogenic decomposition of ferulic acid, a model lignin derivative. Appl. Environ. Microbiol. 39: 436–444.

Hurst, H. M., and N. A. Burges, 1967. Lignin and humic acids. In A. D. McLaren and G. Paterson (eds.), Soil Biochem. 1: 260–286. Marcel Dekker. New York.

Jeffries, T. W., S. Choi, and T. K. Kirk, 1981. Nutritional regulation of lignin degradation by *Phanerochaete chrysosporium*. Appl. Environ. Microbiol. 42: 290–296.

Kaiser, J.-P., and K. Hanselmann, 1983. Fermentative metabolism of substituted monoaromatic compounds by a bacterial community from anaerobic sediments. Arch. Microbiol. 133: 185–194.

Kaplan, D. L., and R. Hartenstein, 1980. Decomposition of lignins by microorganisms. Soil Biol. Biochem. 12:65–75.

Kerr, T. J., R. D. Kerr, and R. Benner, 1983. Isolation of a bacterium capable of degrading peanut hull lignin. Appl. Environ. Microbiol. 46: 1201–1206.

Keyser, P., T. K. Kirk, and J. G. Zeikus, 1978. Ligninolytic enzyme system of *Phanerochaete chrysosporium*: Synthesized in the absence of lignin in response to nitrogen starvation. J. Bacteriol. 135: 790–797.

Kirk, T. K., W. J. Connors, and J. G. Zeikus, 1976. Requirement for a growth substrate during lignin decomposition by two wood-rotting fungi. Appl. Environ. Microbiol. 32: 192–194.

Kirk, T. K., E. Schultz, W. J. Connors, L. F. Lorenz, and J. G. Zeikus, 1978. Influences of culture parameters on lignin metabolism by *Phanerochaete chrysosporium*. Arch. Microbiol. 117: 277–285.

Liu, S.-Y., R. D. Minard, and J.-M. Bollag, 1981. Oligomerization of syringic acid, a lignin derivative, by a phenoloxidase. Soil Sci. Soc. Am. J. 45: 1100–1105.

Maccubbin, A. E., and R. E. Hodson, 1980. Mineralization of detrital lignocelluloses by salt marsh sediment microflora. Appl. Environ. Microbiol. 40: 735–740.

Martin, J. P., K. Haider, and G. Kassim, 1980. Biodegradation and stabilization after 2 years of specific crop, lignin, and polysaccharide carbons in soils. Soil Sci. Soc. Amer. J. 44: 1250–1255.

Milstein, O. A., Y. Vered, A. Sharma, J. Gressel, and H. M. Flowers, 1983. Fungal biodegradation and biotransformation of soluble lignocarbohydrate complexes from straw. Appl. Environ. Microbiol. 46: 55–61.

Norris, D. M. 1980. Degradation of ^{14}C-labeled lignins and ^{14}C-labeled aromatic acids by *Fusarium solani*. Appl. Environ. Microbiol. 40: 376–380.

Odier, E., G. Janin, and B. Monties, 1981. Poplar lignin decomposition by gram-negative aerobic bacteria. Appl. Environ. Microbiol. 41: 337–341.

Odier, E., and B. Monties, 1983. Absence of microbial mineralization of lignin in anaerobic enrichment cultures. Appl. Environ. Microbiol. 46: 661–665.

Ohta, M., T. Higuchi, and S. Iwahara, 1979. Microbial degradation of dehydrodiconiferyl alcohol: A lignin substructure model. Arch. Microbiol. 121: 23–28.

Pettey, T. M., and D. L. Crawford, 1984. Enhancement of lignin degradation in *Streptomyces* spp. by protoplast fussion. Appl. Environ. Microbiol. 47: 439–440.

Reid, I. D. 1983a. Effects of nitrogen supplements on degradation of Aspen wood lignin and carbohydrate components by *Phanerochaete chrysosporium*. Appl. Environ. Microbiol. 45: 830–837.

Reid, I. A. 1983b. Effects of nitrogen sources on cellulose and synthetic lignin degradation by *Phanerochaete chrysosporium*. Appl. Environ. Microbiol. 45: 838–842.

Reid, I. D., and K. A. Seifert, 1980. Lignin degradation by *Phanerochaete chrysosporium* in hyperbaric oxygen. Can. J. Microbiol. 26: 1168–1171.

Sorensen, H. 1962. Decomposition of lignin by soil bacteria and complex formation between autoxidized lignin and organic nitrogen compounds. J. Gen. Microbiol. 27: 21–34.

Stevenson, F. J. 1982. Humus chemistry. John Wiley & Sons, New York, 443 pp.

Sulfita, J. M., and J.-M. Bollag, 1981. Polymerization of phenolic compounds by soil-enzymes complex. Soil Sci. Soc. Amer. J. 45: 297–302.

Taylor, B. F. 1983. Aerobic and anaerobic catabolism of vanillic acid and some other methoxy-aromatic compounds by *Pseudomonas* sp. strain PN-1. Appl. Environ. Microbiol. 46: 1286–1292.

Tien, M., and T. K. Kirk, 1983. Lignin-degrading enzyme from the hymenomycete *Phanerochaete chrysosporium* Burds. Science (Washington, D.C.) 221: 661–663.

Trojanoswki, J., K. Haider, and V. Sundman, 1977. Decomposition of ^{14}C-labeled lignin and phenols by a *Nocardia* sp. Arch. Microbiol. 114: 149–153.

Wicklow, D. T., R. W. Detroy, and B. A. Jessee, 1980. Decomposition of lignocellulose by *Cyathus sterocorus* (Schw.) de Toni NRRL 6437, a "white rot" fungus from cattle dung. Appl. Environ. Microbiol. 40: 169–170.

Wilson, J. D. 1985. Decomposition of (^{14}C)lignocelluloses of *Spartina alterniflora* and a comparison with field experiments. Appl. Environ. Microbiol. 49: 478–484.

Zeikus, J. G. 1978. Fate of lignin and related aromatic substrates in anaerobic environments. In T. K. Kirk, T. Higuchi, and H.-M. Chang (eds.), Lignin Biodegradation: Microbiol. Chem., Potential Appl., (Proc. Int. Semin.) 1: 101–109. CRC, Boca Raton, Florida.

EIGHT

HUMIC AND FULVIC ACIDS: FORMATION AND DECOMPOSITION

For most research involving soil humic substances, operational definitions have been used to differentiate its various fractions; that is, humic acids are considered to be those components removed from soil by extraction with alkaline solutions, primarily sodium hydroxide solutions, and precipitated from these alkaline extracts through acidification to a pH of approximately 1.0. This soil organic fraction includes not only the aromatic components comprising the humic acid core but also a variety of humified plant components, such as proteins, peptides, amino acids, and polysaccharides. Thus in reality, humic acid as usually studied is an extremely heterogeneous soil fraction. Consideration of this molecular heterogeneity is important when evaluating reports of humic acid synthesis and degradation processes. Entirely different results are anticipated from measurements of humified protein decomposition than are encountered when catabolism of the aromatic humic acid nucleus is considered. Much of the biodegradation research conducted to date involves use of humic substrates consisting of degradation resistant aromatic humic acid core plus a variety of humified plant and animal components. Therefore, most of the carbon dioxide collected would originate from the more labile components. Interpretation of data derived from study of such complex substrates is difficult; that is, the use of these complex substrates provides data which is analogous to data that would be collected if the decomposition of lignocellulose were measured as an entity without knowledge of the differential decomposition rates of the lignin (and even the individual components of the lignin) and the cellulose moieties. Also, as with lignocellulose, the two general humic acid components, aromatic core and easily humified metabolized substances, are not metabolized independently. The easily metabolized materials provide carbon, energy, as well as macronutrients to the microorganisms involved in the aromatic polymer decomposition. Thus although the primary objective of this chapter is to evaluate synthesis and decomposition of the complex aromatic component of humic acids, the properties and presence of readily metabolized substrates must be considered from the view of their direct and indirect impact on aromatic decomposition.

Fulvic acid catabolism is closely allied with soil humic acid transformations. Evaluation of both the modern and classical literature comparing humic and fulvic acids and their behavior in soil reveals a difference in opinion concerning the relationship of these two substances. Given that fulvic acids are generally more oxidized, of greater water solubility, and of smaller molecular weight than are humic acids (Stevenson, 1982), fulvic acids may be either precursors of humic acid synthesis or decomposition products. Although a number of reports may be cited in favor of either possibility and substantive arguments are provided opposing the alternatives, it is reasonable to consider that fulvic acids occur both as precursors and products of humic acid transformations. For example, depolymerization of lignin to the level of oligomers produces acid and alkali soluble products which are aromatic in nature, and are of smaller molecular weight and of greater oxidation state than humic acids. But these oligomers may be incorporated into humic acids, thereby operating as a precursor of humic acid while expressing the properties of fulvic acids. Conversely, as humic acids are metabolized by soil microorganisms, cleavage of the covalent linkages forming the complex polymer produces smaller molecular weight substances which are soluble in acid, and are more oxidized. Subsequent metabolism of the soluble product further increases its oxidation state. Thus in this situation, a classical fulvic acid product is produced through decomposition of humic substances. Therefore, we shall consider fulvic acid to be both a precursor and product of humic acid transformation.

The various theories relating to both synthesis and degradation of humic acid will be examined in this chapter. Primary emphasis involves the biological interactions associated with these reactions and the difficulties in elucidating the details of the process.

8.1. SYNTHESIS OF HUMIC SUBSTANCES

As would be anticipated from an evaluation of the overall complexity of humic acids, our lack of specific knowledge of their molecular structure, and the diversity of humic acid precursors, a large number of theories have been proposed to explain humic acid synthesis (Felbeck, 1971; Flaig, 1971). Although each theory, at times, has been vehemently defended, analysis of the conditions in soil suggests that no single theory is adequate to describe the complex reactions leading to accumulation of humic acids. In reality, humic acids are the product of condensation of a variety of plant, animal, and microbial products (Fig. 8.1). This polymerization may result from biological or chemical mechanisms. The process is further complicated by the fact that nascent humic acid is "matured" following its synthesis; that is, the polymer is increased in size due to chemical bonding to one or more humic or fulvic acid polymers in the soil, internal crosslinking between components, and internal oxidation of the molecule through chemical or biological processes. Thus a portion of a single complex humic acid molecule may have originated totally from microbial metabolic products that were polymerized through enzymatic processes. This portion of the molecule thus could be said to origninate through a microbial synthetic pathway and thus support that theory of humic biosynthetic mechanisms. But, this "humic acid fragment" may be incorporated into a portion of a humic acid derived from lignin. The latter fragment would be produced by mechanisms associated with plant biosynthetic pathways. Furthermore, each of the humic acid components, irrespective of their original source, may be transformed (matured) subsequent to their combination into a single molecule through soil chemical and biological mechanisms. Thus no single theory has been proposed to date which totally explains soil humic acid synthesis. But, although it is obvious that none of the theories adequately explain the processes, since the individual theories do provide a convenient tool for dissecting the reactions occurring in soil into easily understood components, our analysis of humic acid synthesis will involve studies of lignin (lignin theory), microbial products (microbial synthesis theory), and simple sugars and amino acids (Maillard Reaction) as precursors of soil humic acids. Other proposed routes of humic acid synthesis involve at least the principles of these pathways if not the actual mechanisms. For example, microbially synthesized humic acid-precursors may originate as products of depolymerization of lignin or be synthesized *de novo* from other plant carbons. Thus whether the carbon structure as assembled by the plant is preserved (lignin theory) or modified through microbial catabolism of the plant carbon and synthesis of aromatic compounds, the reactions are microbially mediated. The individual humic acid precursors may be polymerized either microbiologically or chemically. Two such mechanisms will be examined, enzymatic action by phenol oxidases or peroxidases, and chemically mediated reactions with soil clays.

8.1.1. Lignin as a Humic Acid Precursor

Lignin humification and the properties of this plant component leading to incorporation into humic substances were discussed in Chapter 7. The primary justification of proposing lignin as a humic acid precursor initially related to the basic chemical similarities between the lignin and humic acid (Waksman, 1926). Generally, the early evidence favoring this theory is less than convincing because the parallel properties are not specific to either lignin or humic acids. For examples, support was derived from general similarities in solubilities in alkali and acids as well as organic acids, biodegradation resistance, contents of related functional groups (methoxy, carboxy, etc.), the aromaticity of both compounds, and the acidic nature of the two complex molecules.

Hatcher et al. (1981) recently reexamined the relationship of the aromatic component of the two complex molecules through the study of a variety of humic acids isolated from soils developed in several different climatic zones. They used cross polarization, magic-angle spinning ^{14}C nuclear magnetic resonance (NMR) analytical techniques to quantify the aromatic content of the humic acids. Aromatic contents for the humic acids ranged from 35 to 92 percent, which is in general agreement with values obtained from the conventional procedures

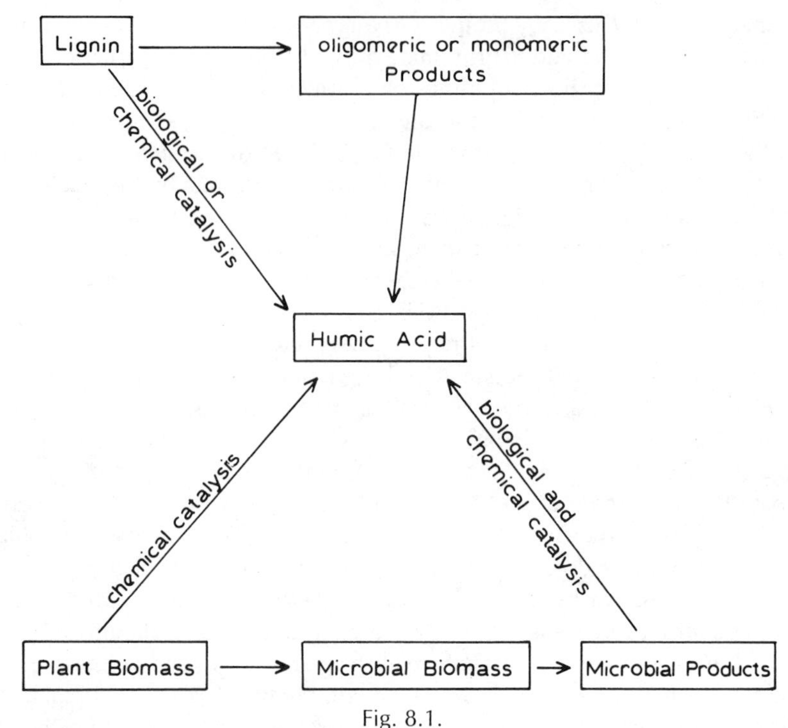

Fig. 8.1.
Soil processes leading to the formation of the aromatic portion of humic acids.

of chemical oxidation/gas chromatographic mass spectrometric studies. The NMR procedures did suggest greater aliphatic contents than are generally detected through chemical oxidation procedures. Even with these general similarities in molecular structure and properties, definitive evidence of the direct incorporation of lignin carbons into humic and fulvic acids was lacking until it became possible to label specific lignin carbons with ^{14}C. It was then shown that humic acid carbons are indeed derived from lignin. This stabilization of lignin carbon by incorporation into humic acid was recently reviewed by Martin and Haider (1978).

Reactions that have been demonstrated to be involved with humification of lignin carbons actually support both a lignin theory of humus origin as well as a microbial theory. As proposed by Waksman, lignin may be incorporated directly into humic molecules. Alternatively, lignin, along with other plant components, may be decomposed to aromatic monomers, which, following modification by the microbial community, are incorporated into humic acids. This reaction sequence forms the basis of a microbially mediated humus synthesis processes depicted by Varadachari and Ghosh (1984). Lignin monomers in a variety of chemical forms have been isolated from sediments (Katase, 1983).

8.1.2. Microbial Humic Acid Synthesis

Microorganisms are involved in humic and fulvic acid synthesis through modification of plant components to humic substance precursors, synthesis of peroxidases and phenol oxidases which catalyze polymerization of phenolic and aromatic monomers into humic acids, and in direct synthesis of humic acid-like polymers. Significance of the latter compounds in humic acid production is derived from the fact that if they are synthesized in soil, they could be incorporated directly into the more complex humic acids comprising the mature humic acid pool. Modification of plant components and subsequent incorporation into humic substances was evaluated in the discussion of humification processes. Enzymatic involvement will be presented in Section 8.2. Thus the discussion in this section will in the most part be limited to microbial synthesis of humic acid-like polymers and the evidence relating these compounds to native soil humic acids.

Humic acid-like products are synthesized by a variety of fungi including *Pisolithus tinctorius* (Tan et al.,1978), *Epicoccum nigrum* (Martin et al., 1967), *Hendersonula toruloidea* (Martin et al. 1972), *Aspergillus sydowi* and *Stachybotrys* spp. (Haider and Martin, 1970), *Aspergillus glaucus* sp. and *Eurotium echinulatum* (Linhares and Martin 1978), and *Streptomyces* spp. (Huntjens, 1972). In each case, a product is synthesized by the microorganism which has a number of biological and chemical properties similar to those of native soil humic acid. For example, when the fungus *Epicoccum nigrum* is grown in a glucose asparagine medium, orsellinic and cresorsellinic acids are synthesized. These metabolic products are modified by decarboxylation, hydroxylation, and oxidation of methyl groups to more than 20 other phenolic compounds which are subsequently linked by the microorganism into humic acid-like products (Haider and Martin, 1967). (These humic acid-like products have also been classified as melanins.) Weak phenol oxidase activity was detected in the fungal mycelium. The authors proposed that upon autolysis of the fungus, the phenol oxidase activity could be released into the growth medium, thereby catalyzing polymerization of the fungal phenolic products. Martin et al. (1967) detected yields of this humic acid-like polymer of 2 to 6 g per 15 liters of growth medium. The fungal product was classed as a humic acid-like polymer based on its similarity to the Leonardite humic acid fraction. Both preparations were relatively biodegradation resistant, and increased soil aggregation and hydraulic conductivity. Chemical similarities included carbon and nitrogen contents, total acidity, exchange capacity, molecular weights, and functional group analysis (carboxyl groups and phenolic hydroxyl groups). Chromatographic analysis of the product of reductive degradation with sodium amalgam revealed the presence of 14 phenols. Also, 30 to 48 percent of the fungal humic acid product was released by proteases. Similar data, including the presence of phenol oxidase, was collected for the humic acid-like product of *Hendersonula toruloidea* (Martin et al., 1972). Tan et al. (1978) examined two products of the ectomycorrhizal fungus, *Pisolithus tinctorius* by acid/base solubility and infrared analysis. They found that a greenish-brown substance produced when the fungus was grown with sucrose resembled humic and fulvic acids, whereas the black substance produced during growth on L-malic and L-succinic acids growth resembled humic acid.

Several strains of streptomycetes, when grown on a glycerol-nitrate medium, synthesized humic acid-like products (Huntjens, 1972). Interestingly, with these organisms, significant portions of the nitrate present in the growth medium was incorporated into the humic product. In one case, 13.7 percent of the nitrogen was isolated in the product. Hydrolysis of the humic acid-like product with 6 N hydrochloric acid released 40 percent of the nitrogen as ammonium and amino acid nitrogen indicating that the microbe had used the nitrate to synthesize amino acids which were subsequently excreted from the cell.

There has been considerable debate concerning the relationship of these microbially synthesized humic compounds to native soil humic acids. With each of the reports cited previously, similarities between the two humic acid products were stressed. This is the general pattern of reports from Martin's laboratory where most of this type of research has been conducted over the past 20 years. In opposition, workers from Schnitzer's laboratory have stressed differences in the substances. For example, Schnitzer et al. (1973) evaluated the chemistry of a number of fungal humic acid-like polymers and compared them with native humic acids. Growth conditions for the microorganisms were the same as those used in Martin's laboratory. The humic acids were characterized by elementary and functional group analyses and by permanganate oxidation of methylated preparations. The major oxidation products from the fungal preparations were (1) aliphatic mono- and dicarboxylic acids, (2) benzencarboxylic acids, (3) phenolic acids, and (4) aromatic compounds containing sulfur and nitrogen. Their data indicated that compared to peat and soil humic acids, the fungal materials contained relatively small quantities of aromatic compounds and more aliphatic compounds per gram of initial material. Based on these differences, it was concluded that "claims by Martin et al. that simple phenols and phenolic acids were significant constituents of fungal humic acids were not confirmed." Thus, the role of these products in soil humic acid synthesis was questioned. In contrast, Martin et al. (1974) using sodium amalgam reduction under nitrogen to decompose their fungal humic acids into aromatic substituents detected 4 to 32 percent simple phenols in the fungal products. They found that yields of such compounds from soil and peat humic acids by their procedures to be 3 to 6 percent. Further evidence of the similarity of the fungal humic acid-like products to soil humic acids was derived from ^{13}C-NMR spectra of fungal products and soil and Leonardite peat humic acids (Ludemann et al., 1982). All of the polymers gave complex spectra with strong signals, shoulders, or plateaus in the area of resonance for aliphatic, peptide, polysaccharide, aromatics, and carboxylic acid groups. The authors concluded that the spectral similarities between the various humic acid spectra supported the hypothesis that fungal humic acids are similar to soil humic acids in their chemical structure.

The significant factor in these studies is that the microorganisms produce not only the chemical precursors necessary for synthesis of a humic acid-like product but that they synthesize enzymes capable of catalyzing their polymerization. No difficulty is encountered in reconciling the molecular difference between these humic acid-like products and native soil humic acids. It is reasonable to conclude that should these products be synthesized in the soil ecosystem, further "maturation" would occur resulting in the conversion of these microbially synthesized materials to products more reminiscent of native soil humic acids. The question at this point is not whether these products can be precursors of soil humic acids or even whether they are in the mixture of complex organic materials extracted from soil and labeled as humic acid, but rather whether the microorganisms that synthesize these products in culture catalyze the same synthetic reactions in soil. Should these humic acid-like polymers be synthesized within the soil matrix by these fungi and other microorganisms, then the conclusion that the carbons contained within their structure are incorporated into soil humic acids is logical.

8.1.3. Other Mechanisms of Humic and Fulvic Acid Synthesis

Throughout the years of study of humic substances, a number of hypotheses explaining the origin of these complex molecules have been advanced. Many proposals have been shortlived, others were merely variants of those preexisting theories. Sufficient background was presented previously to evaluate the relationship of past theories and any new revision that appears. Thus variations on the common themes of plant product or microbially synthesized humic acid already discussed will not be presented. One unique hypothesis that has appeared frequently involves the Maillard reaction; that is, sugar and amine condensation [see Stevenson (1982) for a detailed analysis of the chemical mechanism of this reaction]. In this predominantly chemically mediated process, sugars and amines react to form high molecular weight melanoidins, which are brown to black compounds with some properties similar to soil humic acid. The reaction is used in the food industry to produce the dark brown to black colors in a number of foods. Benzing-Purdie and Ripmeester (1983) compared melanoidins and humic acids from a well-humified fraction of an organic soil through the use of ^{13}C CP-MAS NMR spectral analysis. The melanoidins prepared through the reactions of xylose with glycine, xylose with ammonium sulfate, and glucose with glycine were compared to soil humic acids. The melanoidins studied were those with molecular weights greater than 12,000. The melanoidans formed from xylose reacting with glycine and glucose and glycine were similar to humic acid in that they were highly aliphatic and had the same distribution of carbon types: aliphatic carbons bound to oxygen and nitrogen, aromatic, carboxyl and carbonyl. Glycine and xylose were insignificant substituents of the polymer in that acid hydrolysis yielded no xylose and less than 1 percent glycine. The authors concluded that their data supported the hypothesis that the Maillard reaction contributes to the origin of humic substances in soil and suggest that some of the nitrogen which is in the unknown soil nitrogen fraction may exist as melanoidin type polymers.

8.2. CATALYSTS AND MECHANISMS OF HUMIC ACID POLYMERIZATION

A variety of humic acid-like polymers have been synthesized through oxidative coupling of a number of simple aromatic ring containing compounds. Similarities in susceptibility to biological decomposition between these synthetic humic acid preparations and native humic acid preparations as well as the overall aromatic nature of the compounds lead to the proposal that such mechanisms are involved in native humic acid synthesis. The mechanism of oxidative coupling reactions of aromatic compounds and the microbial interactions involved therein were recently reviewed by Sjoblad and Bollag (1981). Biologically, this reaction is catalyzed by peroxidases (E.C. 1.11.1.7) and polyphenol oxidases (monophenyl monooxygenase, E.C. 1.14.18.1), including that oxidase activity commonly referred to as laccase. Chemically, the reaction is catalyzed by the clay minerals. Both biological and chemical mechanisms may be significant in native humic acid synthesis in soils. The coupling of the aromatic subunits into oligomers involves the production of a free radical (Sjoblad and Bollag, 1981). As shown in Fig. 8.2, the resonance of the electron distribution around the aromatic ring results in chemical structures which may form complexes in the ring ortho or para positions. Thus it is predicted that at least for humic acids formed by this mechanism, ortho and para linkages predominate. This oxidative coupling results in C–C, C–N, C–O, and N–N linkage of aromatic monomers. For example, when vanillic acid is coupled by the laccase from *Rhizoctonia praticola*, dimers with C–O (2-methoxy-6-(2′-methoxy-4′-carboxyphenoxy)-1,4-benzoquinone) and C–C (2-methoxyl-6-(2′-hydroxy-3′-methoxy-5′-carboxyphenyl)-1,4-benzoquinone) linkages were produced (Bollag et al., 1982). A variety of dimer, tetramers, and greater-size oligomeric structures produced by oxidative coupling are described in the review by Sjoblad and Bollag (1981). Occurrence of this polymerization mechanism in soil humic substance synthesis rests upon both the occurrence of these enzymatic activities in soil and the existence of free radicals in soil humic substances. Free radicals are a common soil substituents (Steelink and Tollin, 1967). Phenol oxidase and peroxidase activities are commonly detected in a variety of mineral and organic soils.

Several hypothetical humic acid structures based on the properties of these oxidative coupling reactions have been proposed. One such structure which was developed by Stevenson (1982) is presented in Fig. 8.3. Note the ortho and para linkages between the aromatic rings as dictated by the enzymatic reaction mechanism. Also nitrogen moieties are included as amine groups as well as members of heterocyclic rings. Phenolic hydroxyl groups may also participate in external as well as internal hydrogen bonds. Typical covalent linkages of peptides and sugars are also depicted.

Free radicals and the soil conditions leading to their formation explain not only the synthesis of soil humic substances, but also their stability. The relationship of stability of humic substances and free radicals was demonstrated in a study of effects of soil pH and electrolyte concentration on radical formation

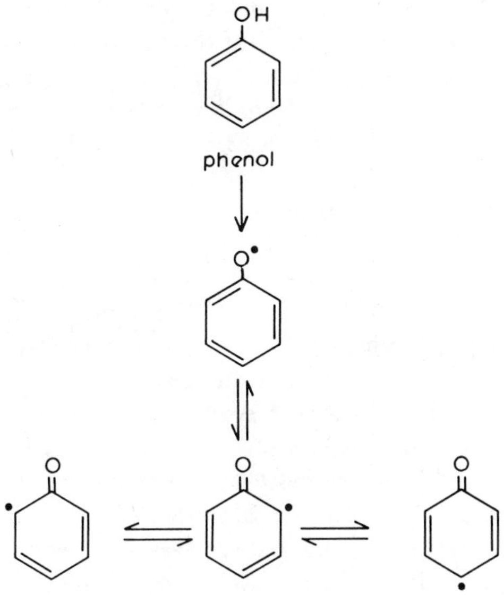

Fig. 8.2.
Free radical formation from phenol and its resonance structures.

(Ghosh and Schnitzer, 1980). Electron spin resonance analysis of humic substances revealed a decrease in free radical concentrations in humic materials with increases in neutral electrolyte concentrations. The free radical concentration and, therefore, the reactivity of the humic materials was lowest between pH 5.0 and 6.5. The authors concluded that this stability in the common pH range of agricultural soils explained, at least in part, the stability of humic materials in soils.

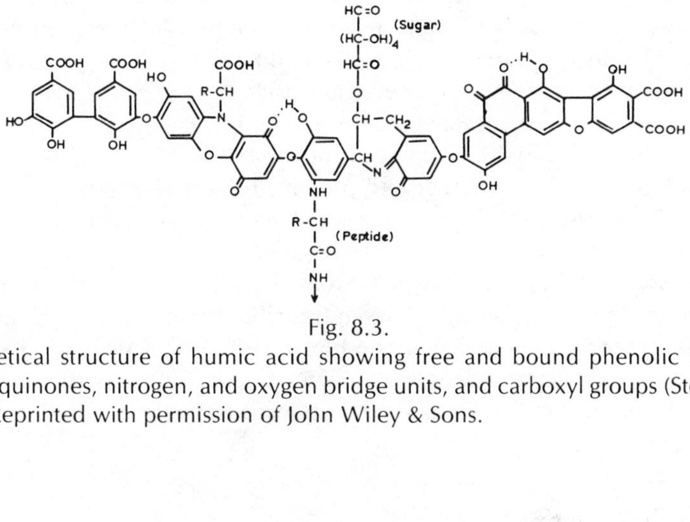

Fig. 8.3.
Hypothetical structure of humic acid showing free and bound phenolic hydroxyl groups, quinones, nitrogen, and oxygen bridge units, and carboxyl groups (Stevenson, 1982). Reprinted with permission of John Wiley & Sons.

Phenol oxidase (laccase) isolated from *Rhizoctonia praticola* catalyzes the polymerization of humus substituents into oligomeric products (Bollag et al., 1980). Orcinol, syringic acid, vanillic acid, and vanillin were oxidatively coupled into oligomers which ranged in size from dimers to pentamers. The decomposition product of a number of herbicides, 2, 4-dichlorophenol, was also incorporated into the oligomers by the phenol oxidase activity. Phenol oxidase oxidative coupling of xenobiotic compounds in soils was proposed as a mechanism for the prolonged persistence of some pesticides in soil. Vanillic acid, a decomposition product of lignin, is also polymerized into mixtures of dimers to pentamers by this fungal phenol oxidase (Bollag et al., 1982). Some selectivity of these enzymatic activities is suggested by the fact that the latter polymerization was also catalyzed by peroxidase and not by tyrosinase.

Metabolic control of laccase activity synthesis in a variety of basidiomycetes, ascomycetes, and deuteromycetes, grown in a sugar-rich liquid medium, was examined by Bollag and Leonowicz (1984). Xylidine stimulated enzyme synthesis in cultures of *Fomes annosus*, *Pholiota mutabilis*, *Pleurotus ostreatus*, and *Trametes versicolor*. No effect of the inducer was observed with cultures of *Rhizoctonia praticola* and *Botrytis cinerea*. Laccase synthesis by *Podospora anserina* was inhibited by the presence of xylidine in the growth medium. Purification of the laccase enzymes produced by these fungi indicated that each fungus synthesized distinct enzymes. Bands of induced enzyme forms on polyacrylamide gels were only detected with the basidiomycetes. The enzyme of *Rhizoctonia praticola* had an optimal pH in the neutral region, whereas activities of all other laccases examined were in the range of pH 3.0 to 5.7. Kinetics of oxidation of methoxyphenolic acids of the enzymes varied with enzyme source.

The synthesis of model humic acid polymers by phenol oxidase and peroxidase were compared by Martin and Haider (1980). With the phenol oxidase (tyrosinase—ICN 101197), yields of humic and fulvic acids were variable. Their reaction mixture contained 4 g of a phenol mixture alone or combined with 1 g of various amino acids, glucosamine, or protein plus 150 to 300 mg of enzyme. Humic product yields with these reaction mixtures ranged from 1.5 to 2.5 g with this enzyme. About 33, 55, 5 to 14, 50, and 100 percent of the ferulic acid; catechol; glycine, lysine, and glucosamine; cysteine; and protein, respectively were incorporated into the humic polymers. In comparison, with peroxidase the yields were more consistent and varied from 2.5 to 3.5 g. Approximately 80 percent of the polymers was recovered as humic acid. With peroxidase about 77, 95, 23 to 34, 63 and essentially 100 percent of the ferulic acid; catechol; glycine, lysine, and glucosamine cysteine; and protein, respectively, were recovered in the humic polymers. Biodegradation resistance of the polymers synthesized by either enzyme was similar to that of native humic acids.

The reaction mechanism of the peroxidase coupling of anilines and phenols involves formation of a positively charged transition state as the rate controlling state in the reaction sequence (Berry and Boyd, 1984). Reactivity was increased by the presence of electron-donating substituents on the aromatic ring. These substituents act to stabilize the positively charged transition state.

Peroxidase is not only involved in *de novo* synthesis of humic acids but it apparently is instrumental in maturation of the humic materials—and perhaps their decomposition (Mangler and Tate, 1982). A peroxidase produced by a fungus isolated from cultivated organic soils of south Florida when incubated in the presence of humic acid increased its E_4/E_6 ratio. The substrate was an acid hydrolyzed humic acid extracted from the organic soil from which the fungus originated. Increases in the E_4/E_6 ratio are indicative of increases in the oxidation state of the humic acids. In this study, the ratio was increased from 4.5 to 8.0.

As was indicated previously, nonbiological processes also catalyze oxidative coupling of aromatics into humic polymers. For example, benzene or phenol when adsorbed on smectite with Fe^{3+} or Cu^{2+} on the exchange complex are polymerized (Mortland and Halloran, 1976). With benzene, the products could not be attributed to coupled benzene rings, whereas extensive coupling of the phenolic rings occurred. Examination of the reaction mixture with electron spin resonance spectroscopy indicated radical formation, as was recorded in the enzymatically catalyzed reactions.

Compounds which are susceptible to chemical oxidation to humic acids can be associated into four chemical groups (Dragum and Helling, 1985). The chemical properties of 93 organic compounds known to be oxidized in soil- or clay-catalyzed reactions into humic acids were analyzed for common chemical properties. The objective of the study was to develop models based on these properties which would be predictive of potential for chemically catalyzed incorporation of xenobiotics into humic acid fractions. All of the compounds examined contained at least one aromatic ring. Many contained fused-ring structures. The compounds were found to fall into four basic chemical groupings: (1) aromatic chemicals that contain electron-withdrawing and weak electron-donating fragments (lower water solubility limit of 200 ppm), (2) aromatic chemicals that contain electron withdrawing and a very strong electron-donating fragment (lower water solubility limit of 112 ppm), (3) aromatic chemicals that contain only electron-donating fragments (lower water solubility limit of 29 ppm), and (4) aromatic chemicals that contain extensive conjugation.

These data indicate a variety of somewhat nonspecific mechanisms for polymerization of aromatic compounds into humic and fulvic acid-like products. The catalysts are either commonly found in soils, as is the case with the clay minerals, or appear to be common components of the enzymatic arsenal of soil microbes. Although it is difficult to demonstrate these reactions *in situ* in a native soil sample and it is even more difficult to analyze quantitatively native soil humic acids for conclusive evidence for activity of these oxidative coupling catalysts, we may reasonably conclude that these processes operate, at least in part, in the synthesis of humic substances in soil ecosystems.

8.3. HUMIC ACID DECOMPOSITION

Humic acid is probably the most formidable, naturally occurring molecule that soil microorganisms are called upon to decompose. Aside from the fact that a number of organisms have been isolated capable of this task, the biodegradation of humic acid is attested to by the fact that an equilibrium concentration characteristic of each specific soil ecosystem occurs. If this complex molecule was totally biodegradation resistant, since it was demonstrated previously that synthesis is catalyzed both biologically and chemically, than an ever increasing quantity of humic acid would be expected to be found in soil. This does not occur.

Examination of any of the hypothetical structures proposed for humic acid (Stevenson, 1982) (Fig. 8.3) allows deduction of the major reasons for the relative biodegradation resistance of humic acids as compared to readily decomposed plant substituents. The list is topped by the observation that humic acid is not comprised of a repeating subunit. Current data indicate that humic acid is a random polymer containing a wide variety of aromatic subunits. Although similar lists of aromatic and phenolic compounds are derived from analyses of humic acids from a variety of sources, the order of polymerization as well as the ratios of the various components must vary between sources. For those substituents arising from exogenously supplied organic matter such as plant debris, it is clear that should the vegetation of a given region lack a particular humic acid precursor, than that material may very well be absent from humic acid isolated from that site. The practical result of such variability in structure is that a variety of enzymatic activities are needed to depolymerize or catabolize humic acids. Energetic requirements for synthesis of this battery of enzymes are the second barrier to rapid, general humic acid decomposition. Not only is the core of the molecule composed of major quantities of aromatic ring containing compounds which require significant energy expenditures before the microbe derives any energy benefit, but synthesis of the large number of enzymes required for the total mineralization of humic acid is energy intensive. Proteins require relatively large expenditures of energy for their synthesis. Thus from the teleological viewpoint, it is more logical for a microbe to catabolize a simple polymer, such as a protein which requires only a few enzymes for carbon and energy recovery, than to develop the complex enzyme systems needed for humic acid mineralization.

The enzyme complex needed to completely mineralize a humic acid molecule necessarily includes those activities associated with decomposition of the associated humified materials, such as proteins and polysaccharides. In fact, as will be discussed in the following text, these activities may in reality provide the bulk of the carbon, nitrogen, and energy return to the growing microbe. Enzymatic activities that are included in this category are proteases, carbohydrases, a variety of hydrolases and so on. This enzyme arsenal is supplemented by monophenol oxidases, peroxidases, and etherases which are most probably synthesized for other cellular functions, but because of their low substrate specificity, are involved in humic acid decomposition. The monophenol oxidases and peroxidases are included in both the synthetic and degradative pathways in that the hydroxylation of aromatic rings is instrumental in both pathways. Indeed, it is likely that the primary mechanism of these enzymes favors decomposition processes, in that the aromatic ring must be hydroxylated before it is split. The polymerization reaction would therefore be deemed to be a side- or secondary-reaction. This hypothesis is supported by the fact that in many studies of these hydroxylating enzymes, the major products of their activity are degradative products rather than oligomers.

The ultimate factor controlling the rate of humic acid decomposition is the energy yielded to the microorganism. The prime reasons for enzyme production by microorganisms, or any living organism, is the requirement of energy for cell synthesis and maintenance. Thus the most efficient microorganism and hence the best competitor, is the one capable of recovering the total energy needed for growth at the least expense. Thus when faced with a formidable structure like humic acid, the more efficient microorganism is the one that utilizes the associated humified proteins and carbohydrates for carbon and energy rather than the complex random polymer of the aromatic backbone, assuming that the pool of nonhumified carbonaceous substrates has been exhausted. If this is not the case, this latter pool would serve as the major carbon and energy reservoir for the microbial community. The microorganisms acquire some energy through metabolism of the aliphatic portions of the molecule and the aromatic side chains, but it is questionable whether these structures occur in sufficient concentrations to provide the microorganisms with enough energy to maintain their cell structure. Thus the microorganism involved in humic substance catabolism would benefit primarily from the slow catabolism of humified substrates, with some direct or perhaps cometabolic attack of the associated aromatics.

In spite of these limitations, a wide variety of microorganisms have been isolated capable of catabolizing humic acid (Table 8.1). It is difficult to determine the true nature of the association of these microbes with humic acid in that an operational definition of humic acid is used for most of the enrichments; that is, the alkaline solubile, acid precipitated product is used in the microbial enrichment cultures. Thus it is highly likely that in most of these cases the microorganisms are using the humified proteins, peptides, amino acids, and polysaccharides as carbon, nitrogen, and energy sources. Under this situation, any catabolism of the aromatic moieties of the humic preparation would be the result of cometabolic processes. A true test of whether a given isolate is metabolizing the central portion of the humic acid molecule and not the adhering proteins, polysaccharides, and so on would be to purify the humic acid preparation through an acid reflux step. This would remove all proteins and carbohydrates.

Organisms from essentially every microbial grouping have been shown to be capable of decomposing humic acid to some degree (Table 8.1). Many of these microorganisms have limited capability to decompose the aromatic components of the humic acid for energy and carbon in that their growth and activity is stimulated by glucose amendment of the culture medium. It is likely that these microorganisms are predominantly metabolizing the humified, readily decomposable plant and animal components of the humic acid preparation. Andreyuk and Gordienko (1978) isolated a number of actinomycetes from lowland peat which not only decomposed humic acids, but simultaneously resynthesized the molecules. A situation that would be anticipated from the dual role of

Table 8.1.
Representative Reports of Humic Acid Degrading Microbes

Microbial Group	Reference
Actinomycetes	Andreyuk and Gordienko, 1978
Associative cultures	Andriiuk et al., 1973
Fungi	Biederbeck and Paul, 1971
Aspergillus versicolor	Duboska, 1980
Penicillium lilacinum	
Penicillium citrinum	
Absidia glauca	
Actinomyctes	Fedorov and Il'ina, 1963
Streptomycetes	Ibrahim and Ibrahim, 1977
Pseudomonas rubigenosa	Mal'tseva et al., 1975
Penicillium frequentans	Mathur and Paul, 1967
Penicillium sp.	Seraya and Dul'gerova, 1977
Actinomyctes	
Nocardia	Sidorenko et al., 1978
	Steinbrenner and Mundstock, 1978

monophenol oxidases and peroxidases. These isolates decreased the content of high molecular weight humic acids in the humic acid mixture and increased their functional group content. Dubovska (1980) found that humic acid decomposition from peat soil by *Aspergillus versicolor*, *Penicillium lilacinum*, *Penicillium citrinum*, and *Absidia glauca* was stimulated by amendment of the culture medium with glucose of sodium nitrate. This preference of soil actinomycetes for glucose as the carbon source was also shown by Fedorov and Il'ina (1963). Although significant use of humic acids by pure cultures of actinomycetes was not observed, respiration data suggested that humic acid was utilized to a small degree by the soil actinomycetes. Humic acid mineralization by oligonitrophilic microorganisms (Mal'tseva et al., 1975) and streptomycetes (Ibrahim and Ibrahim, 1977) was also stimulated by amendment of cultures with an easily metabolized carbon and energy source.

In contrast, *Nocardia corallina* decomposed 23.4 percent of the humic acid amended to a mineral medium as the sole carbon source (Sidorenko et al., 1978). The microbial population increased 15 fold during the incubation. Changes in the hydrogen to carbon ratio of the substrate suggested that the microorganisms were catabolizing the aromatic ring compounds. Analysis of the substrate following six months incubation by infra red analysis indicated increases in CH_2 and methyl groupings. Similarly, *Penicillium frequentans* catabolized the aromatic moiety of humic acid (Paul and Mathur, 1967; Mathur and Paul, 1967). The cleavage did not involve reduction of carboxyl groups, suggesting that one of the steps involved was hydrolysis of the humic acid ether bonds. Gel filtration analysis of the metabolized humic acid preparation revealed that the humic acid fraction with molecular weight of more than 50,000 was degraded to the largest extent.

As was observed with lignin decomposition, catabolism of the aromatic fraction of the humic acid may involve either oxidation of the individual aromatic rings *in situ* on the humic molecule or depolymerization and subsequent oxidation of the aromatic substituents. In either case, the reactions are slow, and generally involve participation of molecular oxygen. Soil humic acid fractions may be divided into a more biodegradation resistant fraction and extremely resistant fractions. The former humic acids exhibit decomposition times in years to decades, whereas the latter exhibit radiocarbon ages of many thousands of years (Stout et al., 1981). Even the more susceptible fractions are decomposed at an extremely slow rate compared to the rates generally associated with readily metabolized plant and animal components. For example, Martin et al. (1982) noted that from 6 to 9 percent of the protein, cysteine, lysine, and glucosamine carbons linked into model humic acid polymers were lost during a one-year incubation period in allophanic soils. Somewhat more rapid rates were detected in a variety of agricultural soils. Between 13 and 24 percent of the same compounds was catabolized in a comparable period in the latter soils. Note that the humic components whose catabolism was measured were the more easily decomposed moieties. Aromatic substances would be of greater biodegradation resistance.

8.4. CONCLUSIONS

Humic acid synthesis pathways could be said to be as complex as the molecule itself. Specific humic acid carbons may arise from plant derived substituents, microbial biomass or products, as well as animal tissues. With the development of modern chemical synthetic procedures, anthropogenic sources must also be added to this list. The actual polymerization of humic acids may be biologically catalyzed through the action of enzymes such as monophenol oxidases or peroxidases, or chemically mediated through common soil components such as the clay fraction. Commonly analyzed and discussed humic acid precursors include the major plant substituent lignin and microbially synthesized humic acid-like components (melanins). In either case, these materials may be incorporated *in toto* into humic acids or be modified, even to the point of total depolymerization, and incorporated. Subsequent to initial synthesis, the nascent humic acid molecule is "matured" through combination with other humic acid molecules and internal oxidation reactions.

As a result of the complexity of the humic acid molecule itself as well as the fact that biological and chemical processes contribute to its formation, humic acid molecules are relatively biodegradation resistant when compared to the more easily decomposed plant and animal components. But, the fact that concentrations of this soil organic matter component do not continuously accumulate in soils where they are formed, and a wide variety of fungi, bacteria, and actinomycete strains have been isolated with the capability of catabolizing at least portions of the humic molecule attest to the conclusion that eventual mineralization is possible. Because of uncertainties associated with the molecular structure as well as its extremely complex structure, the exact mineralization mechanisms of humic acid are unknown. Enzymatic activities that must be involved in the mineralization processes are phenol oxidases, etherases, peroxidases, as well as those enzymatic activities associated with depolymerization of humified proteins and polysaccharides.

REFERENCES

Andreyuk, E. I., and S. A. Gordienko, 1978. Transformation of humic acids by soil actinomycetes. Mikrobiol. Zh. (Kiev) 40: 690-697 (in Russian). Chemical Abstracts 90: 102586w.

Andriiuk, K. I., S. O. Hordienko, I. N. Havrysh, H., I. Konotop, and V. A. Martynenko, 1973. Decomposition of peat humic acids by associative cultures of microorganisms. Mikrobiol. Zh. 35: 554-559 (in Russian).

Benzing-Purdie, L., and J. A. Ripmeester, 1983. Melanoidins and soil organic matter: Evidence of strong similarities revealed by ^{13}C CP-MAS NMR. Soil Sci. Soc. Am. J. 47: 56-61.

Berry, D. F., and S. A. Boyd, 1984. Oxidative coupling of phenols and anilines by peroxidase: Structure-activity relationships. Soil Sci. Soc. Am. J. 48: 565-569.

Biederbeck, V. O., and E. A. Paul, 1971. Fungal degradation of soil humic nitrogen. Argon. Abst. 1971: 80.

Bollag, J.-M., and A. Leonowicz, 1984. Comparative studies of extracellular fungal laccases. Appl. Environ. Microbiol. 48: 849-854.

Bollag, J.-M., S.-Y. Liu, and R. D. Minard, 1980. Cross-coupling of phenolic humus constituents and 2, 4-dichlorophenol. Soil Sci. Soc. Am. J. 44: 52-56.

Bollag, J.-M., S.-Y. Liu, and R. D. Minard, 1982. Enzymatic oligomerization of vanillic acid. Soil Biol. Biochem. 14: 157-163.

Dragun, J., and C. S. Helling, 1985. Physicochemical and structural relationships of organic chemicals undergoing soil- and clay-catalyzed free-radical oxidation. Soil Sci. 139: 100-111.

Dubovska, A., 1980. Decomposition of humus substances by microorganisms. VI. Utilization of carbon from the humus acids of a peaty soil by some micromycetes. Acta Fac. Rerum Nat. Comenianae, Microbiol. 9: 81-111. Chemical Abstracts 94: 153172Y.

Fedorov, M. V., and T. K. Il'ina, 1963. Utilization of humic acid by soil actinomycetes as a sole source of carbon and nitrogen. Microbiology 32: 234-237.

Felbeck, G. T., Jr., 1971. Structural hypotheses of soil humic acids. Soil Sci. 111: 42-48.

Flaig, W., 1971. Organic compounds in soil. Soil Sci. 111: 19-33.

Ghosh, K., and M. Schnitzer, 1980. Effects of pH and neutral electrolyte concentration on free radicals in humic substances. Soil Sci. Soc. Am. J. 44: 975-978.

Haider, K., and J. P. Martin, 1967. Synthesis and transformation of phenolic compounds by *Epicoccum nigrum* in relation to humic acid formation. Soil Sci. Soc. Am. J. 31: 766-772.

Haider, K., and J. P. Martin, 1970. Humic acid-type phenolic polymers from *Aspergillus sydowi* culture medium, *Stachybotrys* spp. cells and autoxidized phenol mixtures. Soil Biol. Biochem. 2: 145–156.

Hatcher, P. G., M. Schnitzer, L. W. Dennis, and G. E. Maciel, 1981. Aromaticity of humic substances in soils. Soil Sci. Soc. Am. J. 45: 1089–1094.

Huntjens, J. L. M., 1972. Amino acid composition of humic acid-like polymers produced by streptomycetes and of humic acids from pasture and arable land. Soil Biol. Biochem. 4: 339–345.

Ibrahim, A. N., and I. A. Ibrahim, 1977. Biodegradation of soil humus by streptomycetes. Agrokim. Talajtan 26: 415–423. (in Hungarian).

Katase, T. 1983. The significance of different forms of *p*-coumaric and ferulic acids in a pond sediment. Soil Sci. 135: 151–155.

Linhares, L. F., and J. P. Martin, 1978. Decomposition in soil of the humic acid-type polymers (melanins) of *Eurotium echinulatum*, *Aspergillus glaucus* sp., and other fungi. Soil Sci. Soc. Am. J. 42: 738–743.

Ludemann, H. D., H. Lentz, and J. P. Martin, 1982. Carbon-13 nuclear magnetic resonance spectra of some fungal melanins and humic acids. Soil Sci. Soc. Am. J. 46: 957–962.

Mal'tseva, N. N., S. A. Gordienko, and V. V. Izzheurova, 1975. Use of humic acids by oligonitrophilous microorganisms. Tr. S'ezda Mikrobiol. Ukr. 4th. 61–62. In D. G. Zatula (ed.), Naukova Dumka, Kiev, USSR. Chemical Abstracts 85: 18897l.

Mangler, J. E., and R. L. Tate III, 1982. Source and role of peroxidase in soil organic matter oxidation in Pahokee muck. Soil Sci. 134: 226–232.

Martin, J. P., and K. Haider. 1978. Microbial degradation and stabilization of carbon-14-labeled lignins, phenols, and phenolic polymers in relation to soil humus formation. In T. K. Kirk, T. Higuchi, and H.-M. Chang (eds.), Lignin Biodegradation: Microbiology, Chemistry, Potential Application (Proc. Int. Symp.) 1: 77–100. CRC, Boca Raton, Florida.

Martin, J. P., and K. Haider, 1980. A comparison of the use of phenolase and peroxidase for the synthesis of model humic acid-type polymers. Soil Sci. Soc. Am. J. 44: 983–988.

Martin, J. P., K. Haider, and C. Saiz-Jimenez, 1974. Sodium amalgam reductive degradation of fungal and model phenolic polymers, soil humic acids, and simple phenolic compounds. Soil Sci. Soc. Am. Proc. 38: 760–764.

Martin, J. P., K. Haider, and D. Wolf, 1972. Synthesis of phenols and phenolic polymers by *Hendersonula toruloidea* in relation to humic acid formation. Soil Sci. Soc. Am. Proc. 36: 311–315.

Martin, J. P., S. J. Richards, and K. Haider, 1967. Properties and decomposition and binding action in soil of "humic acid" synthesized by *Epicoccum nigrum*. Soil Sci. Soc. Am. Proc. 31: 657–662.

Martin, J. P., H. Zunino, P. Peirano, M. Caiozzi, and K. Haider, 1982. Decomposition of ^{14}C-labeled lignins, model humic acid polymers, and fungal melanins in allophanic soils. Soil Biol. Biochem. 14: 289–293.

Mathur, S. P., and E. A. Paul, 1967. Microbial utilization of soil humic acids. Can. J. Microbiol. 13: 573–580.

Mortland, M. M., and L. J. Halloran, 1976. Polymerization of aromatic molecules on smectite. Soil Sci. Soc. Am. J. 40: 367–370.

Paul, E. A., and S. P. Mathur, 1967. Cleavage of humic acids by *Penicillium frequentans*. Plant Soil 27: 297–299.

Schnitzer, M., M. I. Ortiz de Serra, and K. Ivarson, 1973. The chemistry of fungal humic acid-like polymers and of soil humic acids. Soil Sci. Soc. Am. Proc. 37: 229–236.

Seraya, L. I., and A. N. Dul'gerova, 1977. Decomposition of soil humic acid by microorganisms. Mikrobiol. Zh. (Kiev) 39: 643–644. Chemical Abstracts 88: 5435x.

Sidorenko, O. D., V. I. Aristarkhova, and V. A. Chernikov, 1978. Changes in the composition and properties of humic acids after treatment with nocardia microorganisms. Izv. Akad. Nauk SSSR, Ser. Biol. 2: 195–202. (in Russian).

Sjoblad, R. D., and J.-M. Bollag, 1981. Oxidative coupling of aromatic compounds by enzymes from soil microorganisms. In E. A. Paul and J. N. Ladd (eds.), Soil Biochem. 5: 113–152. Marcel Dekker, New York.

Steelink, C., and G. Tollin, 1967. Free radicals in soil. In A. D. McLaren and G. H. Peterson (eds.), Soil Biochem. 1: 147–169. Marcel Dekker, New York.

Steinbrenner, K., and I. Mundstock, 1975. Untersuchungen zum Huminstoffabau durch Nokardien. Arch. Acker-pflanzenbau Bodenkd. 19: 243–255.

Stevenson, F. J. 1982. Humus Chemistry. John Wiley & Sons. New York, 443 pp.

Stout, J. D., K. M. Goh, and T. A. Rafter, 1981. Chemistry and turnover of naturally occurring resistant organic compounds in soil. In E. A. Paul and J. N. Ladd (eds.), Soil Biochem. 5: 1–73. Dekker. New York.

Tan, K. H., P. Sihanonth, and R. L. Todd, 1978. Formation of humic acid like compounds by the ectomycorrhizal fungus, *Pisolithus tinctorius*. Soil Sci. Soc. Am. J. 42: 906–908.

Vardachari, C., and K. Ghosh, 1984. On humus formation. Plant Soil 77: 305–313.

Waksman, S. A. 1926. The origin and nature of the soil organic matter or soil "humus": III. The nature of substances contributing to the formation of humus. Soil Sci. 22: 323–333.

NINE

SOIL ORGANIC MATTER AS A PLANT NUTRIENT RESERVOIR

Soil organic matter is unquestionably the largest pool of plant nutrients in the world's soil. Exceptions are generally localized in time or space and consist primarily of heavily managed soils, such as intensive agricultural systems that receive large inputs of fertilizers. Since nutrient availability is a major controlling factor in biomass productivity and ecosystem stability, understanding processes contributing to nutrient exchanges in the soil organic matter pool are essential for the development of cohesive ecosystem models.

Although substantial quantities of plant nutrients are contained in soil organic matter pools, these nutrients are not directly available to the plant community; that is, they must be converted to mineral forms by the soil microbial community. It is the mineral forms that are readily incorporated into plant biomass. This mineralization process is the rate limiting step in biomass production in many ecosystems, but consideration of this process alone provides an incomplete and, perhaps, erroneous picture of nutrient cycling. The efficiency of plant biomass synthesis and the rate of return of this biomass to the soil ecosystem also controls ecosystem productivity. The processes involving movement of nutrients into plant biomass and their return to the mineral nutrient pools are referred to as biogeochemical cycles.

A commonly studied example of such a cycle, the nitrogen cycle, is depicted in Fig. 9.1. With the nitrogen cycle, the primary soil nitrogen source is fixation of atmospheric dinitrogen. **Nitrogen fixation** (i.e., reduction of dinitrogen with its triple-bonded nitrogen to organic nitrogen) is catalyzed by a variety of bacterial and blue green algal species (**diazotrophs**). This fixed nitrogen is first incorporated into microbial biomass. Thus for free-living diazotrophs, the microbial cell must be mineralized for transfer of the nitrogen to plant biomass to occur. This mineralized nitrogen generally enters plant biomass through the root systems of plants associated with free-living nitrogen fixing microorganisms. In contrast, most symbiotically fixed nitrogen is transferred directly from the bacterial or actinomycete symbiont to the host plant. This fixed nitrogen becomes part of the soil organic matter pool as a result of the demise of the plant, or

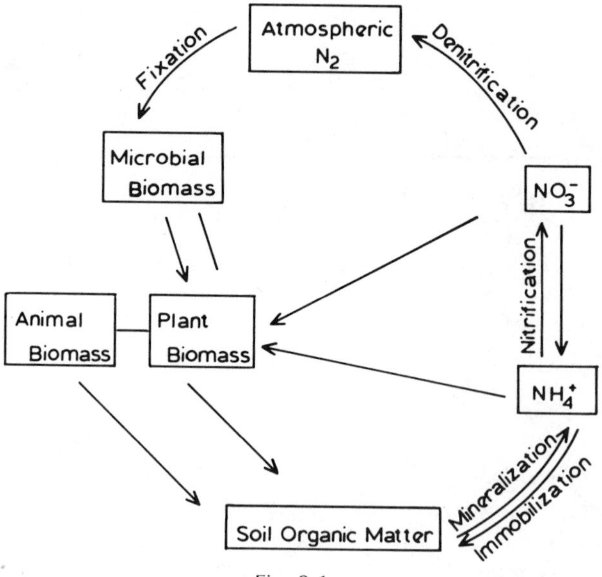

Fig. 9.1.
The nitrogen cycle.

portions of the plant (leaves or root tissue) and the subsequent incorporation of the plant substituents into the soil matrix. Other routes to soil organic matter include transfer as root exudates or immobilization of fixed nitrogen into soil microbial biomass. The nitrogen within the colloidal soil organic matter pool becomes available to the plant community through mineralization to ammonium and nitrate. Note the cyclic nature of the process in that the terminal soil nitrogen product, nitrate, may be biologically reduced to atmospheric dinitrogen.

Many of the steps in biogeochemical cycles involve a series of oxidation state changes of the nutrient atom. These oxidation/reduction reactions provide the primary benefit to the biological mediators of these reactions. In the case of the nitrogen cycle, an example of such transformations is seen in the conversion of organic nitrogen to nitrate. The nitrogen oxidation state nitrogen varies from -3 in ammonium to $+5$ in the terminal oxidation product, nitrate. Throughout the carbon, sulfur, and nitrogen cycles, various oxidative reactions result in liberation of energy for microbial growth. With the reductive processes, the oxidized compounds (e.g., nitrate in denitrification) serve as terminal electron acceptors, thereby allowing the microorganism to grow in anoxic or anaerobic soil sites.

The basic principles of the reactions of the nitrogen cycle also relate to the other biogeochemical cycles. For example, the reactions of the carbon and sulfur cycles are similar to those of the nitrogen cycle in that biologically fixed carbon or sulfur is biochemically transformed through a series of oxidative reactions to carbon dioxide and sulfate. Transformations of the phosphorus cycle are in

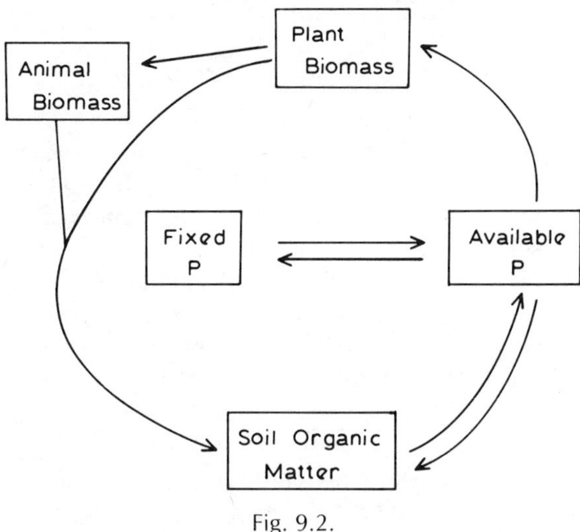

Fig. 9.2.
The phosphorus cycle.

some ways biologically less complex then other cycles in that the oxidation state of the phosphorus atom generally does not change, but the cycle is complicated by an interchange of soil mineral phosphorus between plant available and fixed, or unavailable, forms (Fig. 9.2). Organic phosphorus pools are soil organic matter and plant and animal biomass. Within the mineral matrix of the soil, phosphate is in equilibrium between a water soluble state and precipitates. Similar accumulation of a nutrient in water insoluble forms is noted with the sulfur cycle where sulfide and sulfate are retained in soil minerals. For further details on the biology and chemical properties of these cycles, see Stevenson (1986).

The biochemical reactions involved in the nitrogen cycle and other biogeochemical cycles have been widely discussed and reviewed (e.g., see Alexander, 1977). Most commonly studied cycles are those involving carbon, nitrogen, phosphorus and sulfur transformations. Elementary biology, ecology, and microbiology texts usually contain detailed descriptions of the essentials of the processes. Unfortunately, because the general trend is to present the various biogeochemical cycles as isolated entities, the student frequently is left with the impression that these cycles operate independently; that is, the implication is that it is possible for a functional nitrogen cycle to exist in the absence of a carbon cycle, for example. None of the cycles function independently. Thus to overcome these "subliminal" misconceptions, rather than repeating the already adequate presentation of the biochemical reactions involved with the individual cycles, the interactive nature of the various biogeochemical cycles and the impact of their interactions on total ecosystem structure and function will be examined in this chapter.

9.1. SOIL ORGANIC MATTER AS A PLANT NUTRIENT POOL

The quantity of plant nutrients produced by decomposition of the soil organic matter pool depends on the physical and chemical factors limiting mineralization, the rate of organic matter accretion, and the quantities of nutrients supplied from external sources. The biomass inputs to the soil ecosystem may be balanced with the mineralization rates so that ecosystem exists in equilibrium or there may be a net production of mineral nutrients or a sequestering of these substances. For example, even with abundant biomass accumulation and nutrient surpluses, physical or chemical properties of the ecosystem may retard mineralization to the degree that partially decayed biomass accumulates; that is, the system becomes a net nutrient sink. Physically limited systems include swamps and bogs where mineralization is controlled in part by the anoxia created by the flooded conditions. Drainage of the swamp results in the reduction of soil moisture to a level more favorable to agricultural production. Drainage of swamplands, as occurred in the northern reaches of the Florida Everglades in the early 1900s, results in accelerated mineralization of the accumulated plant debris. The system becomes a net nutrient source rather than a sink. This mineralization may result in nutrient production in quantities far greater than can be incorporated into aboveground biomass. For example, organic matter mineralization in the drained Histosols of the Everglades Agricultural Area produces 1200 to 1400 kg nitrogen per hectare annually (Tate, 1980a). The more common situation in soils involves a state where nitrogen inputs into soil organic matter reserves approximately equal mineralization outputs. In a reasonably closed system, (e.g., climax forests), tight coupling between the forest floor organic matter mineralizing community and plant root nutrient uptake mechanisms is extremely important for nutrient preservation. (Other nutrient retention mechanisms in forests include translocation of nitrogen reserves from leaves prior to leaf fall.) For example, in a study of a series of edaphic climax forests, soil nitrogen mineralization rates (26 to 84 kg/ha/yr) were highly correlated ($r^2 = 0.92$) with net aboveground production (4.1 to 9.5 Mg/ha/yr) (Pastor et al., 1984). As may be expected from evaluation of plant biomass cycling to soil organic matter in forest ecosystems (Chapter 2), positive correlations were also found between soil nitrogen mineralization rates and litter production and nitrogen and phosphorus return in litter. Negative correlations were observed between nitrogen mineralization rates and carbon:nitrogen and carbon:phosphorus ratios and efficiency of phosphorus use in the litter.

In an ideal system, all biomass nitrogen mineralized is directed to new biomass synthesis; realistically some nutrients are diverted to nonproductive pathways. Loss of nitrogen through denitrification or leaching necessarily reduces the quantities of biomass which can be produced. In agricultural systems, major quantities of the nutrients are removed with the harvested crop. Prolonged removal of nutrients from the ecosystem results ultimately in a decline in total plant biomass and system stability unless the nutrient losses are compensated for by exogenously supplied nutrients. In the absence of anthropogenic intervention, this would involve nitrogen fixation, solubolization of mineral nutrient sources, and so on.

In contrast to native sites where biomass productivity is controlled primarily by mineralization processes, due to the large quantities of fertilizers used in most agricultural systems in developed countries, mineralization of soil organic matter has little impact on total ecosystem productivity and crop yield. Although large quantities of nutrients are removed from the soil with harvesting, leaching, and denitrification, sufficient fertilizers are usually added to the soil to counteract the losses. Generally, nutrient inputs from fertilization are several orders of magnitude greater than those from soil organic matter mineralization. (This does not imply that use of nitrogenous fertilizers could not be reduced were mineralization of crop residues considered in making soil test recommendations. This alternative will be discussed in the following text.)

A major portion of the world's soils are managed to recover nutrients retained in soil organic matter for crop production. The most common example of such systems is called subsistence agriculture. The objective of subsistence agriculture is to "mine" soil organic matter pools for crop production. Once these pools are exhausted, the soil is usually allowed to return to its "wild" state and the farmer moves on to a newly cleared site.

The impact of anthropogenic intervention on soil nutrient cycling varies with the nutrient under study. Some soils may receive sufficient quantities of nutrients "naturally" from exogenous sources or from continued input from mineral solubolization to counteract any losses due to ecosystem disturbance. For example, soluble phosphorus pools may be maintained in part by solubolization of rock phosphates. This process is enhanced by products of microbial activity. Acid products produced by the microorganisms during fermentation of organic matter and reduction of soil water pH by biogenically produced carbon dioxide results in slow dissolution of the soil phosphates. Thus in high phosphorus soils, phosphorus fertilization may have minimal impact on plant growth and development. Also, sulfur fertilization in developed countries rarely has led to improved crop yields in the past. This has been more the result of the incorporation of atmospheric sulfur into soil sulfur cycles rather than sulfur produced by mineralization of indigenous organic materials. With improved air quality (i.e., reduced atmospheric sulfur loads), soil organic matter pools in native sites and sulfur fertilization in managed ecosystems may become the primary biomass sulfur sources.

9.1.1. Soil Organic Fractions Yielding Plant Nutrients

Essentially all soil organic matter fractions are sources of mineral nitrogen, sulfur, and phosphate to some extent, but, it is the readily metabolized fractions which are the greatest contributors of plant nutrients to the ecosystem. Thus primary contributors to ecosystem productivity are the pools of decomposing plant debris (including incorporated plant biomass and litter layers), roots and root exudates, and microbial biomass. Nutrient contributions from the humified soil fractions essentially are negligible in most soils. The latter soil organic matter fraction is only an important nutrient source in soils containing or receiving no, or minimal, inputs of easily metabolizable carbon substrates (i.e., fallow or bare soils). Obviously in soils with humic substances providing the bulk of the energy and carbon to the decomposer community, overall biomass production for the ecosystem is limited essentially to microbial cell synthesis since there is no influx of photosynthetically fixed carbon.

Physical manipulation of the soil also affects the availability of nutrients in various soil fractions by altering physical protection of decomposable organic matter pools. Some soil organic matter is unavailable to microbial mineralization by virtue of its location within the soil matrix. Microorganisms are precluded physically from reaching the organic matter. For example, Ladd et al. (1977) noted that wet/dry cycles increased the available, thus the mineralization, of native soil organic matter in a calcareous sandy soil and a calcareous clay soil amended with glucose and wheat straw. This physical protection of soil organic matter and plant debris from mineralizing populations has practical implications relating to soil tillage and the return of plant biomass nitrogen to subsequent crops. Larger surface litter accumulations are found on no-till and old-field agricultural soils than on conventionally tilled soils (Stinner et al., 1984). The greater exposure of the surface litter to temperature and moisture extremes and the physical isolation from the soil microbial community may contribute to accumulation of this litter in that nutrient mineralization appeared to be more rapid in the conventionally tilled soils.

Ratios of nutrients produced by mineralization of soil organic matter and total nutrients contained in this soil fraction vary with ecosystem type. Total quantities of nutrients retained in the soil organic matter necessarily varies directly with total soil organic matter levels. The higher the organic matter content, the greater the potential nutrient yield. But, a range of the carbon : nitrogen : sulfur ratios in soils depending upon ecosystem is observed (Bettany et al., 1973). This ratio ranged from 58 : 6.4 : 1 in arid chernozemic brown soils to 129 : 10.6 : 1 in leached grey wooded soils.

9.1.2. Estimation of Residual Soil Organic Nitrogen

As suggested in the preceding text, an economic problem associated with mineralization of recently incorporated organic matter and the subsequent production of plant nutrients relates to estimation of the quantities of residual soil nitrogen retained in soil biomass and crop residues at the end of the growing season. Mineralization of these crop residues does provide plant nutrients for subsequent crops. Since fixed nitrogen is an expensive fertilizer, any means of reducing quantities added to crops improves the prophet margin. Estimates of nitrogen quantities that must be added to soil for maximal crop yield rely on analyses of the mineral nitrogen. Residual organic nitrogen pools are not commonly evaluated. But, it is known that a portion of the residual soil organic nitrogen will be mineralized during the growing season releasing nutrients which may be used by the crop. If it were possible to estimate the quantities of this mineralized soil organic nitrogen, quantities of fertilizer applied and their associated costs could be reduced. Soil nitrogen mineralization rates are commonly estimated, but correlation of these rates with crop yield is difficult because it is impossible at this time to determine the soil nitrogen fraction that is being mineralized and its contribution to crop yield.

Fox and Piekielek (1984) have found some promising associations between anaerobically mineralized nitrogen and nitrogen availability to corn. In their study, soils from 67 experiments located throughout Pennsylvania were incubated anaerobically for seven days to determine if the nitrogen mineralized under anaerobic conditions correlated better with nitrogen availability indices than did aerobic mineralized nitrogen. The nitrogen mineralized under anaerobic conditions correlated more closely with total soil N content ($r = 0.79$) and boiling 0.01 M calcium chloride extractable N ($r = 0.74$) than did that mineralized aerobically. The authors conclude that the high correlations between the anaerobically mineralized nitrogen and a number of soil nitrogen parameters is promising, but the relationships were not high enough to be used in a routine soil test procedure at this time.

In a related study, Griffin and Laine (1983) examined nitrogen availability in soils amended with cow and poultry manure, sewage sludge, and mycelium. Nitrogen mineralization potential (organic nitrogen substrate at 0 time, N_0) and the mineralization rate constant (k) (see Section 9.2.1 for explanation of the rate equations) were determined from long-term aerobic incubations of soil samples. Field soil temperature and moisture values were determined so that laboratory values could be corrected to approximate field rates. Corn (*Zea mays* L.) was grown on the test plot so that the relationship of plant biomass yield and nitrogen uptake to nitrogen mineralization potential and rate could be determined. Positive relationships provide data necessary for developing soil test procedures for estimating effects on crop growth by mineralization of soil organic fractions. Nitrogen mineralization potential was a "very poor predictor of yield or nitrogen uptake," whereas both yield and nitrogen uptake correlated well with the product of nitrogen mineralization potential and the mineralization rate constant ($N_0 \times k$). Practical use of the procedure is limited by the fact that lengthy incubations were required for accurate estimations of k.

The recent demonstration that soil organic matter could be separated by density fractions with different carbon:nitrogen ratios and mineralization rates (Chichester, 1969; Young and Spycher, 1979; Tiessen and Stewart, 1983; Sollins et al., 1984) suggests a potential means of quickly estimating residual soil nitrogen. The light fraction has been shown to be related to the nascent plant debris, microbial biomass and more humified fractions are associated with heavier particles. Mineralization of the light particles, as a result of their wide carbon:nitrogen ratios results in a net nitrogen immobilization whereas mineralization of the heavy fraction yields a net increase in available soil nitrogen. Models descriptive of the interactions of these two nitrogen pools could be predictors of crop yield.

Although our attention in managed soil sites is directed to aboveground biomass production, in native sites, precipitous declines in soil microbial biomass also may provide significant nutrients for plant production. The microbial community not only participates in mineralization of the nutrients and in the case of phosphorus, nutrient solubilization (e.g., see Hannapel et al., 1964; Alexander, 1977), but they may also be a significant nutrient reservoir. Brookes et al. (1984) found that 3 percent of the phosphorus in arable soils was in microbial biomass, whereas 5 to 24 percent of the soil phosphorus was in this fraction in grassland soils. They found that the decline in biomass phosphorus when a grassland was cultivated over a 20-year period accounted for approximately 50 percent of the total decline of soil organic phosphorus. Reversion of cultivated soil to woodland resulted in a doubling of the soil organic phosphorus over a 100-year period. This was accompanied by an 11-fold increase in microbial biomass phosphorus over the same period.

9.2. INTERACTIONS OF BIOGEOCHEMICAL CYCLES

Soil organic matter could be said to have a "uniform" origin in that it is derived primarily from plant biomass with significant contributions from microbial and animal sources. Thus each biogeochemical cycle has two major pools in common with the other cycles, biomass and soil organic matter. This indicates that mineralization of one biomass elemental component simultaneously results in production of mineral forms of the others. Thus mineralization is a key soil biochemical processes for total ecosystem function and stability because it controls the occurrence and/or rate of a number of other biological processes as well as directly or indirectly contributes to the control of aboveground biomass production. In development of ecosystem nutrient models, carbon, nitrogen, phosphorus, and sulfur mineralization processes may be also categorized based on the driving forces behind the reactions (McGill and Cole, 1981). Nitrogen and sulfur are said to involve biological mineralization, whereas biochemical mineralization processes are associated with phosphorus and sulfur. McGill and Cole (1981) defined biological mineralization as the conversion of organic nitrogen and sulfur to inorganic forms during oxidation of carbon by soil organisms to provide energy for microbial growth and reproduction. The primary driving force of this process is therefore microbial energy requirements. In contrast, biochemical mineralization is the release of phosphorus and sulfur from organic substrates through the action of extracellular enzymes. Primary control of this process is by requirements for the elements themselves for microbial growth, not energy needs. Hence, in the former situation, nitrogen and sulfur are released into the ecosystem gratuitously as a result of the organisms oxidation of carbon for energy production, whereas in the latter case, nutrient supply directly controls the mineralization rate. Thus although mineralization of each element appears to be linked, some differentiation is possible at the level of specific biochemicals mineralized.

Variations in net mineralization kinetics of carbon, nitrogen, phosphate, and sulfur result from both chemical sequestering of mineral nutrients into water insoluble forms within the ecosystem and microbial interactions with the nutrients subsequent to mineralization (primarily immobilization). Incorporation of these sequestered nutrients into plant biomass is controlled by the longevity and mineralization rates of the microbial community and the chemical equilibria reactions involved in the interchange of water insoluble and soluble minerals. Many of the reactions in this partitioning of the nutrients into various environmental pools are nonbiological. Therefore, chemical and physical factors must also be considered when evaluating nutrient movement within an ecosystem. Nutrients for biomass synthesis are derived from either the atmosphere or soil mineral substituents (Fig. 9.3). Carbon and nitrogen originate in the atmosphere as carbon dioxide and dinitrogen, respectively. Thus gaseous diffusion properties become important in many carbon and nitrogen fixation reaction kinetics models. Although major quantities of sulfur are incorporated into biomass from atmospheric sources in industrialized regions of the world, the main pool of both sulfur and phosphorus for biomass synthesis are soil mineral pools as sulfides and sulfates or phosphates, respectively. Soil sources will become more important to plant nutrient as air quality is improved in that atmospheric sulfur sources will most probably be significantly reduced. Net mineralization is therefore the difference between total nutrient mineralized and that immobilized into microbial biomass plus that fixed through soil chemical reactions. Thus the rates of return of of the macronutrients, nitrogen, phosphorus, and sulfur, into mineral pools—hence, the mineralization rates measured in the field—do not reflect the actual mineralization rate. We could term the observed rate as the net mineralization. Simplified net mineralization rate equations for each of the macronutrients may be conceptualized as follows:

—nitrogen:

$$\text{Mineralization}_{\text{net}} = \text{mineralization}_{\text{actual}} - \text{immobilization}$$

—carbon:

$$\text{Mineralization}_{\text{net}} = \text{mineralization}_{\text{actual}} - \text{carbon}_{\text{fixed in carbonates}}$$

—phosphorus:

$$\text{Mineralization}_{\text{net}} = \text{mineralization}_{\text{actual}} - \text{phosphate fixation} - \text{immobilization}$$

—sulfur:

$$\text{Mineralization}_{\text{net}} = \text{mineralization}_{\text{actual}} - \text{sulfide and sulfate fixation} - \text{immobilization}$$

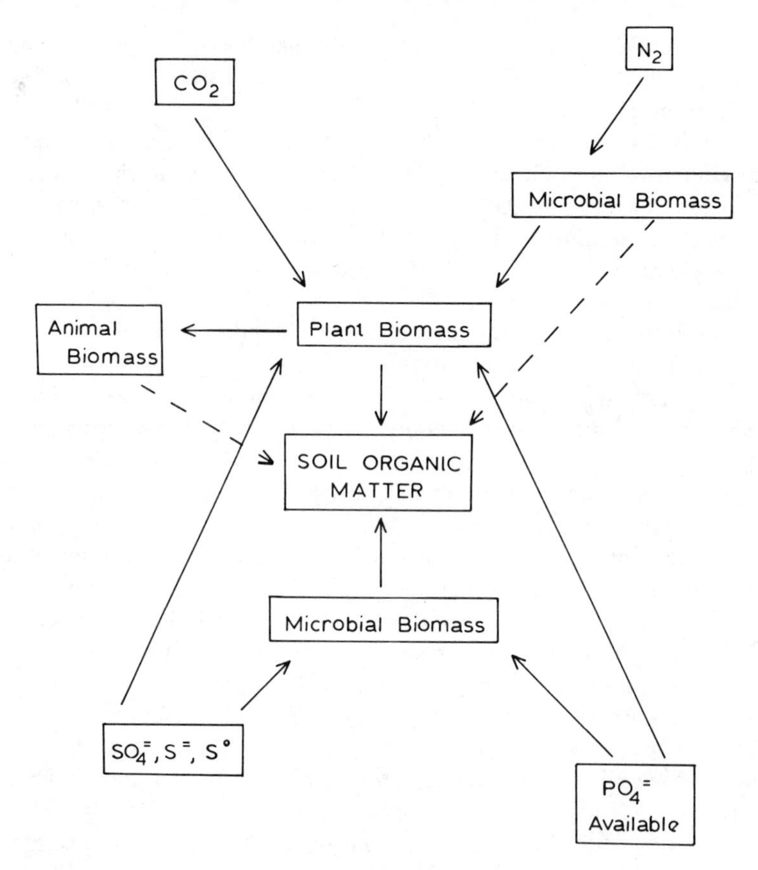

Fig. 9.3.
Plant nutrient sequestering into soil organic matter. A system receiving no external supplies of nitrogen, phosphorus, or sulfur is depicted.

These processes are depicted in Fig. 9.4. It is apparent that although each of these nutrients originates from common sources, plant biomass and soil organic matter, the fates of each nutrient following initial mineralization varies such that the net mineralization rates are quite different from each other. Thus, although the commonality in sources predicts similarities in mineralization rates—not quantities—this can not be detected in the field, In fact the equations presented in the preceding text are an over simplification of the true field situation.

Since mineralization processes are generally evaluated individually, our analysis of the effects of their interactions on total ecosystem function is necessarily limited. Thus to gain the maximum understanding of the processes involved, nitrogen and phosphorus mineralization will be evaluated individually, as examples of the types of reactions involved, with an interpretation of how the observed data relate to processes involving other plant nutrients.

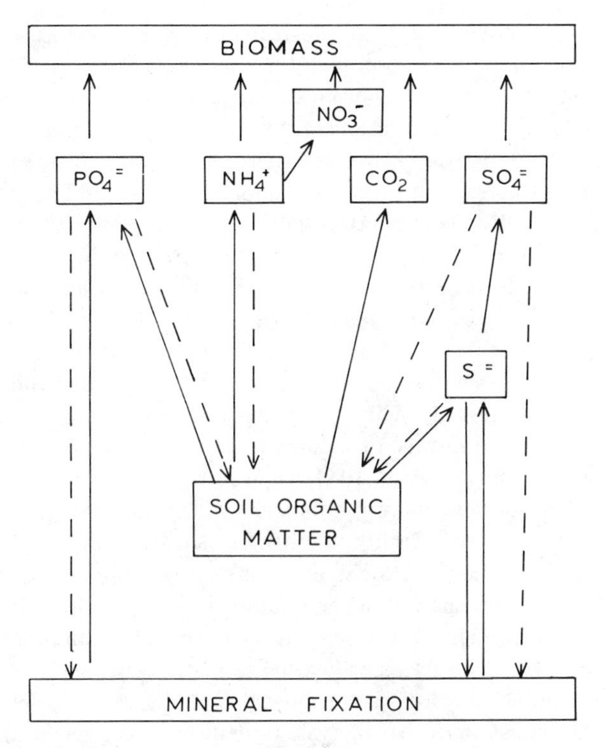

Fig. 9.4.
Soil organic matter as a source of plant nutrients. (Solid arrows, nutrient production pathways; dashed arrows, nutrient immobilization.)

9.2.1. Nitrogen Mineralization

Nitrogen is found both in biological tissues and soil in a variety of easily metabolized forms. Such nitrogen containing compounds as amino acids, peptides, proteins, amino sugars, and nucleic acids are easily isolated from soil (Stevenson, 1982). These compounds occur in soil as biomass substituents, as nonhumified, transient organic matter components, as well as in humified forms. Rapid catabolism of amino acids amended to soils (e.g., see Tate 1980b; Tate, 1985) and mineralization of soil organic nitrogen by soil microorganisms (e.g., see Huntjens, 1972) document the susceptibility of these nitrogen sources to microbial attack. Nitrogen is also incorporated into the more biodegradation resistant structures of humic acids (Stevenson, 1982). These latter compounds are extremely stable. Although they are decomposed slowly, their contribution to the observed nitrogen mineralization rate is minimal.

To gain a better understanding of total ecosystem function and, in the case of managed systems, to avoid the over or under use of nitrogenous fertilizers, it is useful to have a basic understanding of the *in situ* nitrogen mineralization rate.

This includes knowledge of the rate, substrate mineralized, and seasonal and/or diurnal variation in quantities of mineral nitrogen released. Procedures for estimating these values generally involve measurement of mineral nitrogen production in soils incubated under a variety of conditions. This is time-consuming and occassionally the relationship of the results to the actual field situation is questionable. Prediction of soil nitrogen mineralization rates could be greatly simplified were it possible to associate nitrogen mineralization, at least predominantly, with a specific soil nitrogen fraction, such as protein or biomass nitrogen. It would then be possible to estimate the nitrogen mineralization potential of specific soils by assaying the quantity of that fraction present. Combination of these values with basic information on physical and chemical properties of the ecosystem and the effect of these properties on microbial activities would allow development of predictive models. This would save time and reduce the questions of possible changes in the nitrogen fractions prior to assay or during incubation. Recent studies by Juma and Paul (1984) suggest that a specific soil organic nitrogen fraction does not contribute the majority of the soil nitrogen mineralized. They evaluated the extractability of nitrogen from an ^{15}N-labeled Weirdale loam. Comparison of the ^{15}N-enrichment of mineralized nitrogen with extracted nitrogen allowed determination of the relationship of the nitrogen fraction extracted with that mineralized. A variety of nitrogen extraction procedures ranging from water extracts prepared by heating soil/water suspensions under steam pressure to strong acid hydrolysis were used. Quantities of labeled nitrogen in the microbial biomass were also assessed to determine the role of this soil fraction in nitrogen mineralization processes. The authors concluded that extraction of the highly labeled nitrogen pool only explained part of the source of nitrogen mineralized. Their data indicate that soil nitrogen is mineralized from a variety of pools, thus it is only remotely possible that a single extractant could be developed which would approximate the size of mineralization-susceptible nitrogen pool. Microbial biomass was a major source of mineralized nitrogen in that 15 to 25 percent of the nitrogen mineralized over a 12-week period originated from this pool, but 75 to 85 percent of the nitrogen mineralized was still derived from other sources.

A variety of incubation procedures have been developed for estimation of soil nitrogen mineralization potential. Basically, each technique consists of incubation of discrete soil samples under conditions conducive for mineralization to occur. Periodically, mineral nitrogen is extracted from the soil and quantified. In that nitrogen immobilization is not quantified with these procedures, net nitrogen mineralization is detected. (Corrections for nitrogen immobilization require amendment of the soil with ^{15}N-labeled nitrate and measurement of its incorporation into soil organic nitrogen components. Alternatively, ^{15}N-depleted nitrate can be used. In this case, reduction of the levels of ^{15}N nitrogen in the organic fraction is quantified.) Results from two laboratory procedures and one technique that approximates field conditions commonly are reported in the literature. With the laboratory studies, differences relate to the techniques of soil manipulation for extraction. In the method of Stanford and Smith (1972), soils periodically are leached with extractant then returned to the incubation chamber. The disadvantage of any leaching procedure, aside from each laboratory manipulation further increasing the differences between the laboratory sample and true field conditions, is that not only is mineral nitrogen removed from the sample, but so is any water soluble organic nitrogen contained in the soil. This organic nitrogen could contribute to subsequent nitrogen mineralization in the samples. Thus it is believed that with the leaching procedures, the net mineralization rate is underestimated. This difficulty is alleviated by the batch incubation procedure. With these types of studies, discrete soil samples are sacrificed for each analysis; that is, the total soil sample is analyzed and discarded without further incubation. Large numbers of samples must be prepared with the batch procedure, but the loss of soluble nitrogenous substrates is avoided.

A further problem with incubation methods is one that is common to any laboratory procedure, the soil microcosms created in the laboratory may be extremely different from those occurring in the field. Among other modifications, the physical structure of the soil has been altered by sampling and preparation. Also, the incubation temperature and moisture are generally controlled at a reasonably constant value which is not the case in the field. Some of these variations from *in situ* conditions are avoided by the measurement of nitrogen mineralization by the buried bag method. For this procedure, soil samples are collected and sealed into polyethylene bags that are returned to the soil (Fig. 9.5). Soil moisture remains that of the soil when collected and temperature variation is essentially that of the native ecosystem. Criticisms of this buried bag procedure relate to the potential for oxygen depletion and carbon dioxide accumulation in the sealed bags, both of which could alter microbial growth and metabolic rates. Originally, it was felt that soil gases are freely permeable through the polyethylene barrier, but Bremner and Douglas (1971) found that this was a faulty assumption. They concluded that the various films used were not sufficiently permeable to carbon dioxide and oxygen for use in the soil incubation experiments. In their study, polyethylene film with a thickness of 0.5 mil was superior in gas permeability properties to the commonly used 1.0 mil film, but this advantage was off set by other problems with water permeability. Free or even limited exchange of water in the bag with soil moisture alters soil mineral nitrogen content. Other types of plastic film (e.g., teflon, polypropylene, cellulose acetate, and nylon) were less desirable than polyethylene. Others, have found the data collected with the buried bag procedure relates closely to that collected in the laboratory. Smith et al. (1977) found in a study extending over two cropping seasons with eight Oklahoma soils that the field data collected with plastic bags and glass filter tubes, after correction for variations in soil moisture and temperature, differed from laboratory calculated values by <10 ppm nitrogen. Thus they found the procedure to be useful for field nitrogen mineralization quantification. Similarly, Westermann and Crothers (1980) evaluated the buried bag procedure by comparing soil nitrate-nitrogen content of soils incubated in bags with that of the adjacent crop root zone. Measurement of the nitrate concentration in the buried soils and in the root zone allowed calculation of the quantity of

Fig. 9.5.
Measurement of nitrogen mineralization rates with the buried bag procedure.

nitrogen that should have been incorporated into plant tissue. All values were in agreement. Thus the authors also concluded that the buried bag procedure was a useful method for determining soil nitrogen mineralization rate. Considerable differences of opinion exist and will likely persist over the relative value of each of these methods. As with any other scientific techniques, their utility can only be extended to the degree that the experimenter understands their limitations. Thus for studies of nitrogen mineralization, the procedure selected must be that which yields data that best approximates actual field values under the conditions of the ecosystem of interest.

Once values for nitrogen mineralized either in the laboratory or field are collected, a mathematical model must be developed that produces reliable numerical values which can be used to compare mineralization capacities between different soil ecosystems. Quantities of interest are the original nitrogen available for mineralization and its mineralization rate. Traditionally, a first-order model has been used for estimating nitrogen mineralization potentials. This is described mathematically by the following relationship:

$$dN/dt = -kN \tag{1}$$

where N = the concentration of mineralizable organic nitrogen, k is a rate constant, and t = time. Integration of this relationship yield this equation:

$$N_t = N_0 e^{-kt} \tag{2}$$

Two new unknowns introduced into this equation are N_0 (organic nitrogen substrate at time 0) and N_t (organic nitrogen substrate at time t). Since the nitrogen at time t is equivalent to the initial nitrogen minus the nitrogen mineralized (N_m), this equation may be reduced to one containing a single unknown value as follows:

$$N_0 - N_m = N_0 e^{-kt} \tag{3}$$

or

$$N_m = N_0 (1 - e^{-kt}) \tag{4}$$

The unknowns are generally calculated through linear regression of a log transformation of this equation. To develop a valid estimation of N_0 and k, it is necessary to specify the deviation of the values from equation (4). Smith et al. (1980), after assuming that their data followed first order kinetics as described previously, compared values calculated with the traditional least squares fit of the straight line generated by the log transformed data with those derived from a root mean square deviations of the experimental data from a nonlinear least squares equation. They found that the nonlinear least squares equation gave a more accurate estimation of the nitrogen mineralization potential (N_0) and mineralization rate constant (k) than did the traditional calculation method. Similar conclusions were reached by Talpaz et al. (1981) and Juma et al. (1984). Talpaz et al. (1981) indicate that the differences between these two techniques result from the behavior of the error term in the mineralization rate equation. These authors also concluded that the nonlinear regression procedure is preferable to the traditional linear regression analyses of the log transformed data.

The complexity of the soil ecosystem is reflected in the current difficulties of measuring nitrogen mineralization and mathematically representing the results. Although it is reasonably easy to list the biochemical reactions that must be occurring and the various soil organic nitrogen fractions participating in the process, interactions with the chemical and physical environment of the microorganism and the organic matter, especially from the view of microsite variation, field data are generally quite complex and frequently contradictory. Problems are the result of the inability to totally define the site (in many cases, the microsite) and conditions where the microorganisms are active. The importance of nitrogen mineralization to overall ecosystem function and the desire to develop total ecosystem models are stimulating greater effort to understand these processes and thereby increase our capability to predict the impact of environmental perturbations on this key soil process.

9.2.2. Phosphorus Mineralization

Phosphorus is among the most tightly conserved plant nutrients in forest ecosystems. Biological conservation of this nutrient within the forest community is accomplished by close coupling of biological decomposition and uptake processes in surface soils (Wood et al., 1984). Mycorrhizal associations and other rhizosphere microbiological populations as well as the general soil microbial flora contribute to this association of mineralization and assimilation processes. Free-living microbes involved with phosphorus mineralization include a wide variety of bacterial, fungal, and actinomycete species (Alexander, 1977; Molla et al., 1984). Phosphorus conservation in a forested ecosystem is exemplified in data from Hubbard Brook Experimental Forest in central New Hampshire (Wood et al., 1984). In this 55- to 65-year old northern hardwood forest ecosystem, plants assimilated an estimated 12.5 kg soil phosphorus per hectare annually. Of this phosphorus, about 1.5 kg/ha was stored in accumulating biomass and 11.0 kg/ha was returned with aboveground and belowground litter to the soil system annually. This included both root and aboveground litter sources. Within the soil ecosystem, more than 10.0 kg/ha of organic phosphorus was mineralized annually. Another 1.5 to 1.8 kg phosphorus/ha was provided through weathering of primary soil minerals annually. As with this temperate forest, phosphorus also appears to be highly efficiently cycled in tropical forests (Vitousek, 1984). Analysis of fine litter fall suggested that phosphorus, but not nitrogen, availability limited litter fall in a substantial subset of intact tropical forests. Sites on old oxisols and ultisols appeared to be especially low in available phosphorus. This phosphorus cycling is also linked to the fate of other major plant nutrient substituents of plant biomass (litter fall). In an extensive study of nutrient release (nitrogen, phosphorus, sulfur, potassium, manganese, calcium, zinc, iron, magnesium, copper, and sodium) during litter fall decomposition, it was shown that phosphorus levels may have a controlling influence on mineralization rates of other nutrients (Gosz et al., 1973).

A number of simulation models have been developed which are descriptive of phosphorus mobilization and immobilization in a variety of ecosystems. In a model designed to predict plant and decomposer uptakes and turnover rates of principal phosphorus compartments of semiarid grasslands (data was used from the Pawnee (Colorado, U.S.A.) and Matador (Saskatchewan, Canada) sites), it was found that labile inorganic phosphorus pools of each soil layer were replenished primarily from mineralization of labile organic phosphorus as well as leaching of water soluble phosphorus forms from standing dead biomass and litter (Cole et al., 1977). In a model of soil and plant phosphorus processes in agricultural soils, ecosystem phosphorus was divided into pools of stable, active, and labile inorganic phosphorus; fresh organic and stable organic phosphorus; and grain, stover, and root phosphorus (Jones et al., 1984a). Data from 78 soils from the continental United States and Puerto Rico were used to test the model (Sharpley et al., 1984a). Phosphate was extracted from the soil with a variety of extractants (bray 1 P, Olsen P, North Carolina P, and labile P). Organic phosphorus and an index of fertilizer phosphorus sorption were also evaluated. Regression analysis was used to derive equations predictive of soil labile phosphorus, organic phosphorus, and a phosphorus sorption index from soil chemical and physical properties. Labile phosphorus related to extractable phosphorus, whereas organic phosphorus related to total soil nitrogen and pH. As would be anticipated sorption was impacted by soil clay, calcium carbonate, labile phosphate, and base saturation. These soil parameters accounted for 84, 64, and 78 percent of the labile phosphorus, organic phosphorus, and phosphorus sorption variation, respectively. The mathematical relationships derived from this study were tested by evaluation of changes of soil surface organic phosphorus due to 40 years of cultivation (Jones et al., 1984b). Soils evaluated were collected from several areas in the Great Plains (U.S.A.). Concentrations of the soil test phosphorus were accurately simulated by the model over time periods up to 20 years with and without fertilization of the soils. The model was also useful for prediction of fertilizer phosphorus requirements necessary to maintain adequate phosphorous fertility for maize and wheat.

Soil chemical properties do affect soil phosphorus composition (Tiessen et al., 1984). Nine different organic and inorganic soil phosphorus fractions were extracted sequentially from samples of 168 USDA-SCS (U. S. Department of Agriculture–Soil Conservation Service) benchmark soils. Eight soil orders of soil taxonomy were represented in the soil samples. Correlation and regression analyses of the phosphorus distribution and soil chemical analyses data showed that organic matter accumulation depended, in part, on the available forms of phosphorus. Secondary phosphorus forms were related to weathering indicators, such as base saturation. Soil chemical properties and soil taxonomic relationships related to the relative proportions of available, stable, organic, and inorganic phosphorus forms. Mineralization was a major determinant of phosphorus fertility in the more weathered Ultisols, in that 80 percent of the variability in labile phosphorus was accounted for by organic phosphorus levels.

Incorporation of exogenous organic phosphorus containing waste materials also alters phosphorus distribution in soil (Sharpley et al., 1984b). Cattle feedlot waste applied in concentrations of 176 to 1614 Mg/ha to irrigated continuous-grain sorghum [*Sorghum bicolor* (L.) Moench] increased the total, inorganic, organic, and available phosphorus content and decreased the phosphorus adsorption index of the surface soil (0 to 30 cm). Quantities of phosphorus in the surface soils were highly correlated with the total feedlot waste phosphorus applied and time since the last application. With larger feedlot waste applications, the proportion of total phosphorus as inorganic phosphorus (34 to 71 percent) increased. Cessation of amendment of soils with feedlot wastes caused soil organic phosphorus contents decreased to pretreatment levels more rapidly than inorganic phosphorus contents. This indicates that mineralization of organic matter is the principal source of this inorganic phosphorus.

9.3. CONCLUSIONS

The quantities of plant nutrients retained within the soil organic matter fraction in native systems control total ecosystem biomass production are an interactive component in determining the overall nature of plant community that develops on the soil surface. In reality, although this pool of nutrients may be quite substantial, they are unavailable to the growing plants. The soil microbial community must metabolize the organic substances to mineral forms which may then be incorporated into plant biomass. The major plant nutrients, nitrogen, phosphorus, sulfur, and a variety of minerals, are derived from a common soil organic matter source, partially decomposed plant and animal debris, microbial biomass, and humified organic matter. Although a variety of organic components are mineralized, the most significant contributor to mineral nutrient pools is the easily metabolized soil organic matter pool. Thus each of the biogeochemical cycles has two common intersection points, biomass, (microbial, plant, and plant) and soil organic matter. This origin, while suggesting a common mineralization rate, actually produces a variety of net mineralization rates. Appearance of mineral nutrients in soil interstitial water is controlled by the difference between their synthesis rate and immobilization or sequestering within a variety of water insoluble mineral forms. Thus the identity of the ecosystem limiting nutrient, albeit nitrogen or phosphorus, is controlled not only by the quantities available from mineralization, but by the fate of this newly mineralized nutrient in soil and the inputs of nutrients from mineral sources, such as the solubilization of phosphates.

REFERENCES

Alexander, M. 1977. Introduction to soil microbiology. John Wiley & Sons, New York, 467 pp.

Bettany, J. R., J. W. B. Stewart, and E. H. Halstead, 1973. Sulfur fractions and carbon, nitrogen, and sulfur relationships in grassland, forest, and associated transitional soils. Soil Sci. Soc. Am. Proc. 37: 915–918.

Bremner, J. M., and L. A. Douglas, 1971. Use of plastic films for aeration in soil incubation experiments. Soil Biol. Biochem. 3: 289–296.

Brookes, P. C., D. S. Powelson, and D. S. Jenkinson, 1984. Phosphorus in the soil microbial biomass. Soil Biol. Biochem. 16: 169–175.

Chichester, F. W. 1969. Nitrogen in soil organo-mineral sedimentation fractions. Soil Sci. 107: 356–363.

Cole, C. V., G. S. Innis, and J. W. B. Stewart, 1977. Simulation of phosphorus cycling in semiarid grasslands. Ecology 58: 1–15.

Fox, R. H., and W. P. Piekielek, 1984. Relationships among anaerobically mineralized nitrogen, chemical indexes, and nitrogen availability to corn. Soil Sci. Soc. Am. J. 48: 1087–1090.

Gosz, J. R., G. E. Likens, and F. H. Bormann, 1973. Nutrient release from decomposing leaf and branch litter in the Hubbard Brook Forest, New Hampshire. Ecol. Monogr. 43: 173–191.

Griffin, G. F., and A. F. Laine, 1983. Nitrogen mineralization in soils previously amended with organic wastes. Agron. J. 75: 124–129.

Hannapel, R. J., W. H. Fuller, and R. H. Fox, 1964. Phosphorus movement in a calcareous soil: II. Soil microbial activity and organic phosphorus movement. Soil Sci. 97: 421-427.

Huntjens, J. L. M. 1972. Availability of microbial and soil organic nitrogen to a *Pseudomonas* strain and the effect of soil organic matter on the availability of casein nitrogen. Soil Biol. Biochem. 4: 347-368.

Jones, C. A., C. V. Cole, A. N. Sharpley, and J. R. Williams, 1984a. A simplified soil and plant phosphorus model: I. Documentation. Soil Sci. Soc. Am. J. 48: 800-805.

Jones, C. A., A. N. Sharpley, and J. R. Williams, 1984b. A simplified soil and plant phosphorus model: III. Testing. Soil Sci. Soc. Am. J. 48: 810-813.

Juma, N. G., and E. A. Paul, 1984. Mineralizable soil nitrogen: Amounts and extractability ratios. Soil Sci. Soc. Am. J. 48: 76-80.

Juma, N. G., E. A. Paul, and B. Mary, 1984. Kinetic analysis of net nitrogen mineralization in soil. Soil Sci. Soc. Am. J. 48: 753-757.

Ladd, J. N., J. W. Parsons, and M. Amato, 1977. Studies of nitrogen immobilization and mineralization in calcareous soils—II. Mineralization of immobilized nitrogen from soil fractions of different particle size and density. Soil Biol. Biochem. 9: 319-325.

McGill, W. B., and C. V. Cole, 1981. Comparative aspects of cycling of organic C, N, S, and P through soil organic matter. Geoderma 26: 267-286.

Molla, M. A. Z., A. A. Chowdhury, A. Islam, and S. Hoque, 1984. Microbial mineralization of organic phosphate in soil. Plant Soil 78: 393-394.

Pastor, J., J. D. Aber, and C. A. McClaugherty, 1984. Aboveground production and N and P cycling along a nitrogen mineralization gradient on Blackhawk Island, Wisconsin. Ecology 65: 256-268.

Sharpley, A. N., C. A. Jones, C. Gray, and C. V. Cole, 1984a. A simplified soil and plant phosphorus model: II. Prediction of labile, organic, and sorbed phosphorus. Soil Sci. Soc. Am. J. 48: 805-809.

Sharpley, A. N., S. J. Smith, B. A. Stewart, and A. C. Mathers, 1984b. Forms of phosphorus in soil receiving cattle feedlot waste. J. Environ. Qual. 13: 211-215.

Smith, J. L., R. R. Schnabel. B. L. McNeal, and G. S. Campbell, 1980. Potential errors in the first-order model for estimating soil nitrogen mineralization potentials. Soil Sci. Soc. Am. J. 44: 996-1000.

Smith, S. J., L. B. Young, and G. E. Miller, 1977. Evaluation of soil nitrogen mineralization potentials under modified field conditions. Soil Sci. Soc. Am. J. 41: 74-76.

Sollins, P., G. Spycher, and C. A. Glassman, 1984. Net nitrogen mineralization from light- and heavy-fraction forest soil organic matter. Soil Biol. Biochem. 16: 31-37.

Stanford, G., and S. J. Smith, 1972. Nitrogen mineralization potentials of soils. Soil Sci. Soc. Am. Proc. 36: 465-472.

Stinner, B. R., D. A. Crossley, Jr., E. P. Odum, and R. L. Todd, 1984. Nutrient budgets and internal cycling of N, P, K. Ca, and Mg in conventional tillage, no-tillage, and old-field ecosystems on the Georgia piedmont. Ecology 65: 354-369.

Stevenson, F. J. 1982. Humus chemistry. John Wiley & Sons, New York, 443 pp.

Stevenson, F. J. 1986. Cycles of soil. John Wiley & Sons, New York, 380 pp.

Talpaz, H., P. Fine, and B. Bar-Yosef, 1981. On the estimation of N-mineralization parameters from incubation experiments. Soil Sci. Soc. Am. J. 45: 993-996.

Tate, R. L. III 1980a. Microbial oxidation of organic matter of Histosols. In M. Alexander (ed.), Adv. Microbial Ecol. 4: 169-201, Plenum, New York.

Tate, R. L. III 1980b. Effect of several environmental parameters on carbon metabolism in Histosols. Microb. Ecol. 5: 329-336.

Tate, R. L. III 1985. Carbon mineralization in acid, xeric forest soils: Induction of new activities. Appl. Environ. Microbiol. 50: 454-459.

Tiessen, H., and J. W. B. Stewart, 1983. Particle-size fractions and their use in studies of soil organic matter. II. Cultivation effects on organic matter composition in size fractions. Soil Sci. Soc. Am. J. 47: 509-514.

Tiessen, H., J. W. B. Stewart, and C. V. Cole, 1984. Pathways of phosphorus transformations in soils of differing pedogenesis. Soil Sci. Soc. Am. J. 48: 853-858.

Vitousek, P. M. 1984. Litterfall, nutrient cycling, and nutrient limitation in tropical forests. Ecology 65: 285-298.

Westermann, D. T., and S. E. Crothers, 1980. Measuring soil nitrogen mineralization under field conditions. Agron. J. 72: 1009-1012.

Wood, T., F. H. Bormann, and G. K. Voigt, 1984. Phosphorus cycling in a northern hardwood forest: Biological and chemical control. Science (Washington, D.C.) 223: 391-393.

Young, J. L., and G. Spycher, 1979. Water-dispersible soil organic-mineral particles: I. Carbon and nitrogen distribution. Soil Sci. Soc. Am. J. 43: 324-328.

TEN

MINERAL AVAILABILITY AND SOIL ORGANIC MATTER

With the current understanding of the variety of metallic cations contained within soil inorganic particulates, limitation of either soil microbial populations or plant community development by trace mineral concentrations would not be anticipated, But, not all of the minerals, whether stimulatory or inhibitory to plant growth, present in soil are plant available. Even in the presence of high levels of mineral bearing soil particulates, total ecosystem biomass production may be limited or even precluded by either inadequate levels of a variety of metal cations or elevated concentrations of toxic metals in soil solution. Metals, such as iron, calcium, potassium, magnesium, are generally required by plants or microbes in high concentrations (as macronutrients), whereas zinc, manganese, cobolt, molybdenum, nickel, and a number of other trace minerals are needed in small quantities for cellular growth. Hence, these latter metals are referred to as micronutrients. These micro- and macronutrient cations are essential nutrients in that they are substituents of a variety of cellular enzymes, participatory in reactions involving nucleic acid synthesis, and critical for maintenance of cell membrane stability.

For the minerals to be incorporated into cellular biomass, they must be contained in the soil interstitial water. Solubility of these cations is controlled by physical and chemical parameters of the soil, such as pH or redox potential (E_h); by interactions with abiontic organic matter; and by direct and indirect microbial activity. For example, microorganisms may enhance nutrient availability by modifying the soil pH, reducing the soil E_h, synthesis of chelating agents, secreting biological products which form ionic associations with soil cations, or simply through the decomposition of biomass components which contain metallic components, thereby releasing the associated cation.

Micro- and macronutrients obviously provide positive benefits to the plant and microbial communities; but, not all of the soil cations stimulate biomass synthesis. Some minerals have little or no known effect on plant growth and metabolism at the concentrations generally encountered in soil. Others, like mercury, inhibit biomass productivity at extremely low concentrations. Prediction of the impact of various soil minerals on the plant community is complex in that many essential mineral nutrients that are required in small amounts (e.g., manganese) when present in the interstitial water in large concentrations inhibit biomass synthesis. If soil were a simple mixture of the soil mineral complexes and water, analysis of trace mineral effects on ecosystem productivity would involve little more than calculation of the metal water solubility and plant toxicity or nutrient relationship at various metal ion concentrations. Soil systems are complicated by the fact that the inorganic particulate components are intermixed with a variety of microbial, plant, and animal biomass components plus colloidal humic substances and abiontic components, which may combine with various soil metal cations and reduce or augment their plant availability. Metal properties, including availability to the plant community, are then determined to a large degree by the nature of the soil colloidal organic matter fraction and its interactions in the ecosystem. Thus the impact of inhibitory metals on biomass synthesis may be mitigated or augmented as a result of association with soil organic matter. Combination of a metal with an insoluble organic component, such as microbial polysaccharide slimes or humic acids, tend to reduce its availability or associated toxic effects, whereas combination of insoluble minerals with water soluble microbially produced organic chelators necessarily increases the availability of the metal to ecosystem biomass.

Trace minerals are not only supplied through solubilization of native soil minerals, but those which are incorporated into native or exogenously supplied organic matter enter the plant available mineral pools through mineralization of plant, microbial, and animal debris. In managed ecosystems, a third source, anthropogenic supplies, must be added to this list. An appreciation for the magnitude of macronutrient cations contained in aboveground biomass can be gained from a study of the biomass and nutrient contents of a post oak-blackjack oak (*Quercus stellata-Q. marilandica*) forest in central Oklahoma (Johnson and Risser, 1974). Along with large quantities of nitrogen and phosphorus, the biomass was estimated to contain 1,258 kg potassium/ha, 4549 kg calcium/ha, 311 kg magnesium/ha,and 124 kg manganese/ha. The authors concluded that the unusually high contents of calcium in the biomass resulted from high concentrations of this element in the post oak bark. Decomposition of litter from two loblolly pine (*Pinus taeda* L.) stands, 11 and 32 years old, was examined over an eight year period (Jorgensen et al., 1980). After eight years decomposition, 91 percent of the potassium, 67 percent of the calcium, and 79 percent of the magnesium were released from the forest floor materials. This mineralization provided 47, 74, and 65 percent of the potassium, calcium, and magnesium, respectively, assimilated into aboveground biomass annually in the younger stand. In the older stand, 73, 72, and 71 percent of the potassium, calcium, and magnesium were supplied, respectively. Translocation of nutrients from leaves into perennial tissues prior to leaf fall has been shown to be an important nutrient retention mechanism for nitrogen, phosphorus, and potassium, but this mechanism was shown to be less important for sulfur, magnesium, and iron, and unimportant for calcium and zinc in an Aspen-mixed hardwood-spodosol ecosystem of northern Wisconsin (Pastor and Bockheim, 1984). Most data used to evaluate nutrient cycling in forested ecosystems, including the macro- and micronutrients, has been derived from analysis of litter decomposition. As was shown for nitrogen and phosphorus cycling rates, including fine root dynamics in the calculations greatly reduces the time required for cycling of the nutrients from senescent biomass into new plant biomass (Vogt et al., 1983). Mean residence times for potassium, calcium and magnesium in forest floors of young and mature *Abies amabilis* stands were decreased 81 to 90, 52 to 55, and 65 to 85 percent, respectively, by including root dynamics data.

An interesting reaction demonstrating the complexity of metal–biomass interactions is the impact of leaf surface tissue contamination with high concentrations of anthropogenically generated metallic cations over long periods of time. The toxic minerals reduce the rate of return of essential nutrients to the plant community. Direct effects of metal contamination on both the composition of the phylosphere microbial community (Bewley, 1980) and the reduction of the overall biomass decomposition rate (Strojan, 1978) have been shown. Bewley (1980) examined leaf surface (phylosphere) microflora of two groups of mature oak trees (one in the vicinity of a smelting complex thereby contaminated by heavy metals, and the other in a relatively uncontaminated site) and two groups of oak saplings at the uncontaminated site, one of which was sprayed with a mixture of zinc, lead, and cadmium. The latter study served to demonstrate immediate toxic effects of the metal mixture, but did not provide an indication of the long-term adaptation of the microbial community to the presence of these inhibitors. Spraying of the saplings did allow a reduction of many extraneous variables at the smelter site which could affect the data. Direct spraying of the leaves had little effect on total viable populations of bacteria, yeasts, or filamentous fungi (as isolated by leaf washing), but polluted leaves of mature trees near the smelter had fewer bacteria than mature leaves collected form the uncontaminated site. This suggests more of a long-term effect of the pollutant than the controlled experiment was designed to measure. Populations of pigmented yeasts were reduced on both the smelter contaminated and sprayed leaves. The populations of the contaminated mature leaves contained greater numbers of *Aureobasidium pullans* and *Cladosporium* spp. as well as a greater percentage of metal-tolerant fungi than did the uncontaminated leaves. Litter decomposition, as measured through the use of litter bags, was reduced in sassafras leaves and a mixture of chestnut oak-red oak leaves collected within 1 and 6 km of a zinc smelter in Palmerton, Pennsylvania (U.S.A.) as compared to decomposition rates collected from a site 40 km from the smelter (Strojan, 1978). Sassafras average first-year weight losses were 39.3 (40 km), 21.8 (6 km), and 17.5 (1 km) percent. Comparable declines in decomposition were recorded for chestnut/oak leaves collected from the three sites. The reduction in weight loss of the litter had a comparable effect on average quantities of organic matter on the forest floor and the mean thickness of the litter horizons in the three sites. Forest floor organic matter quantities were estimated to be 3.8 kg/m^2 in the control site (40 km), 3.8 kg/m^2 at 6 km and about 8.1 kg/m^2 at 1 km from the smelter. Average thickness of the litter layer declined between the three sites (60 cm at 40 km, 7.0 cm at 6 km, and 12.4 cm at 1 km). This suggests a long-term depression of decomposition, mineral cycling, and subsequent biomass regeneration at the contaminated sites.

An impact of anthropogenic processes also is noted in some managed ecosystems where waste material has been amended to the soil. Considerable research has been conducted in recent years to evaluate the potential for land disposal of sewage sludges. This is attractive in that it provides a reasonably economical means of disposing of this waste by-product while providing material that could *a priori* be predicted to increase the soil organic matter levels. Unfortunately, sludges generally contain high concentrations of metallic cations (Table 10.1). Whereas many of the contaminants would supply macro- and micronutrients for the plant community, many of the nutrients are present in concentrations far exceeding those need for biomass production and, in many cases, are actually present at toxic levels. Ranges of typical values of some common metallic sludge contaminants plus values for two specific sludges are presented in Table 10.1 (Axley et al., 1985; Williams et al., 1984). Note the high variability in metal loadings as exhibited by the range values. The fact that the median value is closer to the lower end of the range suggests that in most instances high metal loadings are exceptions to the generally observed situation. The actual concentrations found in any specific sludge are quite variable and are strongly affected by the nature of the community supplying the sludge. The Oakland sludge is somewhat "typical" of the situation with systems with a significant amount of industrial input, whereas the Pacheco sludge was from essential a domestic supply. Amendment of soils with the Oakland and Pacheco sludges would significantly enhance soil heavy metal levels. For example, in the study of Williams et al. (1984) where these sludges were used, native cadmium, copper, lead and zinc concentrations (mg/kg soil) were 0.29, 31, 89 and 90, and 0.17, 23, 50 and 57 for the soils receiving the Oakland and Pacheco sludges, respectively. Difficulties anticipated from the use of these sludges as soil amendments include the obvious possibility of accumulation of sufficient levels of toxic metals to reduce or to preclude plant growth, bioaccumulation of the metals in plant biomass at undesirable concentrations for human consumption, and migration of the metals from the site of amendment, such as to ground waters.

Behavior of metal ions contained within sludge varies with soil properties and the identity of the metal. For example, amendment of an acid soils (pH $\sim$ 5) with sewage sludge resulted in increased solution concentrations of zinc, manganese, and cadmium (Behel et al., 1983). Copper, nickel and lead ions were present in most solutions at levels below the limits of the analytical procedures used. Cadmium, zinc, and manganese appeared to be present in the soil solution predominantly as free ions. Complexing of these ions to organic soil substituents ranged from 9 to 37 percent, 3 to 22 percent, and 3 to 31 percent of the total cadmium, zinc, and manganese, respectively. With the design of most sludge disposal sites and prudent planning prior to sludge application, the presence of metals in the soil itself is not the environmental problem of greatest concern.

Table 10.1.
Examples of Cation Concentrations in Sewage Sludge

Metal	Typical Values[a]		Oakland Sludge[b]	Pacheco Sludge[b]
	Range	Median		
	mg/kg dry sludge			
Cd	1-3410	15	47	5
Co	1-260	10	—	—
Cu	84-17,000	800	607	188
Ni	2-5300	80	—	—
Pb	13-26,000	500	874	271
Zn	101-49,000	1700	3570	499

[a]Axley et al. (1985)
[b]Calculated from data of Williams et al. (1984)

Prudent planning should include site placement in locations where problems resulting from containment of a waste material with high heavy metal loadings would not be anticipated; that is, as long as the "associated metal pollutants" stay in place, no adverse environmental impacts would occur. Thus greater environmental problems would be associated with the potential for the heavy metal contaminant to migrate or to be carried from the original disposal site.

Considerable research in recent years has involved evaluation of the potential for and kinetics of incorporation of the trace minerals into plant biomass and movement of the toxic metals into ground waters. For example, Williams et al. (1984) examined metal movement in soils receiving sludge amendment over a six-year period. Sewage sludge was mixed into the surface 20 cm of the soil profile annually for six years at rates ranging from 0 to 225 metric tons/h/yr. As would be anticipated if the metals had a greater tendency to remain in the soil than to move within the soil profile, cadmium, copper, lead, and zinc levels increased in the surface soil with sludge rate and with years of addition. Cadmium, copper, and zinc mobility was limited to a depth of 5 cm below the zone of sludge incorporation. Greater mobility was detected with zinc in that significant concentrations were measured at depths 5 to 10 cm below that of maximum application. Thus adsorption of the metals within the soil profile results from both biological and chemical reactions. For example, curium-244 has been demonstrated to react with dissolved and particulate organic matter as well as hydrous oxides (Sibley and Alberts, 1984). Absorption of this transuranic isotope was measured on untreated and extracted sediments in filtered and unfiltered water. Partition coefficients (K^d = radionuclide activity/kg sediment/radionuclide activity/liter solution) ranged from 7×10^3 to 5×10^5. The quantities of curium-244 in the water phase was significantly reduced through the use of ultrafiltered water. This indicates that a significant portion of the curium-244 retained in the water phase was complexed with soluble organic matter. The authors concluded that hydrous oxides may be important for long-term sediment-water partitioning of the curium, but that the initial adsorption was determined by interactions with dissolved and particulate organic matter. Similarly Levi-Minzi (1976) noted that the Langmuir adsorption equation described cadmium adsorption from dilute solutions for 10 soils at 5 and 25°C. The calculated Langmuir adsorption maxima and bonding energy coefficients correlated with cation exchange capacity of the soil and its organic matter content.

From these comments, it is clear that soil mineral components are key in the development of functional plant and microbial communities and that the levels of these minerals, both beneficial and detrimental to the plant community, are controlled to a large part by the association of the soil organic fractions with soil minerals. A number of excellent reviews of this subject have been published in recent years (Page and Bingham, 1973; Huisingh, 1974; Jernelov and Martin, 1975; Gadd and Griffiths, 1978; Tyler, 1981). Primarily, the objective of these presentations was to evaluate the agricultural, general environmental, or public health aspects of trace minerals or heavy metals in soil and water ecosystems. Accordingly, the reader is referred to those reviews for a more detailed evaluation of those topics than will be presented herein. This discussion delineates the interactions between major soil organic matter fractions and the resultant effects on overall ecosystem function. Particular emphasis will be on the abiontic and microbiologically catalyzed metal transformations within the soil itself. To demonstrate potential for indirect aspects of metal-organic matter interactions, the effect of acid rain on soil metal cations will be evaluated.

10.1. CHEMICAL ASSOCIATION OF METALLIC CATIONS WITH ORGANIC MATTER

The impact of soil metal cations on biological processes, both positive and negative, and the general ecosystem adaptations resulting from heavy metal problems depend primarily on the distribution of the metal among a variety of water soluble, insoluble, or colloidal states and the kinetics of the reactions resulting in the transfer of metals between these states (Fig. 10.1). Sequestering of toxic metals into water insoluble forms (soil mineral particulates and particulate organic matter–metal complexes) results in reduction of biomass toxicity, whereas solubolization of the toxicant increases its mobility and plant availability, thereby magnifying the environmental impact. For plant nutrients, desirable reactions, in contrast, are those leading to optimum levels of soluble nutrients. Distribution of the metals between these various compartments depends upon the soil pH, E_h, organic matter levels, and a variety of microbiological interactions. These chemical and physical environmental interactions, although reasonably predictable from our basic knowledge of the chemical processes, are in reality extremely complex because the effect of each of these soil properties on metal distribution among the various soil forms is moderated or enhanced by the

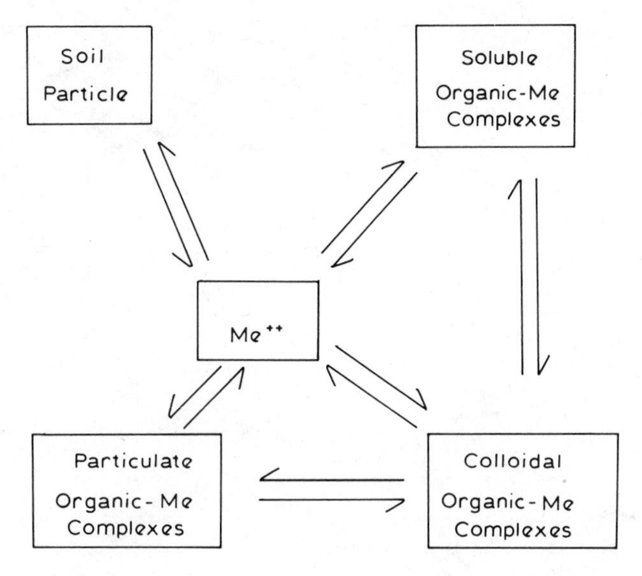

Fig. 10.1.
Conceptual model of metal compartmentalization in soil.

quantities and types of native soil organic matter and, to a large degree, the types of exogenous organic matter entering the soil. For example, pH variation affects the solubility of each metallic species as well as the ionization state of the anionic species on soil organic matter which react with the metal. Fully negatively charged organic components are more likely to combine with the cationic metals than would uncharged species. Also, ionization of humic acid molecules enhances the reaction with metals by expanding the humic molecule through repulsion of the negative charges so that contact between the organic component and the cation occurs more readily. The quantity of cations associated with the humic acid is also affected by the degree of humification.

The actual chemical reactions involved in soil organic matter–metal reactions have been reviewed in detail (see Stevenson, 1982; Flaig et al., 1975). Humic acids contain a variety of anionic reactive groups which may interact with metallic cations. These include a variety of oxygen containing functional groups, such as phenolic hydroxyls and carboxyls, aliphatic carboxyls, and alcoholic hydroxyls. Major humic and fulvic acid components which may react with metallic cations have been identified through the use of the acid preparations with reactive groups chemically "blocked" and, thus, prevented from interacting with solution cations. Schnitzer and Skinner (1965) examined the effect of blocking specific humic acid components on Fe^{3+}, Al^{3+} and Cu^{2+} binding to humic acids. Reactive groups studied were (1) alcoholic plus phenolic hydroxyls, (2) phenolic hydroxyls plus carboxyls, (3) phenolic hydroxyls, (4) both types of hydroxyls plus carboxyls, and (5) carboxyls. Significant blocking of either the acidic carboxyls

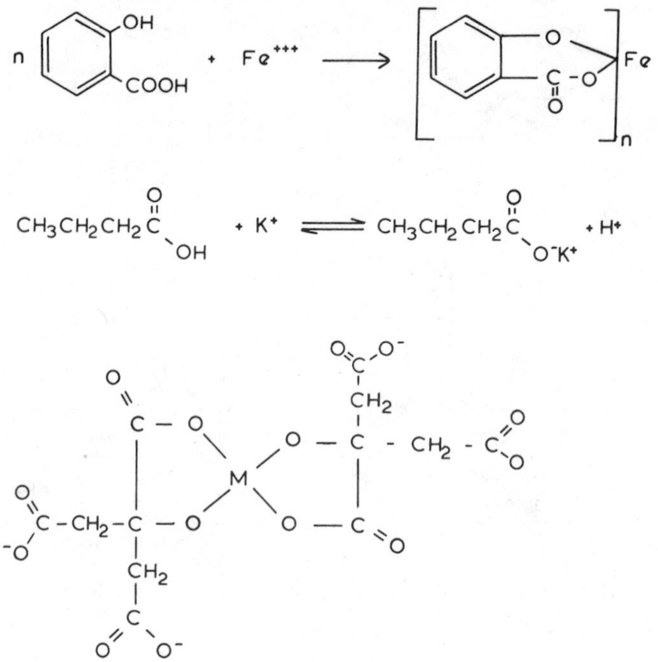

Fig. 10.2.
Metal interactions with soil organic matter components.

or the phenolic hydroxyls resulted in reductions in metal retention. The data indicated that both the phenolic hydroxyls and the acidic carboxyls appeared to react simultaneously with the metal ions by the same type of reaction known to occur between ferric iron and salicylic acid (Fig. 10.2). Minor reactions occurred between less acidic carboxyl groups and the metal ions. Alcoholic hydroxyls did not participate in the reactions. Because these reactive groups increase with degree of humification of organic matter, it is reasonable to conclude that the greater the humification state of an organic matter sample the greater the potential for its reaction with metallic cations.

Boyd et al. (1981) used electron spin resonance spectroscopy to study the mechanism of Cu^{2+} binding to humic acid. They examined copper-humic acid complexes and some adducts with nitrogen donors. Their data indicated that two humic acid oxygen donors were coordinated equatorially to the Cu^{++} center in the copper–humic acid complex. The characterization of the adducts with nitrogen donors yielded data consistent with the formation of a copper chelate in the original copper–humic acid complex with the two equatorial copper–humic acid oxygen bonds occupying the *cis*-position. Similar results were found in evaluating copper binding by humic acids extracted from sewage sludge (Boyd et al., 1983). In the latter study, the copper ions also appeared to form axial bonds with humic acid donor ligands orginating from proteinaceous substances associated with the sludge humic acid fractions. Glycylglycine formed coordinate bonds with copper bound to humic acids. Bloomfield et al. (1976) found that metal associations with partially humified organic matter includes these strong associations plus weaker ionic reactions. In their study or binding of copper, manganese, cobolt, nickel, lead, zinc, and cadmium with aerobically decomposing plant biomass, the metals were associated with both colloidal, humified organic matter and with organic matter in true solution.

Various types of organic matter–metal associations are depicted in Fig. 10.2. Along with the reaction discussed in the previous text, ionic associations and chelation complexes are shown. Solubility of the product is dependent upon the solubility of the organic matter complex. Ionic associations with simple organic acids or chelators, such as the butyric or citric acids associations depicted, results in enhanced solubilization of soil metals. The primary mechanism for the enhancement of dissolution of insoluble minerals is that the equilibrium state existing between that metal complexed in soil minerals and that existent in ionic form is shifted to the right. For example, Fe^{2+} contained within the soil interstitial water may be produced through ionization of ferrous sulfate as described by the following equation:

$$Fe_2(SO_4)_3 \rightarrow Fe^{3+} + SO_4^{2+}$$

Combination of the ferric ion with soil organic matter results in a decrease in the soluble ferric iron. Thus the equilibrium of the ionization reaction is shifted. If the organic matter–metal complex is water soluble, the metal may be translocated to another site within the soil profile. Water soluble organic matter–metal complexes could diffuse or be washed to sites of lower iron concentration. As a result of a shift in the equilibrium or degradation of the organic molecule, the metal could then be released into the soil solution in the ionic form. Subsequently, the metal could be used as a nutrient by the plant or if conditions were favorable, the metal could be precipitated as an iron salt.

In most soil ecosystems, cations are associated with a variety of anionic soil particulates and colloids. These include not only the organic components discussed thus far, but also a wide variety of soil clays. In fact, mineral components may actually react at a faster rate with the cations than does the organic matter. Hodges and Zelazny (1983) found that montmorillonite and kaolinite reacted much faster with aluminum than did peat. But, peat eventually retained more aluminum than did the montmorillonite. Inskeep and Baham (1983) examined the influence of forest litter layer, dried Chicago sludge, and a peat soil as sources of organic ligands on adsorption of cadmium and copper by sodium montmorillonite. pH was varied from 4 to 8.5. The complexing of cadmium with organic ligands in the forest litter was much less than that for copper. Copper adsorption decreased with increased pH. In contrast, cadmium adsorption was affected only slightly by pH variation.

These data reveal a variety of natural soil processes controlling metal mobility and free ion concentrations in soil solution. A variety of reactions with native or exogenously supplied organic matter may occur which serve to reduce toxicant levels or enhance nutrient concentrations. These reactions are affected to a large degree by ecosystem management and naturally occurring microbial processes.

10.2. ORGANIC MATTER INTERACTIONS WITH METALS

From the view of the total ecosystem, soil mineral interactions with soil organic matter components can be divided into two reasonably distinct categories: (1) those reactions which rely upon the living (biotic) soil organic matter fractions, and (2) those resulting from passive interactions with nonliving or abiontic colloidal organic matter. Materials of interest in the latter category are humic and fulvic acids as well as exogenous organic matter entering the soil through natural processes, such as aboveground biomass senescence, or organic debris added to soil sites managed to maximize crop production or land reclamation. A variety of sludges, straws, and wood products may be added to soils to ameliorate reclamation problems (Visser, 1985). Significant differences between the biotic and the abiotic and abiontic interactions affect both short-term and long-term implications of organic matter–metal reactions. Simply mixing organic debris with metal contaminated soil certainly has positive short-term effects, but this procedure may not provide a continuing solution to soil metal problems. For example, difficulties may arise if insufficient organic matter is mixed with the soil in that the chemically reactive sites may become saturated before toxic metal levels are reduced sufficiently to relieve the inhibitory effects. Also, biodegradation of the organic matter could eventually reduce or eliminate the benefits of the combination of the organic matter with heavy metals.

Practical aspects of these interactions are exemplified by work with oil shale waste reclamation. A major limitation in the reclamation of oil shale wastes is the removal of heavy metals remaining in the waste slag following oil recovery processes (Wildung and Garland, 1985). These metals are present in the shale waste in concentrations that are unfavorable for the development of the microbial populations necessary for establishment of active biogeochemical cycles. Thus aboveground biomass production is inhibited or even precluded. An immediate solution for the problem is to add an exogenously produced organic matter to complex the heavy metals. This could, if managed properly, reduce metal loading to the extent that viable microbial and plant populations develop. Encouragement of microbial community development is essential for long-term ecosystem stability. A favorable reclamation management plan is to encourage development of indigenous biomass production and stable microbial populations. This may involve use of some externally supplied organic materials, but it must also include optimization of other environmental factors limiting microbial activity. Through management of the microbial populations, a stable soil ecosystem is developed which would not only alleviate immediate metal problems but also yield a stable, aesthetically pleasing aboveground plant community (Tate, 1985).

10.2.1. Abiotic Organic Matter Interactions

Metal binding in soils is controlled in large part by the soil cation exchange capacity. The organic matter contribution to the cation exchange capacity is derived primarily from soil humic and fulvic acids (abiotic component) and, to a small degree, from abiontic materials. (This includes not only the extracellular enzymes for which the term was coined, but also those cellular components which were originally functional in or around the microbial cell but, due to the demise of the cell, have become separated from living biomass and stabilized into colloidal soil organic matter.) Anionic components of abiontic substances result predominantly from ionization of carboxyl groups, such as those contained in the uronic acids frequently found in microbial capsules and slimes and proteins. Since these materials are generally water insoluble and closely allied with microbial cells, their impact is primarily localized in microsites. In most soils, humic and fulvic acids cation exchange capacity far exceeds any derived from abiontic sources. Exceptions include sandy soils containing little or no organic matter. In these soils, the polysaccharides secreted by microbial colonies to cement the cells to sand grains may become significant. Whereas humic materials are reasonably stable to biological metabolism, abiontic materials are constantly being catabolized and resynthesized in the ecosystem. Thus their contribution to total cation exchange capacity is variable. This is contrasted against the essentially constant background of the more stable soil organic matter components.

Although the predominant reactions between humic acids and metals are simple ionic bonding or chelation reactions and generally do not alter the redox state of the metal, humic acids have been shown to catalyze the reduction of a variety of metal ions. For example, elemental mercury is formed in aqueous solutions of mercuric ion and humic acids (Alberts et al., 1974). The reaction rate is first order and varies with solution pH. The reaction mechanism involves the direct interaction of the metallic cations plus free radical electrons of humic acid. Similarly, Ghosh et al. (1983) when comparing interactions between iron, copper, and manganese with humic acid concluded that these cations could also be reduced chemically by the humic acids.

These chemical reactions are rapid (in the order of a few seconds) and produce tightly linked metal-organic matter complexes. Bunzl et al. (1976) in their study of lead, copper, cadmium, zinc, and calcium by peat measured half times for adsorption and desorption of the metal ions in the range of 5 to 15 seconds. The desorption rate is specific for each individual metallic ion. Humic acid complexes of copper and lead are considerably more stable than those for cadmium (Stevenson, 1976). Since the chemical components of humic acid which react with the metal are common to all humic acid preparations (differences involve quantities of the reactive groups present), little variation between the binding capacities of various humic acids occurs (Stevenson, 1976). The stability constants for various metal complexes with humic and fulvic acids may be derived through graphical solutions to the following equation:

$$v = \Sigma\, n_j k_j(M)/(1 + k_j(M))$$

where v equals sites bound/macromolecule concentration, k_j is the stability constant for binding at class j, n_j is the number of binding sites of class j, and (M) is the free metal ion concentration (Fitch and Stevenson, 1984). Graphical solutions include Scatchard ($v/(M)$ verses v), reciprocal ($1/v$ verses (M)), and double reciprocal ($1/v$ verses $1/(M)$) plots. A procedure similar to Hill plots has also been used, but Fitch and Stevenson found this procedure to be less reliable than the preceding methods.

The distribution of metal cations between sorbed and solution states is important both from the view of plant nutrition and from ecosystem "pollution" problems. Sorption capacities are generally estimated by mixing known concentrations of the test metal ion in water with soil. Following an incubation period, the soil is extracted with an appropriate extractant and the quantity of the metal in the extract is quantified through a method, such as atomic adsorption spectroscopy. Differences between quantities added to the soil and those extracted are considered to have been sorbed to the soil minerals and organic colloids. Zunino and Martin (1977) used dialysis with repeated changes of distilled water to measure the maximum binding ability of a variety of humic acid-like polymers and soil and peat humic acids. Their data were described by a linear Langmuir plot. Others have found the Freundlich sorption isotherm equation provided a better fit of the data (e.g., see Elrashidi and O'Connor, 1982). An interesting result of such studies is the observation that chelation of the metal prior to or subsequent to soil amendment may greatly decrease its sorption. Elrashidi and O'Connor (1982) found that zinc sorption by a variety of soils was significantly decreased by the presence of ethylenediaminetetraacetic acid (EDTA) in the solution prior to soil amendment or in the soil solution after mixing of the zinc solution with the test soil. James and Bartlett (1983) observed that citric acid, diethylenetriaminepentaacetic acid (DTPA), fulvic acids, and water soluble organic matter from air-dried soil kept Cr(III) in solution and prevented its immediate removal by soil. Cr-citrate remained soluble for at least one year in limed surface soils maintained at field moisture. These data are significant in that many of the organic materials tested are common substituents of sewage sludges, animal manures, and industrial waste waters, as is chromium.

The physical effect of the association of metal ions with various abiontic soil organic matter fractions on cation availability and impact on the biotic component of the ecosystem varies with the type of organic matter involved and the nature of the interaction. Organic components may be reasonably water insoluble, (humic acids or humin), or readily water soluble (e.g., fulvic acids and a variety of small molecular weight microbial products). Stability of the metal-organic matter associations depends on the type of interaction between the metal and the organic component—ionic bonding is much weaker than that found with chelators. Thus a variety of physical products result from organic matter-cation interactions ranging from water insoluble tightly bound metal complexes (metals chelated to humic acids) to weak associations (ionic bonding between metals and simple organic acids, such as acetic acid). Tightly bound, water insoluble humic acid-metal complexes remove trace minerals or heavy metals from soil solution, whereas the ionic-simple organic acid associations result in enhancement of metal mobility and/or plant availability.

As has already been alluded to, each of the reactions involving metal complexes in soil are reversible. Reaction kinetics thus may be shifted into the direction of ionization through removal of the cation from solution that is, through reduction of the solution concentration. This may occur through interactions with soil organic matter or from incorporation into living biomass. Reduction of free ion concentrations in soil may result in both reversal of the binding reactions to soil organic components as well as increased solubilization of primary soil minerals.

The overall effect of binding of the metal with various soil organic matter complexes on total ecosystem function is dependent primarily upon whether the free cation is stimulatory or inhibitory to primary production and the solubility of the metal-organic matter complex. Significance is attributed to water insoluble metal complexes when the cation is an obligatory plant nutrient or a toxicant. Removal of aluminum ions from solution through interaction with soil organic matter in acid soils is desirable, whereas reduction of trace minerals by similar reactions reduces biomass production. In contrast to the situation with water insoluble metal-organic matter complexes, the environmental impact of water soluble complexes may extend far beyond their site of formation. Water soluble toxic metals, including radionuclides, are more likely to migrate from the site of deposition into ground waters or recreational waters than are the insoluble complexes. Thus the environmental impact of the water soluble organic matter-metal complex is of major concern. Evidence for such mobility include the natural phenomenon of podzolization (Stevenson, 1982) and water pollution under land fills.

10.2.2. Biological Interactions with Metals

Two types of biological interactions between soil organic matter and metallic cations are of interest in this study: (1) those that alter the mobility of the metal within the soil profile by changing the distribution between water soluble and insoluble forms and/or the equilibrium between metal solubility states, and (2) processes which chemically change the metal ion itself (alkylation or reduction). Each reaction may have both biological and chemical components. For example, Zamani et al. (1984) evaluated zinc immobilization by soil microorganisms in Rubicon sand (an Entic Haplorthod, pH 5.9). A zinc solution was continually perfused through columns of native (biologically active) and gamma-irradiated (sterilized) soils. At steady-state, approximately 75 percent of the zinc was bound to the soil through chemical and physical mechanisms in the sterile soil columns. Amendment of the sterile soil with a carbon and energy source plus a microbial inoculum increased the zinc immobilization an additional 20.5 percent after a 72-hour incubation period. (Note the long time delay for the reaction in comparison to the essentially immediate reaction times discussed previously for the chemically mediated reactions.) Nutrient amendment in the absence of microbial inocula had no effect on the quantity of zinc initially immobilized; that is, the increase in metal immobilization resulted from microbially mediated processes, not simply from the amendment with carbon. The effect of the microbial populations on zinc immobilization did vary with soil type. No biological enhancement of metal binding was found with either a Brookston clay loam (Typic Argiaquolls, pH 7.5) and Haughton muck (Typic Medisaprist, pH 6.0). It is reasonable to expect that in soils with a reasonably high cation exchange capacity that sufficient reactive sites would exist to bind the metallic ions prior to microbial catabolism of a carbon source. In these soils, no significant effect of microbial activity on short-term metal binding would be expected. This does not suggest that the metals chemically bound within the soil mineral and/or organic fractions are unavailable to the microbial community. Microbial alteration of bound metal ions following their initial chemical immobilization is likely. For example, under suitable conditions, the microbial populations could alkylate, reduce the appropriate cations, or immobilize the metal cation within its biomass.

Microbial interactions with soil metals may also directly result in increased metal mobility. The alkylation reactions discussed in the following text are a prominent example of such an effect. In this case, the metal may be transformed from a water insoluble or slightly soluble form to a volatile alkyl metallic compound. An active microbial population may also enhance microbial mobility through simple alteration of the soil physical environment. This could result from a localized pH or redox potential variation in soil microsites or a general alteration of these parameters throughout the entire soil mass. Redox potential modification may result from depletion of electron acceptor, such as molecular oxygen, nitrate, nitrite, or sulfate, whereas examples of mechanisms for pH alteration include carbon dioxide or organic acid synthesis. A more generalized ecosystem pH modification occurs due to microbial synthesis of strong acids, as is noted with acid mine drainage. In that case, *Thiobacillus* species oxidize sulfide ions to sulfate (Mills, 1985). An example of enhanced metal mobility resulting from microbial activity involves cadmium leaching from a silt loam soil (Ausmus et al., 1977). Increased cadmium leaching rates were correlated with dehydrogenase activity which was used as a measure of microbial respiration. An interesting technique was used in this study to sample small volumes of soil water from within the soil column. Cellulose acetate hollow fibers were placed in the soil columns for rapid collection of soil interstitial water for the metal ion analysis.

These biological effects on metal mobility in soil result from direct microbial interactions with metal ions (reduction, alkylation, incorporation into cellular substituents, or catabolism of the carbon substrate, thereby releasing the associated metal ions), from indirect solubolization through reaction with a product of microbial metabolism (carbon dioxide or siderphore production) or from the simple microbially mediated alteration of the physical and chemical environment. In any case, the kinetics of the binding (immobilization) or mobilization reaction resemble those of biological processes and the reactions are abolished by sterilization of the soils. Each of these mechanisms will be discussed in greater detail in the following text.

10.2.2.1. Indirect Interactions

A major forensic science maxim is also applicable to the relationship of the soil biotic community and environs; that is, the presence of a living organism in an ecosystem results in some degree of physical and/or chemical change in the site. Many of these modifications are minor, probably even unquantifiable with current instrumental limitations. Others result in major and, in many cases, irreversible alterations of the ecosystem state. Shifts in the equilibrium status of metallic cations in soil interstitial water are one of these highly significant effects of the presence and respiration of soil microbial flora. Microorganisms may indirectly enhance or reduce metal availability through (1) modification of the site pH or E_h, (2) inorganic and organic acid synthesis, (3) excretion of chelators, (4) catabolism of metal-organic matter complexes, and (5) accumulation of extracellular slimes. An interaction between microbial populations and their surroundings resulting in a shift in metal availability is defined as being an indirect product of microbial activity if modification of the metal, its solubility, or its association with organic matter is not the primary basis for the reaction being catalyzed by the microorganism. For example, oxidation of potassium acetate to carbon dioxide results in release of the potassium. This is an indirect result of the microbial activity in that the primary "objective" of the microorganism was the recovery of carbon and energy.

The solubility and the oxidation state of many metals are controlled, in part, by the soil pH and redox potential (E_h). Under extremely acidic conditions, metals, such as iron, manganese, zinc, and copper, are highly soluble. As the pH increases, metal hydroxides forms which precipitate. The exact pH where precipitation commences varies with each element. In reality, a characteristic gradient of solubilities based on soil pH exists for each metal. Since each of the foregoing minerals is required for plant growth, to some degree, it could be concluded that acidic pH values are not detrimental to plant growth and development. Since there are extensive pools of acidic soils world wide, this situation would be fortuitous. Unfortunately this is not the true situation. As the soil pH declines, these micronutrients may actually be solubilized to levels which are toxic to plant growth. Also, solubility of such toxic metals as lead, cadmium, zinc and lead (Gerritse and van Driel, 1984), and aluminum are also increased under acidic conditions.

Because of this major effect of acidic conditions, any microbial process that yields acids, be they organic or mineral, has a major effect on localized mineral solubility. The simple production of carbon dioxide through respiratory processes causes a localized pH decline. This results from the equilibrium relationship between carbon dioxide, bicarbonate, and carbonate in soil water. More dramatic variation in soil pH is derived from synthesis of strong mineral acids, for example, the oxidation of sulfide by *Thiobacillus* sp. to sulfate. Not only are large quantities of minerals solubolized as a result of sulfuric acid production, but extensive environmental damage can result both at the site of primary acid production and down stream (Mills, 1985).

Whereas these mineral acids alter mineral distribution predominantly through a pH effect, organic acids may solubolize minerals through simple ionization reactions or by chelation. Data of Duff et al. (1963) provide a clear example how one organic acid, 2-ketogluconic acid, can solubilize soil minerals (Table 10.2). A motile, short, gram-negative, rod shaped bacterium was isolated which dissolved natural phosphates and silicates when grown on a glucose medium. The active compound was unequivocally identified as 2-ketogluconic acid. This acid formed a chelate with the various cations which was sufficiently stable to allow separation of the metal-chelate from the acid in a pH 3 chromatographic developing solvent. A wide variety of microbial by-products that may alter soil pH or chelate metals are released into the soil surrounding the microbial colony. Among these are a number of iron chelating compounds, siderophores, which are excreted by a diverse variety of bacteria ranging from common enterics (e.g., van Tiel-Menkveld et al., 1982; Payne et al., 1983), free-living nitrogen fixing bacteria (Fekete et al., 1983; Page and Huyer, 1984), and blue green bacteria (Murphy et al., 1976). Siderophores are apparently active in soil. For example, they have been proposed as a mechanism explaining disease-suppressive soils (Kloepper et al., 1980). Amendment of either fluorescent *Pseudomonas* strain B10, isolated from a take-all suppressive soil, or its siderophore, pseudobactin, to *Fusarium*-wilt and take-all conducive soils resulted in the conversion of the soils to suppressive soils. The mechanism of suppression is apparently the result of efficient complexing of Fe^{3+} by microbial siderophores. This hypothesis is supported by the fact that amendment of the suppressive soils with Fe^{3+} converted them to conducive soils. Cationic, neutral, and anionic secondary hydroxamic acid siderophores have been found in soil extracts (Akers, 1983). In a comparison of the abilities of hydroxamic, synthetic, and other natural organic acids to chelate iron, Cline et al. (1982) concluded that hydroxamate siderophores are of great importance to plant nutrition in natural ecosystems, including agronomic soils.

The oxidation state (hence the plant availability) of many metals varies with changes in E_h. The actual reduction of minerals, such as iron, manganese, or copper, is catalyzed directly by soil microbes or, as indicated previously, chemically by soil humic acids. Reduction of the metal oxidation state is favored by a lowering of the soil (or microsite) redox potential through general microbial me-

Table 10.2.
Solubolization of Metals from Soil by a Bacterial Isolate Growing in a Glucose Medium[a]

Soil (Depth)	Cation Solubilized			
	Ca	Mg	Fe	Al
	μg/ml			
Peniel heugh soil				
0–22.5 cm	190	85	305	190
55–65 cm	250	140	165	200
Aberdeen soil				
top soil	30	20	—	—
Rhum soil				
7.5–15 cm	nil	85	—	—
22.4–30 cm	nil	125	—	—

[a]Duffet et al., 1963.

tabolism under conditions where free oxygen diffusion into the soil is limited (e.g., in flooded or water logged soils). Creation of anoxic conditions in the presence of nitrate or nitrite results in redox potentials of about 150 to 200 mv. Depletion of the nitrogen oxides pools and development of conditions favorable for carbon dioxide and sulfate reduction results in redox potentials of −200 to −1000 mv. Under these highly reducing conditions, reduction of metals with varying oxidation states is favored.

A simple mechanism of increasing cation levels in soil solution is microbial catabolism of the anionic carbon compound to which they are associated. This is an indirect effect of microbial activity in that the primary benefit to the microbial population results from the use of the carbon contained in the metal-organic matter complex for a carbon and energy source. In most soils, anionic organic acids and chelators are found in combination with a variety of metallic cations. Mineralization of the organic component to carbon dioxide leaves the cation free in solution. Data exemplifying this process is provided by work of Firestone and Tiedje (1975) in the study of metal nitrilotriacetate (NTA) complex catabolism by a *Pseudomonas* species. The bacterial species was shown to degrade calcium, manganese, magnesium, copper, zinc, cadmium, iron, and sodium chelates of NTA at equal rates when the appropriate metal concentrations were sufficiently low to avoid toxicity problems. The metals did not accumulate in the microbial cells. The authors postulated that an envelop-associated component, most likely a transport protein involved in binding, was responsible for dissociation of the metal from the NTA. Similarly, a variety of soil bacterial species including *Pseudomonas*, *Bacillus*, *Serratia*, *Acinetobacter*, *Klebsiella*, *Nocardia*, and *Streptomyces*, precipitate iron from the iron salt of ferric ammonium citrate through the metabolism of the citrate (Alexander, 1977). Microorganisms are involved in a variety of direct and indirect conversions of metal cations from soluble to insoluble states. For review of this topic in greater detail, see Alexander (1977).

Bioaccumulation of metals in capsular or slime layers and along cell walls can also serve as a passive mechanism for removal of metallic cations from soil solution. This bioaccumulation simply may involve ionic association with anionic bacterial slime, capsule layers, or cell walls. Common bacterial species, *Caulobacter* sp., *Micrococcus* sp., *Pseudomonas fluorescens*, *Mycobacterium phlei*, *Escherichia coli*, *Klebsiella pneumoniae*, and *Corynebacterium pseudodiphtheriticum*, precipitate iron from solution resulting in encrusting of the microbial cell (Macrae and Edwards, 1972).

10.2.2.2. Direct Microbial Interactions with Metals

Microbiologists, botanists, and biologists in general have traditionally shown a primary interest in the effect of metal cations on cellular processes. This includes both beneficial and detrimental effects. Biological processes directly affecting chemical and physical properties of soil metals are (1) valence (oxidation state) changes, (2) substitutions, (3) methylations, and (4) transalkylations. The cell catalyzing the various reactions may gain energy benefits, a sink for excess reducing power, or relief from metal toxicity. Capability to catalyze each individual reaction is limited to a select group of microbial species interacting with a few metallic ions. These limitations are determined in part by the metabolic capabilities of the microorganism (especially the energetics of the reaction) and by the metallic ion chemistry. For example, metals, such as iron, copper, mercury, and manganese, may exist in a variety of oxidation states. Although the reactions may occur spontaneously, the oxidation and/or reduction of these metals may be directly biologically catalyzed at a rate which is generally significantly greater than the spontaneous chemical transformations. Substitution involves the replacement of one metal with another in a metal complex or living organisms whereas transalkylation is the substitution of one alkyl group associated with a metal for a second. Methylation is the bonding of methyl groups to a variety of metal receptors. Of primary interest for this discussion are those reactions which alter metal mobility and/or plant availability and have the potential to catalyze major changes in the distribution of soil metals between the inorganic- and organic matter-associated metal pools. Of the preceding microbial-metal interactions, valence changes and methylation processes are most significant from this viewpoint and, hence, will discussed herein. For a more detailed review of each of these biologically catalyzed reactions with soil metals than can be presented here, see Jernelov and Martin (1975).

Among the properties of a metal which are altered as a result of changes in metal valence state are its water solubility, plant and microbial availability, and toxicity. Of necessity, these properties are interactive in that a water insoluble metal complex or compound has little or no toxic or nutrient effect on the biological community. But as microorganisms grow, they may oxidize or reduce the water insoluble minerals thereby converting them to a chemical form with greater water solubility and resultant nutrient availability or toxicity. With simpler systems than soil, such as that occurring within microbial cells, prediction of the effect of various metallic ions on cellular processes is relatively easy. As was stated by Jernelov and Martin (1975) "the valence state of a metal ion normally occurring in an organism can be assumed to be that which is least toxic to the organism." This is not the situation with the complex soil ecosystem. Instead we may assume that in soil, the valence state of a metal ion occurring most commonly is that which is most favored by the properties; chemical, physical, and biological, of the soil site. Thus the microbial population, or portions thereof, may be sensitive, resistant, or unaffected by the prevailing metal cations in solution. For those ions which are toxic, the microbe may: (1) develop resistance to the toxic ion; (2) reduce the metal concentration in solution by altering its oxidation state to one which is less toxic; (3) reduce the overall concentration of the metal *in situ* by increasing its mobility, such as through methylation reactions; or (4) not grow in the presence of the offending metal. All four adaptations are found in soil. Microbial survival in this complex situation thus requires a variety of adaptive mechanisms. If low levels of the toxic metal ion are present the microbe may detoxify the metal or develop resistance, whereas with large metal cation concentrations, detoxification may not be effective in reducing the toxicity to manageable levels. Then resistance is of primary importance. [See Robinson and Tuovinen (1984) for further discussion of mechanisms of microbial resistance and detoxification of metallic cations.] Along with the reduction of toxic effects, the microorganism may also gain energy from oxidation of these metals, or use them as an electron acceptor (Alexander, 1977). In this situation, the microbial population is accelerating a reaction that may occur spontaneously, or be indirectly favored through microbial reactions which create a reducing atmosphere. This capability to alter the oxidation state of metals and metal oxides is apparently widely dispersed among the soil microflora. For example, Bautista and Alexander (1972) concluded that that capacity to reduce selenate, selenite, tellurate, tellurite, vanadate, molybdate, molybdenum trioxide, arsenate, chlorate, and MnO_2 was commonly found in soil microorganisms.

Considerable attention has been given in recent years to natural occurrence of methylation processes. Metals which may be biologically methylated include selenium, tellurium, mercury, and arsenic. Primary interest in evaluation of methylation kinetics in a variety of ecosystems has not been the result of concern for the effects on primary plant and microbial productivity, but rather it has related to those processes leading to increased human exposure and/or toxicity. With methylation, solubility of the metal in biomass tissues is altered so that bioaccumulation is encouraged. Thus accumulation of these human toxins in food chain plants and animals is accentuated. Methylated mercury, arsenic, and selenium products have been isolated from soils, sewage, and natural waters. These processes may be considered beneficial from the view of soil biomass production in that each of the metallic ions are biotoxic. Thus the volatility and mobility resulting from methylation would result in reduction of interstitial water concentrations. This has been proposed as one means of reduction of toxic metal cation levels in oil shale wastes (Wildung and Garland, 1985).

10.3. EFFECT OF SOIL METALS ON SOIL ORGANIC MATTER METABOLISM

Heavy metal problems are frequently observed or identified at the ecosystem level (failure of plant growth, yellowing or stripping of leaves, selection of resistant plants), but perhaps of greater interest are those less visible effects of moderate metal levels resulting in less obvious changes in overall biomass productivity. These effects may be the result of direct inhibition of plant growth or they may be derived from the interaction of metal ions with essential soil microbiological processes. For example, a critical step in the nitrogen cycle is mineralization of organic nitrogen. Reduction of the nitrogen mineralization rate by toxic metals could indirectly result in less plant biomass productivity. Thus, soil organic matter-microbial-metal interactions are essential to overall ecosystem productivity and stability. Although metallic cations may either stimulate (as nutrients) or inhibit cellular activity, reactions of interest are generally those involving inhibition of microbial activity. Microbiological trace mineral requirements are as a rule met by solubilization of complex soil minerals. The occurrence of toxic levels of heavy metals in native, undisturbed, sites has been observed, but most problems with heavy metals in soils result from anthropogenic intervention. Sites down wind from smelters, those receiving industrial sludges, as well as areas impacted by acid mine drainage are prime examples of candidates for heavy metal problems.

The interaction of heavy metals with the soil microbial community is complex and highly variable. The impact of the metal on microbial activity may be moderated or enhanced by properties of the chemical and physical environment in which the microbe resides. As indicated above soil pH, E_h, clay content, and organic matter levels may increase or decrease metal ion availability. Thus considering the principles of comparative biochemistry and the vast biochemical literature describing the effect of heavy metals at a variety of concentrations on cellular metabolism, it could be assumed that prediction of the results of contamination of soil with heavy metals on microbial activity would be relatively easy. But, as a result of heterogeneity of soil and the variety of inorganic and organic reactions involving metallic ions in soil, this is not possible. Concentrations of the metal that are known to be toxic in culture may be ineffective in soil due to reduction of the effective concentrations of the metal in the soil interstitial water. Thus models of soil microbes and metal interactions must include terms relating the physical and chemical properties of the site (and microsite) to free-metal concentrations.

This impact of soil on inhibition of microbial respiration by a variety of heavy metals is exemplified in the study conducted by Firestone and Tiedje (1975) involving nitrilotriacetate (NTA) catabolism by a *Pseudomonas* species which was discussed previously. NTA–metal complexes were readily decomposed with low concentrations of the substrate. The limiting factor relating to substrate concentration was the quantity of metal ion liberated due to catabolism of the carbonaceous portion of the substrate. No inhibition of microbial respiration occurred until the concentration of liberated cation reached toxic levels. Calcium-, iron-, manganese-, and sodium-salts were not inhibitory and their catabolism resulted in equivalent respiration rates. Catabolism of nickel, cadmium, and mercury NTA salts resulted in total inhibition of bacterial respiration. Copper and zinc salts gave intermediate results (Fig. 10.3). Amendment of the bacterial growth medium with soil resulted in a reduction of the inhibitory effect with many of the cationined for nickel, cadmium, copper, and zinc. The percent recovery varied with the metal cation with greatest recovery occurring for the copper and zinc salts and least for nickel salts. The total inhibition or respiration by mercury-NTA was not reduced by amendment of the growth medium with soil. A similar inhibition of carbon mineralization by liberation of associated cationic metals was suggested by results of a study of metal salts of a variety of microbial and plant polysaccharides (Martin et al., 1966) Decomposition of iron, aluminum, zinc, and copper salts or complexes of a variety of microbial and plant polysaccharides in Greenfield sandy loam was assessed. The microbial and plant polysaccharides contained between 6 and 83 percent uronic acid units. Decomposition of all the polysaccharides was inhibited by at least some of the metal cations examined. For example, zinc and aluminum had little influence on *Azotobacter chroococcum* polysaccharide decomposition, whereas both copper and iron reduced decomposition of this polysaccharide by 50 percent. These studies exemplify the difficulties encountered by microorganisms in soil gaining their energy from carbon mineralization, the extreme variability of the process, and the ameliorative capacity of soil.

As was exemplified by the data of Firestone and Tiedje (1975) some metal ions, such as mercury, are toxic to microbial activities at extremely low concentrations. Others only cause a partial inhibition of microbial activities at commonly encountered soil levels. This partial inhibition of microbial activities of particular interest from the view of ecosystem stability in that the probability of a differential sensitivity of various microbial populations exists and surviving populations may actually develop mechanisms to overcome the chronic metal toxicity. Thus in the short-run, microbial activity and potentially total ecosystem productivity may be reduced, but over long periods of time, the biomass production rate may approach or reach rates comparable to those existent prior to imposition of the metal limitations.

Critical soil microbial functions inhibited in part or totally by trace metals include methanogenesis, sulfate reduction, carbon dioxide evolution, and microbial biomass production (Capone et al., 1983), nitrogen fixation (Wickliff et al., 1980), and nitrification (Tyler et al., 1974; Liang and Tabatabai, 1978). Capone et al. (1983) found that as a result of the differential sensitivity of various carbon catabolism pathways in a *Spartina alterniflora* salt marsh, carbon flow could be altered through heavy metal contamination. Immediate and long-term

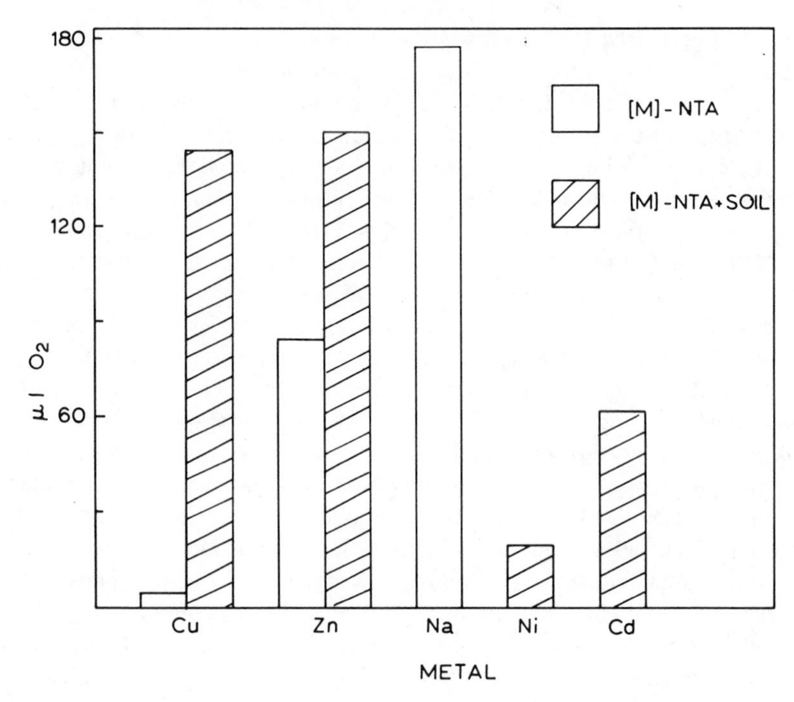

Fig. 10.3.
Effect of amendment of bacterial growth medium with soil on respiration of NTA–metal complexes (Firestone and Tiedje, 1975).

inhibition of methanogenesis was detected in sediments amended with methyl mercury, mercuric sulfide, and sodium arsenite. Methanogenesis was initially inhibited with mercury, lead, nickel, cadmium and copper (all as chlorides), and zinc sulfate, potassium chromate and potassium dichromate. But the initial inhibition period was followed by a period of overall stimulation. Carbon dioxide production was less sensitive to metal amendment than was methanogenesis. Sulfate reduction was inhibited by all metals tested in the short term and over the long term by all but ferric chloride and nickel chloride. Microbial biomass was decreased by amendments with ferrous chloride, potassium dichromate, zinc sulfate, cadmium chloride, and cupric chloride, but was not affected by lead chloride, mercuric chloride, or nickel chloride. The initial suppression of the methane synthesis rate was alleviated by precipitation, complexation, or transformation of the metal. The authors proposed that the stimulation of methanogenesis resulted from the long-term inhibition of competing sulfate reducing bacterial populations.

Nitrogen fixation, an essential process for introduction of fixed nitrogen into nonmanaged ecosystems, is sensitive to trace minerals [see Tate (1985) for more detailed discussion of the *Rhizobium* legume system in acid soils.] For example, nitrogen fixation by the red alder (*Alnus rubra* Bong.) is strongly inhibited by cadmium (Wickliff et al., 1980). This nitrogen fixation is the result of a symbiotic association between the alder tree and an endophytic actinomycete, *Frankia*. Treatment of red alder seedlings for 11 weeks with 0.545 to 135 μM cadmium chloride in nitrogen free solution resulted in a decrease in *in situ* nitrogenase activity of 25 to 89 percent. Total number of nodules per plant decreased between 29 and 74 percent. The data suggested that cadmium was inhibiting nitrogenase activity, growth, nodulation, and nitrate reductase.

Nitrification, oxidation of ammonium to nitrite and nitrate, is also sensitive to a variety of trace minerals. Low concentrations of cadmium, or lead had no effect on nitrification in a mull soil, but as the metal concentration was increased, so did inhibition (Tyler et al., 1974). Liang and Tabatabai (1978) studied the effect of 19 trace elements on nitrification in several soils. Greater than 50 percent inhibition of nitrification of ammonium to nitrite was found with 5 μmoles/g Ag (I), Hg(II), Cd(II), Ni(II), As(III), Cr(III), B(III), Al(III), Se(IV), and Mo(VI). Less than 25 percent inhibition was found for Mn (II) and Pb(II). Inhibition by Co(II), Cu(II), Sn(II), Fe(II), Zn(II), Fe(III), V(IV), and W(VI) was intermediate between these values. *Nitrobacter* populations were less sensitive to trace elements in that inhibition by several of the metals was only detected in one of the soils. Actual percent inhibition for each metal varied with the soil examined. This would be expected in that clay, pH, and organic matter alters the ratio between bound (biologically unavailable) and water soluble (available) metals. The soil pH ranged from 5.8 to 7.8, organic matter from 2.58 to 5.45 percent, and clay from 39 to 50 percent. A similar observation of this variation between soils was reported by Doelman and Haanstra (1979) in their study of the effect of lead amendment of soil respiration and dehydrogenase activity. A concentration of 1500 μg lead/g soil resulted in approximately a 50-percent reduction of respiration in two sandy soil samples. Comparable lead levels in a clay soil reduced respiration by only 15 percent, whereas concentrations as high as 7500 μg/g soil had no effect on respiration of a peat soil. Similar results were observed with dehydrogenase activity.

Chemical speciation was shown through use of computer modeling to explain these types of differences between soils (Lighthart et al., 1983). The chemical speciation of cadmium and copper in two soils was simulated with GEOCHEM, a equilibrium thermodynamic computer model. When the quantity of "free" metal actually present in interstitial soil water was determined, it was found that the percent inhibition of soil respiration occurred at metal concentrations predicted from studies of axenic bacterial cultures. Thus the apparent resistance of the microbial populations in some soils to relatively high metal ion levels is in reality an artifact of the reduction of "free" metal in interstitial water through adsorption or complexation with soil colloids.

Although the discussion thus far has related to soil microbial processes, the actual site of inhibition of these metabolic processes generally involves specific metabolic enzymatic activity. Combination of the trace metal with sulfhydral groups or other reactive protein substituents results in alteration of the protein tertiary structure, hence its ability to interact with its substrate. Other inhibitory interactions include the replacement of a metal cofactor in an enzyme with one that inactivates the enzyme. For example, amendment of soil with lead has been shown to inhibit enzyme synthesis in soil (Cole, 1977). The net synthesis of amylase and α-glucosidase in starch or maltose amended soil is reduced 75 and 50 percent, respectively, by the concurrent or subsequent amendment of the soils with 2 mg lead/g soil. Invertase synthesis was transiently reduced after the lead amendment. In most cases, the enzyme synthesis rate returned to rates of soils not receiving lead amendment within 24 to 48 hours. This recovery follows changes in amylase producing bacterial populations. Immediately following amendment of the soil with lead the population of lead sensitive amylase producing bacteria declined. The subsequent increase in amylase synthesis was accompanied by increases in the amylase synthesizing bacterial populations. The amount of inhibition of enzyme synthesis was related both to the quantity and form of lead amended to the soil.

In most circumstances, any long-term inhibition of soil microbial activity through metal contamination would be considered detrimental to total ecosystem development. But, one example can be presented where it has been suggested that such contamination actually may result in preservation of soil organic matter. Mathur et al. (1979) proposed that copper (<100 kg/ha) may be used to reduce the subsidence of some organic soils (See Chapter 2 for a discussion of Histosol subsidence). They found that carbon dioxide evolution, hence oxidation of the soil organic matter and the resultant soil subsidence, correlated with extractable copper contents. As copper levels increased, carbon dioxide evolution declined. Subsequent investigation demonstrated reduced activities of a variety of soil enzymes which also correlated with residual fertilizer copper levels (Mathur and Sanderson, 1980).

Microbes which survive inhibitory, but not lethal, levels of trace minerals may react in a number of ways to overcome the metal toxicity. These organisms may:

1. **Increase the synthesis rates of those enzymes inactivated with the metals.** This could be a rapid response for those enzymatic activities that are inducible unless the microbes were already producing the enzymatic activity at the maximal metabolic capacity. In that case, increased enzyme synthesis would have to result from selection of individuals capable of greater enzyme synthesis than the general population.
2. **Increase the synthesis of metal inactivating polymers.** In many instances, toxic cations are kept from contact with the microbial cell itself through combination with anionic polymers surrounding the cell, such as microbial capsules and slime layers. Any increase in synthesis of these protective substances, either as the result of induction of new synthesis capacity of selection of "over-producing" variants, would increase the resistance of the microbial cell to the toxicant. As with changes in enzyme synthesis rate, this could be a rapid process were induction of enzyme synthesis involved or a more gradual adaptation if selection of microbial variations occurs.
3. **Selection of microbial variants capable of secreting metal removing polymers.** Although the general microbial population in question may not be

capable of producing metal complexing polymers, some individuals may exist with genetic defects causing over synthesis of some cationic cell polymer. This normally detrimental mutation would become a beneficial survival factor.

4. **Selection of more metal resistant enzymes.** The enzyme active site and tertiary structure is critical for enzymatic activity. Many enzymes are inactivated through combination of the metal with sulfhydral groups within the protein amino acid chain. Substitution of a sulfhydral group with a nonreactive group without significantly altering the protein tertiary structure would increase the protein resistance to metal mediated denaturation. Similarly, modification of the tertiary structure through amino acid substitution in a manner that decreases vulnerability of the metal reactive groups while preserving configurations necessary for enzymatic activity may also provide for metal resistance.
5. **Enter a resting stage or migrate to unaffected area.** The microbe may avoid the toxicant in time or location. Many toxic conditions are transitory in nature. Thus microbes capable of becoming in active through formation of resting structures or via reduce metabolic activity may minimize the impact of a toxic metal on cellular activity. Alternatively, the toxicant may be localized. Then, mobile microbes may increase its survival potential through migration to a nonaffected site.

No single hypothesis explains total survival of microorganisms under conditions that, at least superficially, would appear to be prohibitive to growth and activity. A number of mechanisms may be used by different species within the microbial populations. Actually, it is highly likely that because of the diversity of function of each individual organism (i.e., the variety of enzymatic activities necessary for cell maintenance and growth), two or more of these mechanisms are responsible for survival of the individuals in the populations.

Evidence that the microbial population does indeed respond and adapt to toxic metal conditions in soils is provided by observation of the selection of metal resistant populations in metal contaminated sites and the capability to isolate axenic cultures of metal resistant or tolerant bacteria. Distribution of fungal species along a gradient of copper and zinc contamination around a brass mill has been studied (Nordgren et al., 1983). The soil was a conifer mor soil and maximum copper and zinc concentrations were both approximately 20,000 μg/g dry soil. The frequency of isolation of the genera *Penicillium* and *Oidiodendron* decreased from about 30 and 20 percent, respectively in uncontaminated soil to only a few percent in the highly contaminated soils. *Mortierella* was most frequently isolated from the moderately polluted sites. As with the preceding species, population densities decreased in the highest pollution soils. *Geomyces* and *Paecilomyces* increased in abundance as the metal pollution of the soil increased. Statistical analysis of the data indicated that organic matter content and soil moisture were of little influence on the isolation frequency compared to the heavy metal concentrations. Duxbury and Bicknell (1983) were able to describe bacterial population responses in natural and metal-polluted soils in part by a single exponential equation and *in toto* by the sum of two exponential functions. Bacterial populations from both the soils could be divided into two subgroups, one with a wide range of metal tolerance and the other with a more limited tolerance. Their data suggested that gram-negative bacteria that were multiple drug resistant were the more metal-tolerant. Exceptions related to bacteria tentatively identified as coryneforms that were isolated with nickel supplemented medium. The authors concluded that in general, gram-negative bacteria have greater metal tolerance than to gram-positive bacterial. The multiple drug resistance trait which is carried on bacterial plasmids may related to a major mechanism of metal resistance in bacteria. Both Haefeli et al. (1984) and Bopp et al. (1983) isolated *Pseudomonas* strains whose metal resistance was conferred through possession of a plasmid. Haefeli (1984) isolated a silver resistant strain of *Pseudomonas stutzeri* from a silver mine. The organism contained three plasmids, the largest of which (PKK1) carried the genes specific for silver resistance. Similarly, Bopp et al. (1983) found that chromate resistance of *Pseudomonas fluorescens* was carried on plasmid PLHB1.

10.4. ACID RAIN AND SOIL METAL PROBLEMS

Each of the examples of soil or ecosystem problems resulting from heavy metal pollution or trace mineral nutrient limitations discussed above were reasonably easy to attribute directly to metal–organic matter interactions. A more pervasive metal–organic matter interaction affecting large areas of the landscape is acid rain. The most obvious trait of the rain obviously relates to its pH. The direct effect of the pH factor is most commonly observed with the acidification of lakes and weakly buffered soils. As a result of the adaptability of soil microbial populations to acidic conditions, rain acidity may be the primary cause of alterations in the soil environment, but the primary mediatory of the changes in the biological processes actually may be derived from the changes in mobility trace elements, especially aluminum and related minerals due to the altered soil pH. Hence, acid rain research provides a good example of the complexity and magnitude of environmental problems associated with heavy metals. [For a more detailed review of this subject, see McFee (1980).]

Much of the early interest and concern with acidic precipitation resulted from the pioneering work of Likens and his associates. The basis for regional if not world wide concern for this problem were summarized by Likens and Bormann (1974). They noted that the acidity of precipitation of the northeastern United States averaged about 4 with values as low as 2.1 being recorded. The increase in acidity of precipitation in the region of their study corresponded with augmented use of natural gas and with the installation of particle-removal devises in tall smoke stacks.

Since most societal and economic interest is associated with aesthetic properties of native ecosystems and yields of cropped systems, initial questions of the problems of acid rain related to aboveground changes in the ecosystem. Irving (1983) reviewed the literature reporting effects of acid rain on crop yields. With the majority of the crops species studied, either in field or controlled experiments, overall growth or yield was not affected by simulated acid rain. For those crops whose productivity or growth patterns were affected by the acidic precipitation, both positive and negative responses were recorded. In agricultural systems, aboveground manifestations of acid rain could be considered to be a short-term problem in that the aboveground and, at times, the belowground plant biomass is removed during harvest. A more lasting effect may be observed in native ecosystems. Stability of these sites is dependent upon maintenance of a steady state equilibrium between organic matter inputs and losses. Thus any reduction of the aboveground biomass productivity would be reflected in an eventual depression of colloidal soil organic matter levels. Of parallel interest, from the view of predicting long-term ecosystem stability, is the changes in soil microbial processes resulting from acid precipitation. A severe or even a slight reduction of biogeochemical cycling rates may have lasting effects upon overall ecosystem properties and composition.

Changes in microbial activity could result from either alteration of the soil pH or increased toxic metal mobility. The most severe situation would be if increased acidity resulted in a reduction of the soil pH to the point that a vital microbial process is precluded. This type of reaction may be anticipated with some acid mine drainage problems. This has not been observed to date in soil ecosystems receiving acidic precipiation. More generally with the rainfall problem soil, pH is expected to vary within the range of adaptability of the microbial population. Of greater concern is the potential for accumulation or solubilization of toxic metals, such as aluminum. Thus it could be predicted that reaction rates of microbially catalyzed processes would relate more to trace mineral problems than to direct acidity. Trace mineral mobility could be "buffered" by adsorption on soil clays and organic matter. Therefore, considerable knowledge of the total ecosystem is required to predict the effect of acid rain on basic soil biological properties. Of increasing concern, is the effect of the acid rain on the microsite. There may exist sufficient buffer capacity in the overall soil profile to preclude occurrence of any major modification of microbial activity by the rain, but localized conditions may be such that the microbial activity is reduced. This could be postulated to occur with calcareous soils containing occlusions of partially decomposed organic matter. Overall soil pH would be highly buffered by the carbonate minerals, but should the precipitation reach microsites with large quantities of organic matter with minimal or no carbonates, the localized pH may be reduced sufficiently to inhibit microbial mineralization processes. The assumption underlying metal problems with acid rain is that those minerals normally present in the soil are mobilized as a result of the precipitation induced soil acidification. It must also be remembered that the rainfall itself may carry significant concentrations of trace metals with it which would affect microbial activity (Duce et al., 1975).

A quick perusal of the literature reveals a variety of conclusions relating to the nature of microbial adaptation to acid precipitation. Many papers have been published indicating little or no changes of microbial parameters, but, there are at least as many showing significant, at times major, reductions in key soil microbial processes. For example, Wainwright (1980) exposed samples of soil from a woodland site to high concentrations of heavy atmospheric pollution for a one-year period. The soil pH declined from 4.2 to 3.7; no changes in microbial population density, respiration rate, nitrification rate, or the activity of a variety of soil enzymes were recorded. It should be observed that the soil used was more acidic before amendment than are most agricultural soils and that the decline in soil pH was rather minor. Thus the indigenous microbial populations were most likely already adapted to the acidic soil conditions. Strayer and Alexander (1981a; 1981b) treated soil columns prepared from a forest soil with simulated acid rain (pH 3.2 to 4.1) continuously (100 cm) or intermittently twice weekly over a 19-week period. They found that nitrification in the upper 1.0 to 1.5 cm of the soil columns was inhibited by the continuous exposure to the acid precipitation. The degree of inhibition related to the acidity of the simulated rain. Partial inhibition of nitrification of added ammonium was detected in the intermittently amended soils. No effect by intermittent "rainfall" on nitrification of native ammonium was observed. Carbon mineralization, as estimated by glucose mineralization rates, was reduced by the acid rain treatments. Killham et al., (1983) in their study of a Sierran forest soil (pH 6.4) found maximal effect of acid rain (pH 2.0, 3.0, 4.0, and 5.6) in the surface soils. Only the pH 2.0 rain caused significant inhibition of both respiration and enzyme activities. Some enzyme activities were unchanged, others increased or decreased. Thus the authors concluded that individual microbial processes have differential sensitivities to acid rain and that changes in nitrogen cycle processes are the major mechanism associated with acid rain impact on overall soil microbial activity.

Interpretation of acid precipitation data from the view of predicting the impact on overall or even individual microbial properties is extremely difficult. This results from the extreme capability of microorganisms to adapt to toxic conditions or to moderate these conditions so the toxicity is reduced or eliminated. Studies of individual microbial species or populations are even more difficult in that in soil many microbes are inactive at any given sample time. Thus the active microorganisms may be inhibited, but no effect of the toxicant is detected because most of the individuals of that particular species in the site are in a resting stage. Thus a 10-percent killing of the active population that only comprises 10 percent of the total population results in an overall change of 1 percent. This complexity is increased in situations where the population of interest may enter a resting stage. Thus an even greater number of the individuals *in situ* would still be culturable but inactive.

This suggests that process microbiology may be more useful in studies of acid rain on soil ecosystems. This snyecological research (Tate, 1986) is conducted under the assumption that in reality it is not the identity of the individuals catalyzing a particular process that is important, but rather the magnitude of the process is of primary significance. Thus if a population shift occurs as a result of environmental perturbation but the biogeochemical processes are occurring at the predisturbance level, then the effect of the environmental change would be considered negligible even though the microbial species that are active, are different.

This adaptation of individual microbial populations and within species catalyzing individual processes presents another complication for acid rain research data interpretation. Most microbiological experiments involve amendment of soil samples or eluting soil columns with high levels of acidic solutions. This may be as a single large acid amendment or several, generally large, amendments over several weeks. Periods of several months appear to be the norm for "long-term" experiments; but, native ecosystems develop over thousands of years. Thus even environmental studies of a few decades are "short term" studies in the overall picture. Development of acid soils (Krug and Frink, 1983) and microbial adaptation to acidic conditions are natural processes. For example, the highly leached, acidic soils of the New Jersey Pinelands contain very active mineralizing populations capable of returning nutrients to the active aboveground forest (Tate, 1985). Thus although acid rain is viewed as an acid problem that in soil is probably mediated through mobilization of acid soluble minerals, for determination of long-term impact of this problem, microbial adaptation to acid and toxic minerals must be considered.

10.5. CONCLUSIONS

Trace minerals are a major controlling factor of ecosystem productivity and stability. Although a variety of abiotic phenomena control the distribution of metal cations between plant availability or unavailable forms or water soluble or insoluble complexes, metal transformations and migration between various pools are both enhanced and inhibited through interactions with abiotic and biotic soil organic matter components. Many metal transformations are enzymatically mediated at rates far in excess of chemically catalyzed rates. Obvious aboveground manifestations of problems with trace minerals involve plant death, population shifts, reduced productivity, and discoloration of leaves. But, the impact of metal availability on total ecosystem function may be much more subtle. Low levels of toxic trace metals may reduce soil enzyme activities, thereby slowing growth limiting biogeochemical cycles. Such less obvious metal interactions may have economic impact through reductions of crop yields from maximum potential, cause a slow decline of native ecosystems, or even be an underlying cause of unexplained aboveground population shifts. Transient changes in soil pH may cause obvious declines in soil microbial processes as the result of direct pH interactions with soil microbes or through mobilization of toxic soil metals. These transient declines in microbial activity may not truly reflect long-term problems in that the microbial community is highly adaptable to changes in pH and toxic substances. Experiments must be designed to accommodate this microbial adaptability.

REFERENCES

Akers, H. A. 1983. Multiple hydroxamic acid microbial iron chelators (siderophores) in soils. Soil Sci. 135: 156–159.

Alberts, J. J., J. E. Schindler, R. W. Miller, and D. E. Nutter, Jr., 1974. Elemental mercury evolution mediated by humic acid. Science (Washington D.C.) 184: 895–896.

Alexander, M. 1977. Introduction to soil microbiology. John Wiley & Sons. New York, 467 pp.

Ausmus, B. S., D. R. Jackson, and G. J. Dodson, 1977. Assessment of microbial effects on cadmium-109 movement through soil solumns. Pedobiologia 17: 183–188.

Axley, J. H., J. G. Babish, D. E. Baker, et al., 1985. Criteria and recommendations for land application of sludges in the northeast. Bulletin 851. The Pennsylvania State University. Pennsylvania Agricultural Experiment Station. 94 pp.

Bautista, E. M., and M. Alexander, 1972. Reduction of inorganic compounds by soil microorganisms. Soil Sci. Soc. Am. Proc. 36: 918–920.

Behel, D., Jr., D. W. Nelson, and L. E. Sommers, 1983. Assessment of heavy metal equilibria in sewage-sludge treated soil. J. Environ. Qual. 12: 181–186.

Bewley, R. J. F. 1980. Effects of heavy metal pollution on oak leaf microorganisms. Appl. Environ. Microbiol. 40: 1053–1059.

Bloomfield, C., W. I. Kelson, and G. Pruden, 1976. Reactions between metals and humified organic matter. J. Soil Sci. 27: 16–31.

Bopp, L. H., A. M. Chakrabarty, and H. L. Ehrlich, 1983. Chromate resistance plasmid in *Pseudomonas fluorescens*. J. Bacteriol. 155: 1105–1109.

Boyd, S. A., L. E. Sommers, D. W. Nelson, and D. X. West, 1981. The mechanism of copper (II) binding by humic acid: An electron spin resonance study of a copper (II)-humic acid complex and some adducts with nitrogen donors. Soil Sci. Soc. Am. J. 45: 745–749.

Boyd, S. A., L. E. Sommers, D. W. Nelson, and D. X. West, 1983. Copper (II) binding by humic acid extracted from sewage sludge: An electron spin resonance study. Soil Sci. Soc. Am. J. 47: 43–46.

Bunzl, K. W., W. Schmidt, and B. Sansoni. 1976. Kinetics of ion exchange in soil organic matter. IV. Adsorption and desorption of Pb^{2+}, Cu^{2+}, Cd^{2+}, Zn^{2+}, and Ca^{2+} by peat. J. Soil Sci. 27: 32–41.

Capone, D. G., D. D. Reese, and R. P. Kliene, 1983. Effects of metals on methanogenesis, sulfate reduction, carbon dioxide evolution, and microbial biomass in anoxic salt marsh sediments. Appl. Environ. Microbiol. 45: 1586–1591.

Cline, G. R., P. E. Powell, P. J. Szaniszlo, and C. P. P. Reid, 1982. Comparison of the abilities of hydroxyamic, synthetic, and other natural organic acids to chelate iron and other ions in nutrient solution. Soil Sci. Soc. Am. J. 46: 1158–1164.

Cole, M. A. 1977. Lead inhibition of enzyme synthesis in soil. Appl. Environ. Microbiol. 33: 262–268.

Doelman, P., and L. Haanstra, 1979. Effect of lead on soil respiration and dehydrogenase activity. Soil Biol. Biochem. 11: 475–479.

Duce, R. A., G. L. Hoffman, and W. H. Zoller, 1975. Atmospheric trace metals at remote northern and southern hemisphere sites: Pollution or natural. Science (Washington, D.C.) 187: 59-61.

Duff, R. B., D. M. Webley, and R. O. Scott, 1963. Solubolization of minerals and related materials by 2-ketogluconic acid-producing bacteria. Soil Sci. 95: 105-114.

Duxbury, T., and B. Bicknell, 1983. Metal-tolerant bacterial populations from natural and metal-polluted soils. Soil Biol. Biochem. 15: 243-250.

Elrashidi, M. A., and G. A. O'Connor, 1982. Influence of solution composition on sorption of zinc by soils. Soil Sci. Soc. Am. J. 46: 1153-1158.

Fekete, F. A., J. T. Spence, and T. Emery, 1983. Siderophores produced by nitrogen-fixing *Azotobacter vinelandii* OP in iron-limited continuous culture. Appl. Environ. Microbiol. 46: 1297-1300.

Firestone, M. K. and J. M. Tiedje, 1975. Biodegradation of metal-nitrilotriacetate complexes by a *Pseudomonas* species: Mechanism of reaction. Appl. Environ. Microbiol. 29: 758-764.

Fitch, A., and F. J. Stevenson, 1984. Comparison of models for determining stability constants of metal complexes with humic substances. Soil Sci. Soc. Am. J. 48: 1044-1050.

Flaig, W., H. Beutelspacher, and E. Rietz, 1975. Chemical composition and physical properties of humic substances. In J. E. Gieseking (ed.), Soil Components 1: 1-211. Springer Verlag, New York.

Gadd, G. M., and A. J. Griffiths, 1978. Microorganisms and heavy metal toxicity. Microbial Ecol. 4: 303-317.

Gerritse, R. B., and W. van Driel, 1984. The relationship between adsorption of trace metals, organic matter, and pH in temperate soils. J. Environ. Qual. 13: 197-204.

Ghosh, K., A. Chattopadhyay, and C. Varadachari, 1983. Electron exchange behaviors of humic substances with iron, copper, and manganese. Soil Sci. 135: 193-196.

Haefeli, C., C. Franklin, and K. Hardy, 1984. Plasmid-determined resistance in *Pseudomonas stutzeri* isolated from a silver mine. J. Bacteriol. 158: 389-392.

Hodges, S. C., and L. W. Zelazny, 1983. Interactions of dilute, hydrolyzed aluminum solutions with clays, peat, and resin. Soil Sci. Soc. Am. J. 47: 206-212.

Huisingh, D. 1974. Heavy metals: Implications for agriculture. Ann. Rev. Phytopathol. 12: 375-388.

Inskeep, W. R., and J. Baham, 1983. Competitive complexation of Cd(II) and Cu(II) by water-soluble organic ligands and Na-montmorillonite. Soil Sci. Soc. Am. J. 47: 1109-1115.

Irving, P. M. 1983. Acidic precipitation effects on crops: A review and analysis of research. J. Environ. Qual. 12: 442-453.

James, B. R., and R. J. Bartlett, 1983. Behavior of chromium in soils: V. Fate of organically complexed Cr(II) added to soil. J. Environ. Qual. 12: 169-172.

Jernelov, A., and A.-L. Martin, 1975. Ecological implications of metal metabolism by microorganisms. Ann. Rev. Microbiol. 29: 61-77.

Johnson, F. L., and P. G. Risser, 1974. Biomass, annual net primary production, and dynamics of six mineral elements in a post oak-blackjack oak forest. Ecology 55: 1246-1258.

Jorgensen, J. R., C. G. Wells, and L. J. Metz, 1980. Nutrient changes in decomposing loblolly pine forest floor. Soil Sci. Soc. Am. J. 44: 1307-1314.

Killham, K., M. K. Firestone, and J. G. McColl, 1983. Acid rain and soil microbial activity: Effects and their mechanism. J. Environ. Qual. 12: 133-137.

Kloepper, J. W., J. Leong, M. Teintze, and M. N. Schroth, 1980. *Pseudomonas* siderophores: A mechanism explaining disease-suppressive soils. Curr. Microbiol. 4: 317-320.

Krug, E. C., and C. R. Frink, 1983. Acid rain on acid soil: A new perspective. Science (Washington, D.C.) 221: 520-525.

Levi-Minzi, R., G. F. Soldatini, and R. Riffaldi, 1976. Cadmium adsorption by soils. J. Soil Sci. 27: 10–15.

Liang, C. N., and M. A. Tabatabai, 1978. Effects of trace elements on nitrification in soils. J. Environ. Qual. 7: 291–293.

Lighthart, B., J. Baham, and V. V. Volk, 1983. Microbial respiration and chemical speciation in metal-amended soils. J. Environ. Qual. 12: 543–548.

Likens, G. E., and F. H. Bormann, 1974. Acid rain: A serious regional environmental problem. Science (Washington, D.C.) 184: 1176–1179.

Macrae, I. C., and J. F. Edwards. 1972. Adsorption of coloidal iron by bacteria. Appl. Microbiol. 24: 819–823.

Martin, J. P., J. O. Ervin, and R. A. Shepherd, 1966. Decomposition of the iron, aluminum, zinc, and copper salts or complexes of some microbial and plant polysaccharides in soil. Soil Sci. Soc. Am. J. 30: 196–200.

Mathur, S. P., H. A. Hamilton, and M. P. Levesque, 1979. The mitigating effect of residual fertilizer copper on the decomposition of an organic soil *in situ*. Soil Sci. Soc. Am. J. 43: 200–203.

Mathur, S. P., and R. B. Sanderson, 1980. The partial inactivation of degradative soil enzymes by residual fertilizer copper in Histosols. Soil Sci. Soc. Am. J. 44: 750–755.

McFee, W. W., 1980. Effects of atmospheric pollutants on soils. Environ. Sci. Res. 17: 307–323.

Mills, A. L. 1985. Acid mine waste drainage: Microbial impact on the recovery of soil and water ecosystems. pp. 35–81. In R. L. Tate III and D. A. Klein (eds.), Soil reclamation processes. Marcel Dekker, New York.

Murphy, T. P., D. R. S. Lean, and C. Nalewajko, 1976. Blue-green algae: Their excretion of iron-selective chelators enables them to dominate other algae. Science (Washington, D.C.) 192: 900–902.

Nordgren, A., E. Baath, and B. Soderstrom, 1983. Microfungi and microbial activity along a heavy metal gradient. Appl. Environ. Microbiol. 45: 1829–1837.

Page, A. L., and F. T. Bingham, 1973. Cadmium residues in the environment. Residue Rev. 48: 1–44.

Page, W. J., and M. Huyer, 1984. Derepression of the *Azotobacter vinelandii* siderophore system, using iron-containing minerals to limit iron repletion. J. Bacteriol. 158: 496–502.

Pastor, J., and J. G. Bockheim, 1984. Distribution and cycling of nutrients in an aspen-mixed-hardwood-spodosol ecosystem in northern Wisconsin. Ecology 65: 339–353.

Payne, S. M., D. W. Niessel, S. S. Peixotto, and K. M. Lawlor, 1983. Expression of hydroxamate and phenolate siderophores by *Shigella flexneri*. J. Bacteriol. 155: 949–955.

Robinson, J. B., and O. H. Tuovinen, 1984. Mechanisms of microbial resistance and detoxification of mercury and organomercury compounds: Physiological, biochemical, and genetic analyses. Microbiol. Rev. 48: 95–124.

Schnitzer, M., and S. I. M. Skinner, 1965. Organo-metalic interactions in soils: 4. Carboxyl and hydroxyl groups in organic matter and metal retention. Soil Sci. 99: 278–284.

Sibley, T. H., and J. J. Alberts, 1984. Adsorption of curium-244: Effects of dissolved organics and sediment extractions. J. Environ. Qual. 13: 553–556.

Stevenson, F. J. 1976. Stability constants of Cu^{2+}, Pb^{2+}, and Cd^{2+} complexes with humic acids. Soil Sci. Soc. Am. J. 40: 665–672.

Stevenson, F. J. 1982. Humus chemistry. John Wiley & Sons, New York, 443 pp.

Strayer, R. F., and M. Alexander, 1981a. Effects of simulated acid rain on glucose mineralization and some physicochemical properties of forest soils. J. Environ. Qual. 10: 460–465.

Strayer, R. R., and M. Alexander, 1981b. Effect of simulated acid rain on nitrification and nitrogen mineralization in forest soils. J. Environ. Qual. 10: 547–551.

Strojan, C. J. 1978. Forest leaf litter decomposition in the vicinity of a zinc smelter. Oecologia (Berlin) 32: 203-212.

Tate, R. L. III 1985. Microorganisms, ecosystem disturbance and soil-formation processes. pp. 1-33. In R. L. Tate III and D. A. Klein (eds.), Soil reclamation processes. Marcel Dekker, New York.

Tate, R. L. III 1985. Carbon mineralization in acid, xeric forest soils: Induction of new activities. Appl. Environ. Microbiol. 50: 454-459.

Tate, R. L. III. 1986. Importance of autecology in microbial ecology. In R. L. Tate III (ed.), Microbial Autecology. A Method for Environmental Studies. In Press.

Tyler, G. 1981. Heavy metals in soil biology and biochemistry. In E. A. Paul and J. N. Ladd (eds.), Soil biochemistry 5: 371-414. Marcel Dekker, New York.

Tyler, G., B. Mornsjo, and B. Nilsson, 1974. Effects of cadmium, lead, and sodium salts on nitrification in a mull soil. Plant Soil 40: 237-242.

van Tiel-Menkveld, G. J., J. M. Mentjox-Vervuurt, B. Oudega, and F. K. DeGraaf, 1982. Siderophore production by *Enterobacter cloacae* and a common receptor protein for the uptake of aerobactin and cloacin DF 13. J. Bacteriol. 150: 490-497.

Visser, S. 1985. Management of microbial processes in surface mined land reclamation in Western Canada. pp. 203-241. In R. L. Tate III and D. A. Klein (eds.), Soil reclamation processes. Marcel Dekker, New York.

Vogt, K. A., C. C. Grier, C. E. Meier, and M. R. Keyes, 1983. Organic matter and nutrient dynamics in forest floors of young and mature *Aibes amabilis* stands in western Washington, as affected by fine-root input. Ecol. Monogr. 52: 139-157.

Wainwright, M. 1980. Effect of exposure to atmospheric pollution on microbial activity in soil. Plant Soil 55: 199-204.

Wickliff, C., H. J. Evans, K. R. Carter, and S. A. Russell, 1980. Cadmium effects on the nitrogen fixation system of red alter. J. Environ. Qual. 9: 180-184.

Wildung, R. E., and T. R. Garland, 1985. Microbial development on oil shale wastes: Influence on geochemistry. pp. 107-139. In R. L. Tate III and D. A. Klein (eds.), Soil reclamation processes. Marcel Dekker, New York.

Williams, D. E., J. Vlamis, A. H. Pukite, and J. E. Corey, 1984. Metal movement in sludge-treated soils after six years of sludge addition: 1. Cadmium, copper, lead, and zinc. Soil Sci. 137: 351-359.

Zamani, B., B. D. Knezek, and F. B. Dazzo, 1984. Biological immobilization of zinc and manganese in soil. J. Environ. Qual. 13: 269-273.

Zunino, H., and J. P. Martin, 1977. Metal binding organic macromolecules in soil. 2. Characterization of the maximum binding ability of the macromolecules. Soil Sci. 123: 188-202.

ELEVEN

ORGANIC MATTER AND SOIL PHYSICAL STRUCTURE

Even when our interest in an ecosystem exceeds that of the casual observer, we rarely consider more than its aboveground aspects when characterizing or describing it. A brief perusal of the variety of scientific journals dedicated to ecological reports reveals a preponderance of interest in plant communities and their succession, climatic effects on these communities, and animal populations and their interactions. But, in reality, each of these plant and animal populations literally relies on the existence of a favorable soil structure. With optimal soil structure, a productive, vigorous biotic community develops, but if the structure is destroyed, the living portion of the system concurrently is disrupted. For example, those familiar with desert ecosystems generally can readily expound on the long-term damage that results from vehicular traffic across the desert floor. Destruction of the minimal soil structure associated with desert soils and the interactive biotic communities of desert crusts creates scars in the landscape that endure for decades to centuries. Thus we can conclude that preservation of an optimal soil structure is essential for continued productivity, if not the survival, of the total ecosystem.

Soil structure is essentially a field term descriptive of the gross physical characteristics of a soil profile, including measurement of overall aggregation or arrangement of the soil solids, plus discontinuities or variations in color or composition along the soil profile. Total ecosystem characteristics, especially those relating to the types of biotic communities present and their productivity, are directly affected by soil structure in that this soil property influences such basic soil characteristics as water movement, heat transfer, aeration, bulk density, and porosity. Thus to develop a complete picture of the factors controlling plant community development in both virgin and managed ecosystems, the ecologist must have a sound understanding of soil structure and its associated properties.

Major soil biological processes, including biogeochemical cycling, are also dramatically impacted by soil structural properties. These processes may be stimulated, inhibited, or even precluded at extremes of any of the soil physical state variables listed previously, that is, reduced water movement, high bulk density, anoxic conditions, and so on. For example, aerobic microbial processes as well as growth of plant species or strains whose development require well-aerated soil for root production are severely limited in high moisture containing, heavy clay soils in which all, or most, aggregate development has been destroyed through improper management. Such soils would be characterized by high bulk densities and low oxygen tensions. Water logged, heavy soils are an extreme example of the impact of poor soil structure on soil biological processes and resultant plant community development.

High soil bulk densities that are characteristic of some soils with poor structures may also impede plant root development (Schuurman, 1965; McSweeney and Hansen, 1984) as well as limit aerobic microbial activity. Schuurman (1965) attributed the reduction of oat root development as compaction increased simply to mechanical resistance to root penetration of the soil matrix. In situations where a compacted subsoil was overlaid by a loose top soil, as is frequently found in agricultural soils with a well developed plow pan, penetration of the dense subsoil by the plant roots was precluded presumably by insufficient osmotic pressure in the roots. Not only is soil compaction a problem in agricultural systems, but it is of major importance in reclamation systems and disturbed, native sites, such as along the skid trails in forest soils (Froelich et al., 1985). In a study of means to overcome the impedance of plant community development in compacted minesoils, McSweeney and Hansen (1984) found that specific soil structures were associated with particular mining and reclamation practices. The optimal soil structure for root development consisted of rounded aggregates loosely compressed together, that fall within the size classes currently used for blocklike and polyhedral aggregates. Specific mining procedures were found to favor formation of this soil structure.

Examples of limitations of plant community development by an unstable soil structure aside from soil compaction are found in desert or beach sand dune ecosystems. Because of the lack of cohesiveness among the soil particulates, roots and associated microbial communities generally are disrupted severely by shifting sand particles. Also, because of the low water holding capacity of these soils, biomass production may be limited by the xeric (or dry) conditions. These are the sorts of problems resulting from vehicular traffic on desert soils. Desert crusts are disrupted, thereby reducing natural biogeochemical cycling and moisture retention processes. Thus from this evaluation of sandy soils and that of the clay soils, it can be seen that structureless soils are characterized as loose sandy soils or as massive, irregular, featureless, soil units as might occur with the clay soil alluded to previously.

Development of a favorable soil structure conducive for overall ecosystem stability is a slow, complex process. A number of physical, chemical, and biological factors contribute to the development of soil structure. These include wetting and drying or freezing and thawing cycles, physical activity of plant roots or soil fauna, interactions between charged colloidal particles and sorbed cations, tillage or management practices, and decay of organic matter and microbial activity. Plant roots actually interact with the soil physical particle associations at two levels either to increase soil granule formation or to disrupt this granulation. Direct, positive interactions between soil minerals and plant roots result from the adhesion of soil particles to the roots or with fibrous root systems through trapping of soil particles among the root mass. A favorable, indirect impact of root activity on soil forming processes is derived from the stimulation of soil and rhizosphere microflora by organic carbon substrates entering the soil as root exudates and sloughed root cells. These plant photosynthetic products provide nutrient and energy sources for the microorganisms. As will be discussed later in this chapter, growth of the microbial community, both fungal and bacterial, encourages adhesion of soil physical particles into granules. In contrast to these soil building processes, soil granulation may be reduced by physical penetration of the aggregate itself by the root.

Management procedures designed to optimize the physical, chemical, and biological conditions for improvement of soil structure generally show minimal impact on the soil structure in the short term. Soil structural development may require decades or even centuries to reach the steady-state condition characteristic for the ecosystem (Jenny, 1980). For example, in a study of minesoil development in southeastern Montana, Schafer et al. (1980) found that organic carbon content of the surface 10 cm of the soil profile reached levels found in adjacent undisturbed soils within 30 years, whereas structure development in this soil required 10 to 50 years. Formation processes of subsurface soils were much slower. In the 20 to 50 cm depth, organic matter was estimated to require at least 400 years to reach equilibrium. Structural development below the 10-cm depth was estimated to require 50 to 200 years. The slow maturation of soil is apparently accompanied by changes in the molecular weight distribution of the soil organic matter (Goh and Williams, 1979). In a study of 10 different top soils representative of three soil chronosequences, the proportion of both large molecules (molecular weight > 200,000) and intermediate size molecules (molecular weight = 10,000 to 200,000) in the 0.1 M $Na_4P_2O_7$ and 0.5 M NaOH extractable organic matter increased with increasing soil development. Thus studies of soil management regimes for improvement of soil structure must be long-term experiments with unfortunately a large degree of extrapolation (or perhaps modeling) of the data.

As is demonstrated by a brief examination of essentially any basic soil science text, soil structural analysis is extremely complex both in variety and causative factors. Among the vast array of experiments involving analysis of soil structure are simple descriptions of the specific structure(s) typical of various soil series and the interrelationship of these structures and variations therein on total ecosystem development and productivity. This common evaluation of soil horizon development includes such specialized topics as gley formation, podzolization, and formation of water stable aggregates. Each of these processes depends not only on the presence of organic matter in the soil profile, but also on the active biological metabolism of this organic matter.

Soil structural development and/or modification may be a direct or indirect product of microbial activity. For example, soil bacteria through the production of capsular or slime layer polysaccharides directly encourage development of water stable aggregates. Indirect modification of soil structural properties attributable to the microbe–organic matter interaction is exemplified by gley formation. Gley formation results from a combination of physical, chemical, and biological components. Among the prerequisites for gley formation are (1) at least transiently flooded conditions, (2) presence of microorganisms capable of reducing oxidized sulfur compounds to sulfide, (3) a source of oxidized sulfur, such as sulfate, (4) iron, and (5) an organic carbon substrate to supply carbon and energy to the microbial populations involved. Sulfide is generated biologically via sulfur reduction. Although gley formation is complex and many of the factors involved are as yet little understood, the basic process may be summarized in the following way. Simple aliphatic acids, commonly found in the soil organic fraction, are used as a carbon and energy source during this anaerobic, reductive process. As the microorganisms oxidize the carbonaceous substrates, sulfate is used as a terminal electron acceptor. The product of this reduction is sulfide. This biologically produced sulfide combines chemically with iron forming the sulfide precipitate commonly associated with gley soils.

A basic principle underlying organic matter contribution to soil development is that the greatest impact on soil physical structure by organic matter is derived from the dynamic interactions of various organic matter pools. The presence of the organic matter is not as important to soil structural development as are the biological and chemical modifications of this organic matter. The maximum benefit of soil organic matter on soil structural development therefore results from the interactions of the organic matter–microorganism complex. We could mix an organic residue, such as a grain straw, into a soil and observe some physical benefits of the presence of this crop debris on soil properties. But, in the absence of microbial catabolism of that organic matter, little impact on overall soil structure development would be detected. As microorganisms catabolize soil organic matter, the direct and indirect products thereof are clumping of soil colloidal particles (water stable aggregate formation), dissolution of primary minerals, and stimulation the production of new mineral associations; that is, the general soil structure is altered. This effect of metabolic status of the organic matter was exemplified by work of McCalla (1942) where he evaluated water stable structural development of a loessial subsoil left unamended, or amended with colloidal organic matter (humus suspension) or sawdust. Stability of the soil structure was evaluated with and without a protective cover of straw. Sawdust, a substance which is decomposed slowly in soil (see Chapter 5), did not improve the soil structure, whereas a water stable structure developed in the humus amended soils. The author concluded that optimal soil management would be initially to use the crop residue on the soil surface as a protective mulch. After partial decomposition has occurred, the mulch should be mixed into the soil to facilitate water stable aggregate formation.

As an appreciation of the complexities of soil organic matter and microbial reactions in soil and the nuances of variation in soil structure is gained, it becomes apparent that the topic is much too complex to completely cover in a treatise of this nature. Thus discussion of the topic herein will be limited to a cursory evaluation of such processes as podzolization, gley formation, and mineral dissolution combined with a more detailed study of soil aggregation. This will provide a basic understanding of the organic matter and microbial processes associated with soil physical structural development which may be used for analysis of other soil physical properties.

11.1. BIOLOGICAL REACTIONS CONTRIBUTING TO STRUCTURAL DEVELOPMENT

Many soil maturation processes occur essentially in the absence of measurable quantities of soil organic matter or of significant microbial activities. For example, formation of laterite soils or a variety of clay minerals is attributed primarily to leaching and weathering phenomena. But, most soil structural changes are the direct result of, or the rates of the reactions are greatly accelerated by the presence of soil organic matter plus an active soil microbial community. The best means to envision the microbial contribution to soil structure is to examine the major reactions occurring in soil and determine the microbial products and/or processes which could accelerate or inhibit their occurrence. As was indicated previously, the primary soil formation phenomena to be discussed are podzolization, mineral dissolution, formation of mineral inclusions, and soil aggregation. Podzolization, or the formation of spodosols, results from the mobilization and transport of large quantities of iron and aluminum into the subsoil. Generally, the soil horizon consists of an organic-rich litter layer on the soil surface underlain by a light-colored eluvial horizon. This light colored layer is characterized by having experienced greater iron and aluminum leaching than loss of silicates. Below the light colored eluvial layer is a dark-colored illuvial horizon in which iron, aluminum, and organic matter have accumulated (Stevenson, 1982). Definitions of the other soil forming processes listed are self evident.

The primary source of the fixed carbon, thereby the driving force for the soil biologically catalyzed reactions instrumental in altering basic soil structure, is the aboveground plant community. But, the manifestation of the reaction of these aboveground photosynthetic products and the soil minerals is, for the most part, mediated by the soil microbial community. Thus it could be concluded that the energy for many soil formation processes originates as photosynthetically fixed carbon (solar energy). This linking of higher plant and microbial communities is obvious when soil horizon development (podzolization) is examined in typical forest soils. In many forest ecosystems, with highly productive aboveground communities, a large litter layer develops. Soluble decomposition products of this litter layer are leached into the soil surface where they may interact directly with soil minerals thereby enhancing or reducing their water solubility and subsequent mobility within the soil profile (see Chapter 10 for greater discussion of metal mobility). Alternatively, these organic metabolic products may be metabolically modified by the soil microbial community prior to interaction with the mineral components. Thus the microbial community may mediate soil formation processes at two distinct physical locations: the litter layer, and the underlying soil mineral layer. Origination of soil formation organic compounds in the litter layer results in the nature of the plant community having a direct impact upon the rate and type of soil structure developed. For example, Hanrion et al. (1975) noted that some litters induce mull formation whereas more acidic litters induce mor or moder formation. These workers found that the nature of the soluble organic products of Beech litter also varied with soil type and litter age. Older litters were found to produce leachates with reduced amino nitrogen and water soluble carbon than mull leachates.

Accumulation of an organic layer on the soil surface is not unique to forest soils. In a grassland, a thatch layer frequently develops on the soil surface which enriches for microbial species capable of decomposing plant structural carbon. This is not unlike the situation in the forest litter layer. Again, biodegradation products are leached into the soil surface, which in this ecosystem, is comprised largely of the rhizosphere, where further metabolic transformations may occur. The highly fibrous nature of the grass roots also provides a source of microbial carbon and energy through sloughing of root cells, death and decay of roots themselves, and through production of root exudates. But, it must be stressed that in grassland soils, podzolization does not occur. In this soil, a highly granulated, organic matter rich A-horizon generally develops.

With these undisturbed ecosystems, that is, forests and grasslands, actual mixing of the plant debris within the soil profile is limited to leaching of water soluble, metabolic products into the soil profile or to admixture of litter and soil surface particles through the activity of soil animals, especially earthworms and ants. In contrast, with managed sites, such as agricultural systems, aboveground crop debris is frequently mixed into the surface plow layer at the end of the crop growing season. The predominant effect of this mixing is an acceleration of biomass decomposition processes. Plant biomass is mineralized at rates far in excess of those typical of surface litter layers (Chapter 5). Thus benefits to soil aggregate formation are more transient in the agricultural soils compared to the forest or grasslands and processes requiring leaching of organic acids or chelates into the soil profile are essentially nonexistent.

Evaluation of these four major soil formation process (gley formation, podzolization, mineral dissolution, and aggregation), reveals eight primary biologically catalyzed reactions which enhance and/or mediate soil maturation processes. These are:

1. Catabolism of colloidal soil organic matter
2. Modification of the soil pH
3. Synthesis of chelators

4. Alteration of the soil redox potential
5. Oxidation or reduction of soil cations and anions
6. Synthesis of polysacchrides
7. Physical mincing of organic debris
8. Production of cell or mycelial biomass

Most of these processes are predominantly linked with the microbial community, although physical mincing of the organic debris involves the soil animal population, such as earthworms or ants. This activity was discussed in part in chapter 3. Details of these biological processes from the view of the benefits to the microorganisms themselves were discussed in Chapters 3, 5, and 10. Our discussion here will be limited to a determination of the role of these processes in modification of soil physical structure.

Catabolism of soil organic matter not only has a significant, direct impact on soil physical properties, but there are also many indirect consequences of this catabolism. Microbial metabolism may indirectly cause changes in the soil pH and redox potential, yield metabolic intermediates which may have chelation properties, and alter cation solubility through reduction with electrons derived from oxidation of carbonaceous organic matter. The predominant direct effect of biological metabolism of soil organic matter on soil structure results from the mineralization of those organic elements which link inorganic soil substituents. As will be discussed in the following text in relationship to soil aggregate formation, microbially synthesized polysaccharides are instrumental in cementing mineral colloids together to form the water stable aggregates typically associated with good soil structure. These polysaccharides are biodegradable. Thus microbial populations exist in soil which are capable of catabolizing these polysaccharides thereby disrupting the aggregate structure.

Microbial polysaccharide catabolism provides a good example of the potential for microbial metabolism of soil organic matter to produce the indirect effects on soil structure listed previously. As is discussed in essentially all elementary biochemistry text books, polysaccharides are catabolized aerobically via depolymerization reactions to the respective monomers. These monomers may then be oxidized to carbon dioxide and water or fermented to a variety of acidic and alcoholic products under aerobic and oxygen limiting conditions (Fig. 11.1). Simple oxidation of the monomers results in a reduction, at least at the microsite level, of free oxygen levels. Under conditions where this oxygen is not replenished at rates comparable to its utilization, the oxidation potential of the microsite, and, perhaps if there is sufficient microbial activity, of the macrosite, is reduced. As discussed in Chapter 10, this change in redox potential alters the solubility of a variety of soil cations. As the redox potential of the soil is reduced, total catabolism of the carbohydrate to carbon dioxide and water may not be possible. In this situation, many microorganisms are capable of deriving a portion of the metabolic energy contained in the carbohydrate through partial catabolism of the substrate to a variety of organic intermediates; that is, fermenta-

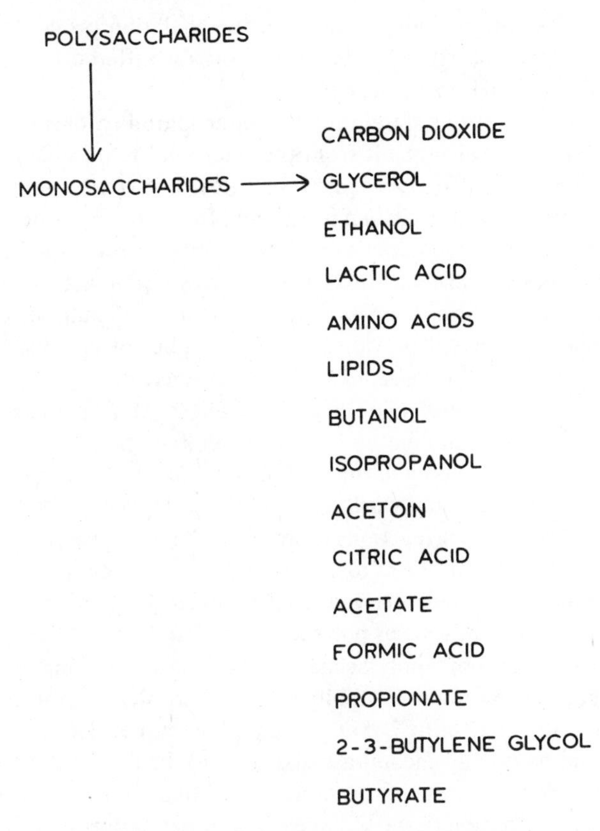

Fig. 11.1.
Potential products of glucose metabolism.

tion products. Under extremely reducing conditions, complete mineralization of carbohydrates results in methane and carbon dioxide production (methanogenesis).

Soil pH is reduced, at least at the microsite, by both incomplete catabolism of the substrate as well as by complete mineralization of the carbohydrate. With total oxidation of the carbohydrate, the carbon dioxide produced may react in water to produce weakly acidic carbonic acid, whereas many of the organic intermediates which may accumulate under oxygen limiting conditions are also weak acids. For example, acetic acid is a common product of oxygen limited carbohydrate metabolism. Some of these organic acids may participate in mineral mobilization as a result of either their acidic or their chelation properties. This may be observed if citric acid is produced as a by-product of carbohydrate catabolism.

Thus mineral dissolution could result from the chelation properties of organic acids accumulated in interstitial water even in highly buffered soils where a measurable reduction of pH does not occur.

Modification of the redox state, thus the water solubility, of soil cations under anoxic conditions may also result from the microbial requirement for electron acceptors during catabolism of organic substrates. With our example of polysaccharide catabolism, once the oxygen is depleted from the microsite, the microorganism requires an electron acceptor for the electrons produced through oxidation of the carbonaceous substrate. A variety of common soil substituents may serve this purpose—nitrate, sulfate (a source of sulfur in sulfide production in gley formation as discussed previously), or a number of cations with variable valences (Fe, Mn, etc.). As the valence of such cations as iron, copper, and manganese is altered, their water solubility also changes. With increased solubility, the cation may become mobile and leach to another point in the soil profile where the redox potential is such that the metal is precipitated again. Chelation complexes, organic acids, and variation in redox state have all been proposed as mediators in podzolization reactions (Davies, 1971; Stevenson, 1982).

Other biological contributions to soil formation which are primarily attributable to microbial activity are derived from biomass production and associated processes. These are synthesis of polysaccharides and production of cellular or mycelial biomass. Microbial polysaccharides and mycelia are major contributors to soil aggregate formation and stabilization. Microbial polysaccharides could be said to provide the "glue" that cements soil mineral particles together, whereas mycelia physically encompass (like a rope) the particles to increase their physical association. Because of the small size of bacterial cells, it is difficult to envision their contribution to the linkage of soil mineral particles, but since they do become bound to clay particles (Marshall, 1969; Marshall, 1971; Stotzky, 1966), the potential exists for at least a minimal role in soil aggregate formation for this major soil biological component.

Water movement throughout the soil profile may be severely restricted through development of high soil bacterial population densities. This could occur at disposal sites for easily metabolizable waste materials, such as is involved in the lagooning of cattle feedlot waste. In this process, sufficient stimulation of the bacterial populations may occur that their cell mass could physically plug soil pores, thereby retarding water movement through the soil profile. This phenomenon was recently quantified in soil columns by Frankenberger and Troeh (1982). They leached saturated soil columns with methanol, propanol, or distilled water for 2000 hours and measured changes in the saturated hydraulic conductivity (K_{sat}). With all three treatments the K_{sat} decreased with time with greater reductions detected in the alcohol treatments than occurred in the distilled water amended soil columns. Initial reductions of K_{sat} were attributed to loss of unstable aggregates and expansion of clays. But the final portion of the declining conductivity-time curves was most probably the result of microbial plugging of soil pores. Pores were postulated to be plugged by microbial growth products and cells. A beneficial aspect of this microbial blockage of soil pores in waste lagooning sites is the reduction of movement of toxic products through the soil profile to ground water. Bacterial blockage of soil pores could also be associated with septic systems in heavy clay soils, land fills, and so on. Furthermore, plugging of pores in rocks by microbial cells may be of major concern in microbial-enhanced oil recovery processes (Jang et al., 1983).

Each of the eight biological processes listed previously occur to some extent in any soil where biological activity is not precluded. The reactions involve active interaction of the biological cell with soil organic and mineral components. Thus it must again be emphasized that for maximal soil structural development, the soil microbial community needs to be actively involved with metabolism of the soil organic matter. Little benefit is derived from the static mixing of organic matter with soil.

These various biological processes are involved to different degrees in the soil formation processes discussed previously. Although overall stability of any ecosystem depends totally upon maintenance of an optimal soil structure, these soil formation reactions are also of major importance in highly managed soils, for example, agricultural soils and reclaimed minelands. In each specific ecosystem, not only aesthetic value, but also economic considerations rely upon the development of an understanding of the soil formation kinetics and the microbial processes affecting this process. Microbial problems associated with mineland reclamation have been recently reviewed (Tate and Klein, 1985). To provide an example of the interactions between the biotic and abiotic soil components for optimization of soil structural development and the management problems associated with developing soils, soil aggregate formation will be discussed herein. Although formation of water stable aggregates does not involve all of the biological reactions presented above, many of them are essential for development of an optimal soil aggregate structure. Thus the principles involved in water stable aggregate formation are applicable to more general reactions associated with soil structural development.

11.2. WATER STABLE AGGREGATE FORMATION

The physical properties of soil water stable aggregates of interest in a study of soil organic matter relate to the importance of biological products in clumping of soil minerals and the stability of these biochemicals to biodegradation and soil management processes. The latter topic is important in predicting long-term benefits from utilization of management procedures which enhance water stable aggregate formation. Biological products may interact with soil minerals on two levels to encourage clumping of soil particles. The organic product may serve as the "glue" which cements the mineral matter together or microbial biomass, fungal mycelia in particular, may be envisioned as forming a "bag-like" structure in which soil particles are trapped or formed into a granular structure. The agronomic, biological, and soil aspects of water stable aggregate formation have been extensively reviewed (Harris et al., 1966; Martin, 1971; Tisdall and Oades,

Table 11.1.
Correlation of Aggregate Stability with Various Soil Organic Matter Fractions[a]

Organic Fraction	Correlation Coefficient (Range)
Total organic matter	0.7439–0.8674
Carbohydrates	0.7106–0.8906
Pyrophosphate extract[b]	0.2253–0.7034
Sodium hydroxide extract 1[b]	0.7463–0.8431
Sodium hydroxide extract 2[b]	0.5175–0.8625

[a]Chaney and Swift, 1984.
[b]Sequentially extracted.

1982). Therefore, this discussion will be limited to an evaluation of the actual biological processes involved in aggregation and, to a limited degree, of the practical management decisions impinging on optimization of soil structure.

11.2.1. Biochemicals Involved in Aggregate Formation

Removal of the soil organic component results in an immediate loss of water stable aggregation. For example, Chaney and Swift (1984) noted significant correlations between aggregate stability and a variety of soil organic matter fractions in several British agricultural soils. Soils were catagorized into six soil groups with a total of 120 soils being analyzed. Correlations between total organic matter and aggregate stability ranged from 0.7439 to 0.8674 with a correlation coefficient of 0.6630 for all soils combined (Table 11.1). Correlations were all significant at the 5-percent level or greater with most being significant at the 1-percent level. High correlations were detected for all soil organic matter fractions. Only one value was found to be not significant (0.2253 for pyrophosphate extract of the Winton soil). This lack of correlation results from the fact that the Winton soil is weakly structured. The correlations between carbohydrate content and aggregate stability are lower than may be expected because of the fact that the polysaccharide values included significant amounts of undecomposed plant structural components which most likely were inactive in terms of aggregate formation and stabilization. Highly significant linear regression equations were developed for two of the soil series evaluated. The following relationships between aggregate stability and organic matter were found for the Steriling and Humbie series, respectively:

$$\text{Aggregate Stability} = 24(\text{Organic Matter}) + 24$$

$$\text{Aggregate Stability} = 30(\text{Organic Matter}) + 49$$

The y-intercepts of 24 and 29 indicate the importance of other factors aside from organic matter in aggregate stability.

Chaney and Swift (1984) also found a highly significant correlation (0.895) between aggregate stability and total nitrogen in the 120 soils evaluated. This high correlation supports the conclusion that readily decomposable organic matter is important in aggregate stability because a major percentage of the soil organic nitrogen is rapidly metabolized by the soil microflora.

The relationship between biodegradation sensitivity of soil organic additives and increases in water stable aggregate formation has been understood for several decades. Martin and Waksman (1941) showed that aggregation effectiveness of soil amendments was dependent on the quantities of readily decomposable constituents contained therein. They noted greater aggregation in soils amended with straw than with farmyard manure. Both soil amendments were more effective in increasing aggregate formation than was lignin. Martin (1942) increased this list of effective soil amendments. The compounds studied listed in descending order of effectiveness were sucrose, cornstalks, straw + clover hay = leaves + timothy hay, salt hay = straw = manure, peat + timothy hay. This relationship of biodegradation susceptibility of organic matter and its impact on soil aggregation was also shown in work by Gilmour et al. (1948). They found that inoculation of soil which had been amended with freshly ground alfalfa with a variety of fungal species increased the degree of aggregation. Inoculation of unamended soils had no effect on aggregation. Effectiveness of the fungal inocula varied with species, type of organic matter added to the soil and the physical composition of the soil.

Griffiths and Jones (1965) also assessed the impact of soil amendments on water stable aggregate formation in soils. A Lincolnshire arable soil and a garden soil were amended with dried, ground fungal mycelia, chitin, glucose, or cellulose, or left unamended (Table 11.2). The data indicate that both the chemical nature and biodegradation susceptibility of the amendment affected aggregation and that the magnitude of the effect of amendment varied with soil type; that is, variations in ratios of soil clays, silts, and sands alter the capacity for the particulates to become associated into aggregates. Cellulose, the most biodegradation resistant substrates used in this study required the greatest time to stimulate aggregation (36 weeks) and with the arable soil was least effective. Maximum impact resulted from soil amendment with fungal mycelia. All substrates were added to the soil at 4 percent (w/w) levels. Thus interpretation of the relationship between the effects of the various amendments is difficult. The higher stimulation by the fungal mycelium compared to the glucose could result either from variation in total carbon added to the soil or from the fact that the fungal mycelia contained all the macronutrients, vitamins, and so on, required for soil microbial metabolism whereas the glucose amendment did not. Martin et al. (1959) also found good stimulation of soil aggregation through amendment with mycelia of a variety of fungal species. Carbon dioxide evolution resulting from mycelia degradation varied as follows: light colored fungi $>$ dark colored fungi

Table 11.2.
Effect of Soil Amendment on Water Stable Aggregate Formation[a]

Amendment	Lincolnshire Arable Soil	Garden Soil
	Maximum Aggregate Stability	
Fungal mycelium	>300 (1)[b]	>300 (1)
Chitin	196 (3)	284 (3)
Glucose	134 (3)	>300 (1)
Grass	142 (36)	153 (3)
Cellulose	29 (36)	230 (36)
Control	5	4

[a]Calculated from Griffiths and Jones, 1965.
[b]Values are weeks of incubation to reach maximum aggregate stability.

= benzene-alcohol extracted light colored fungi > plant residues > benzene-alcohol extracted dark colored fungi. The benzene-alcohol extract was used to evaluate the impact of myalin on decomposition and soil aggregation. Considerable variation in decomposition rates was observed within each of the five groups. Soil aggregation was influenced by the five extraction groups approximately over the same range. Generally, both light colored and myalin containing fungi affected aggregation to the same degree. Pure cultures of *Epicoccum purpurascens* were outstanding in influencing aggregation. Other effective fungi species were *Aspergillus versicolor*, *Diplodina* sp., *Pyrenochaeta* sp., *Stachybotrys atra*, *Stemphylium consortiale*, and *Volutella* sp.

The importance of microbial metabolism of these amendments in augmentation of water stable aggregate formation was demonstrated directly by Harris et al. (1963) and suggested by results of analysis of sugar residues following periodate oxidation of soil sugars (Cheshire et al., 1983). Harris et al. (1963) examined the formation of water stable aggregates in sucrose amended soil incubated either anaerobically or aerobically. Previous work had suggested that anaerobic conditions were contra-indicative of aggregation. Harris et al. (1963) clearly demonstrated that anaerobic conditions were amenable and suggested that failure to observe granulation in previous studies most likely resulted from manipulation of the soil as it was being flooded to induce anaerobic conditions. Soil aggregate stability rests upon both the rate of synthesis of the responsible organic intermediates and their stability or biodegradation resistance. Because of the metabolic limitations imposed by the anaerobic conditions, partially metabolized organic intermediates accumulate in anaerobic soils and these compounds are less likely to be further metabolized then they would be under an oxygen containing atmosphere. Thus aggregate water stability develops at a more rapid rate in the absence of free oxygen and is more durable (Fig. 11.2).

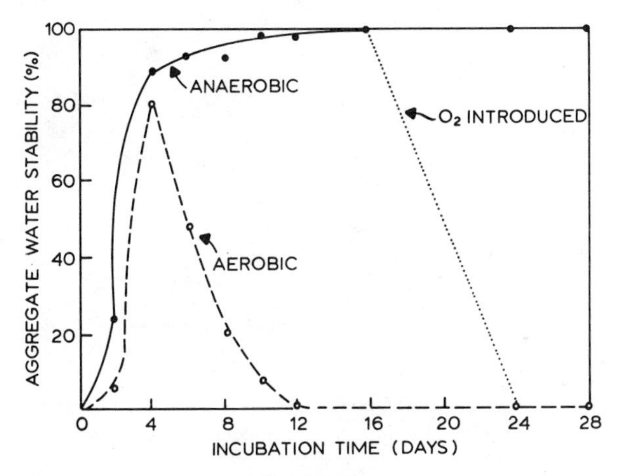

Fig. 11.2.
Changes in water stable aggregates formed in soil amended with sucrose and incubated aerobically or anaerobically. Reproduced from SOIL SCIENCE SOCIETY OF AMERICA PROCEEDINGS, Volume 27, 1963, pp. 542–545 by permission of the Soil Science Society of America.

The anaerobic microorganisms apparently accumulated a stable product which accelerates water stable aggregate formations.

Tisdall and Oades (1982) developed a conceptual picture of the relationship of various plant components and water stable aggregate formation and stability (Fig. 11.3). Readily metabolized polysaccharides, as represented by glucose, cause a rapid increase in granulation followed by as nearly a rapid loss of this structure. In comparison, the more persistent polysaccharide, cellulose, causes a slow accumulation of stable aggregates. The difference in the magnitude of the effect of the two types of polysaccharides likely results from the different quantities of easily metabolizable substrate present in the system at any given time. With easily metabolized polymers, such as amylase, the polymer is rapidly converted to glucose. Thus high concentrations of the microbial substrate, glucose, are available to the microbial community shortly after amendment of the soil with the polymer. In contrast, a variety of enzymes are required to depolymerize cellulose. This is a relatively slow process and is generally the rate limiting step in cellulose mineralization. Thus at any given time, only small quantities of the cellulose derived glucose is available to the microbial population. Ryegrass amendment, which would reflect the conditions encountered with admixture of most perennial crop debris into soil results in a slow but significant increase in aggregate levels. These water stable aggregates formed due to ryegrass amendment are reasonably stable.

Data discussed to this point indicate that the specific organic compound(s) involved in linking soil particulates into granules are at least in part products of the soil biological community and, to some degree, biodegradable. A number of

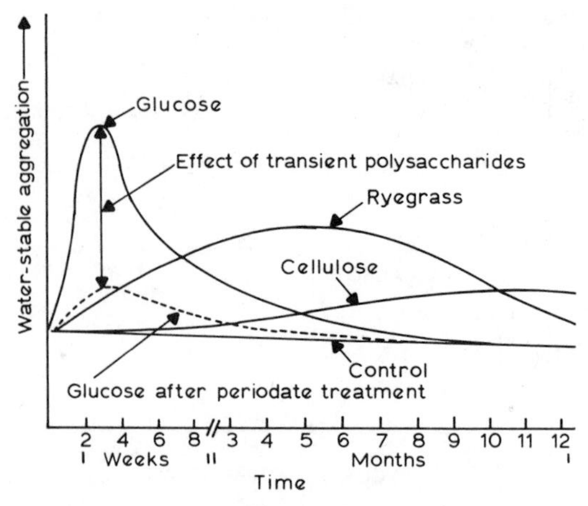

Fig. 11.3.
Alteration of water stable aggregates in soil receiving organic amendments. (Oades, 1982). Reprinted with permission of Tisdall and Pergamon Press.

candidates for these compounds have been proposed, but two classes of compounds, polysaccharides and humic acids, appear to be responsible for the vast majority of the organic matter contribution to soil aggregation properties. This conclusion is based upon results of evaluation of the effects of periodate treatment of soil organic matter on aggregate stability and on the known interactions between humic acids and clay particles.

Periodate disrupts polysaccharide structures through cleavage of carbon–carbon bonds if both contain hydroxyl groups, thereby oxidizing the adjacent hydroxyl groups to aldehydes. If three adjacent carbon atoms with hydroxyl groups interact with periodate, the central carbon atom is released as formate (Fig. 11.4). Measurement of aggregation in soils treated with periodate allows development of models relating soil carbohydrates to granulation. In an example of this type of study, Cheshire et al. (1983) evaluated the effect of duration of periodate treatment on soil carbohydrates and soil structure. A well-structured soil from a field planted to grass was used for the study. Soil samples were untreated or treated with periodate for time periods up to 1176 hours, then the degree of aggregation, total reducing sugar levels, and identity of sugar residues were analyzed. As would be anticipated, increased periodate oxidation decreased the proportion of microaggregates greater than 45 μm and carbohydrate contents. After six hours of periodate treatment, 70 percent of the soil sugars remained unoxidized. After 48 hours, this was reduced to about 45 percent. The residual carbohydrate was enriched in sugars commonly found in plant materials; that is, glucose, arabinose, and xylose. This suggests that microbially synthesized polysaccharides were oxidized preferentially by the periodate. Stability of these sugar residues may be explained structurally; 1 $\rightarrow$ 3 linkages are theoretically resistant to short-term, low concentration periodate oxidation, or physically, oc-

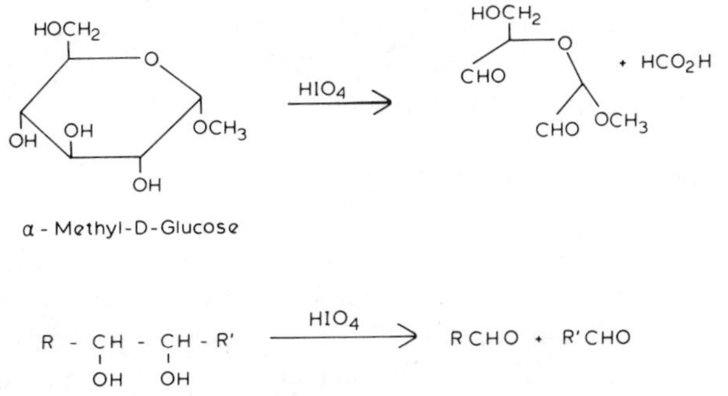

$$\text{α-Methyl-D-Glucose} \xrightarrow{HIO_4} \text{(CHO)(CHO)(OCH}_3\text{) dialdehyde} + HCO_2H$$

$$R-\underset{OH}{CH}-\underset{OH}{CH}-R' \xrightarrow{HIO_4} RCHO + R'CHO$$

Fig. 11.4.
Reactions of periodate with saccharides.

clusion of sugar residues in mineral associations could preclude periodate oxidation. This resistance to periodate oxidation means that some of the earlier data suggesting that polysaccharides are not instrumental in aggregate stability needs to be reexamined in that it has been common to use a six hour incubation period with low periodate concentrations.

Cheshire et al. (1983) also noted that 40-percent disaggregation of the soil resulted from a treatment with sodium chloride-tetraborate. Since during this treatment only 10-percent reduction in soil sugar occurred, another explanation for aggregate disruption is necessary. The authors attributed this disaggregation most probably to result from an electrolytic effect of sodium ions on the soil clays.

In an extension of this study to 15 soils from seven soil series under a variety of cropping systems, Cheshire et al. (1984) found that 15 to 20 percent of the soil carbohydrate fraction was resistant to periodate treatment, even after prolonged reaction times. This residual carbohydrate contained a relatively higher proportion of glucose, arabinose, and xylose than did the susceptible polysaccharides. Monosaccharides which were typical of those derived from microbial sources were preferentially oxidized by the periodate. These residues included mannose, galactose, rhamnose, and fucose. The authors concluded that "over the range measured, aggregate stability was therefore related to the presence of carbohydrate predominantly from microbial sources."

Other soil organic substituents postulated to lead to aggregate formation are the soil humic acids. Humic acids have been shown to react with clays in a manner conducive to the formation of microaggregates (Greenland, 1971; Edwards and Bremner, 1967; Stevenson, 1982). Edwards and Bremner (1967) found that difficult to disperse particles in soils of high-base status were microaggregates ($<259\ \mu$ diameter) consisting largely of clay and humified organic matter. Soil was dispersed by sonic vibration and cation exchange resin techniques. Treatment of the aggregates to replace polyvalent metals by monovalent metals or to destroy the organic matter resulted in a reduction of the quantity of energy needed to disrupt the microaggregate structure. Thus these microaggregates were proposed to result from the linkage of electrically neutral clay minerals and organic particles by polyvalent cations on exchange sites.

11.2.2. Management and Soil Aggregate Structure

Three distinct biological structural levels (plant roots, microscopic fungal mycelia and bacterial cells, and macromolecules) are generally accepted as contributing to soil aggregate formation (Fig. 11.5). Each of these particle organic matter interactions is affected differently by soil management procedures. Processes which encourage plant root development necessarily augment temporary granulation to the maximum levels allowed by the soil physical components. Maximum input of higher plant growth is observed in mature grasslands soils. Conversely, destruction of the plant community, such as bringing grasslands into commercial production, causes a loss in these direct plant-derived benefits. Microbial interactions are stimulated both by increases in root production and exudates from these roots and through the saprophytic attack of senescent plant and animal debris. Exudates from these microbial structures as well as saprophytic conversion of their cell structure to polysaccharides or humic acids contributes to increased aggregation by macromolecules. Of these three levels of aggregate-biomass interactions, the latter is least affected by management procedures. It is relatively easy to visualize destruction of plant root systems and fungal mycelia by working of soil for cropping and through the presence of vehicular traffic. The macromolecular level of interactions where individual biochemicals and soil particles interact is such that physical disruption of structure is unlikely to result from soil physical manipulation.

In an early study, van Bavel and Schaller (1951) found that the nature of the plant community, that is, crop with agricultural systems, has a major impact on soil aggregation. In field plots on Marshall soil located in Clarinda, Iowa, aggregation was approximately twice as great after nearly 20 years of a corn–oats–meadow rotation than it was under continuous corn. Conversion of sod crops to corn resulted in a rapid decline in aggregation, whereas the opposite reaction was noted when cornfields were planted to grass. As would be expected, the best corn yields were found in the better aggregated soils.

Cropping of soils results in a decline in levels of water stable aggregations of virgin soils (Oades, 1984; Powers and Skidmore, 1984). Both studies found that manipulation of soils for crop production resulted in substantial declines in aggregation. Even mildly manipulated soils were found to differ considerably from parallel samples of nondisturbed soils. The greatest differences involved disruption of the large associations derived from the action of plant roots and soil fungi.

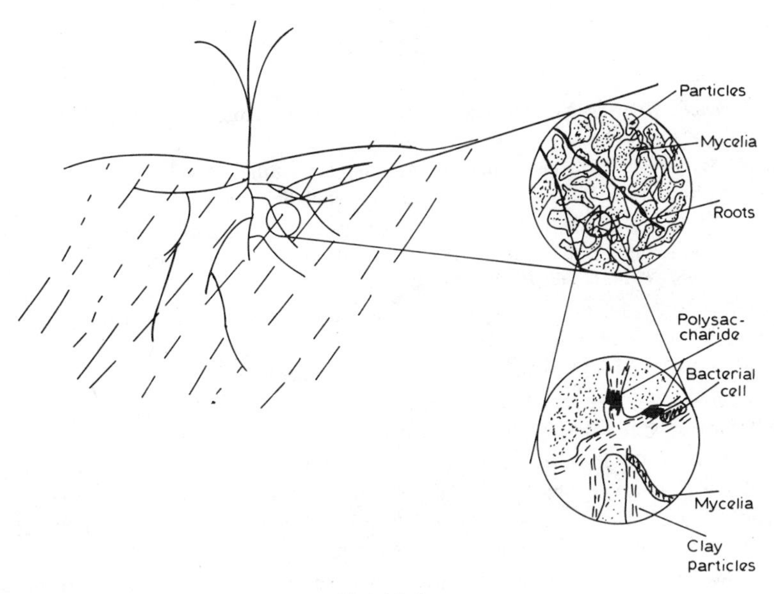

Fig. 11.5.
Plant, microbial, and biochemical interactions leading to water stable aggregate formation.

11.3. CONCLUSIONS

Soil structural development is controlled to a large degree by the levels of organic matter contained therein. Whereas some structural benefits are derived from the presence of organic matter in soil, by far the greatest gain accrues from the microbial metabolism of organic matter entering the soil system. Thus it can be concluded that energy and the fixed carbon associated with soil structural development are derived from solar energy and that the microbial community is the major catalyst of the reactions. Microbial processes instrumental in modification of soil mineral particles and their associations include catabolism of colloidal soil organic matter, modification of soil pH, chelator synthesis, alteration of soil redox potential, direct oxidation or reduction of metalic cations, synthesis of polysaccharides or biomass, and physical mincing of organic debris. Among the processes affected by these microbial community-organic matter actions are mineral dissolution, cation mobility, aggregate formation, and anion levels. Encouragement or disruption of these organic matter-microbial community interactions has a controlling effect on overall ecosystem function, especially those processes associated with agricultural production and land reclamation.

REFERENCES

Chaney, K. and R. S. Swift, 1984. The influence of organic matter on aggregate stability in some British soils. J. Soil Sci. 35: 223-230.

Cheshire, M. V., G. P. Sparling, and C. M. Mundie, 1983. Effect of periodate treatment of soil on carbohydrate constituents and soil aggregation. J. Soil Sci. 34: 105-112.

Cheshire, M. V., G. P. Sparling, and C. M. Mundie, 1984. Influence of soil type, crop and air drying on residual carbohydrate content and aggregate stability after treatment with periodate and tetraborate. Plant Soil 76: 339-347.

Davies, R. I. 1971. Relation of polyphenols to decomposition of organic matter and to pedogenic processes. Soil Sci. 111: 80-85.

Edwards, A. P., and J. M. Bremner, 1967. Microaggregates in soil. J. Soil Sci. 18: 64-73.

Frankenberger, W. T., Jr., and F. R. Troeh, 1982. Bacterial utilization of simple alcohols and their influence on saturated hydraulic conductivity. Soil Sci. Soc. Am. J. 46: 535-538.

Froelich, H. A., D. W. R. Miles, and R. W. Robbins, 1985. Soil bulk density recovery on compacted skid trails in central Idaho. Soil Sci. Soc. Am. J. 49: 1015-1017.

Gilmour, C. M., O. N. Allen, and E. Truog, 1948. Soil aggregation as influenced by the growth of mold species, kind of soil, and organic matter. Soil Sci. Soc. Am. Proc. 13: 292-296.

Goh, K. M., and M. R. Williams, 1979. Changes in molecular weight distribution of soil organic matter during soil development. J. Soil Sci. 30: 747-755.

Greenland, D. J. 1971. Interactions between humic and fulvic acids and clays. Soil Sci. 111: 34-41.

Griffiths, E., and D. Jones, 1965. Microbiological aspects of soil structure. I. Relationships between organic amendments, microbial colonization, and changes in aggregate stability. Plant Soil. 23: 17-33.

Hanrion, M., F. Toutain, and S. Bruckert, 1975. Etude des composes organiques hydrosolubles present dans un sol brun acide et dans podzol sous hetre. I. Evolution comparee. Oecol. Plant 10: 169-185. (in French).

Harris, R. F., O. N. Allen, G. Chesters, and O. J. Attoe, 1963. Evaluation of microbial activity in soil aggregate stabilization and degradation by use of artificial aggregates. Soil Sci. Soc. Am. Proc. 27: 542-546.

Harris, R. F., G. Chesters, and O. N. Allen, 1966. Dynamics of soil aggregation. Adv. Agron. 18: 107-169.

Jang, L. K., P. L. Chang, J. E. Findley, and T. F. Yen, 1983. Selection of bacteria with favorable transport properties through porous rock for the application of microbial enhanced oil recovery. Appl. Environ. Microbiol. 46: 1066-1072.

Jenny, H. 1980. The soil resource. Springer-Verlag, New York, 377 pp.

Marshall, K. C. 1969. Orientation of clay particles sorbed on bacteria possessing different ionogenic surfaces. Biochem. Biophys. Acta 193: 472-474.

Marshall, K. C. 1971. Sorptive interactions between soil particles and microorganisms. In A. D. McLaren and J. Skujins (eds.), Soil Biochem. 2: 409-445. Marcel Dekker, New York.

Martin, J. P. 1942. The effect of composts and compost materials upon the aggregation of the silt and clay particles of Collington sandy loam. Soil Sci. Soc. Am. Proc. 7: 218-222.

Martin, J. P. 1971. Decomposition and binding action of polysaccharides in soil. Soil Biol. Biochem. 3: 33-41.

Martin, J. P., J. O. Ervin, and R. A. Shepherd, 1959. Decomposition and aggregating effect of fungus cell material in soil. Soil Sci. Soc. Am. Proc. 23: 217-220.

Martin, J. P., and S. A. Waksman, 1941. Influence of microorganisms on soil aggregation and erosion: II. Soil Sci. 52: 381-394.

McCalla, T. M. 1942. Influence of biological products on soil structure and infiltration. Soil Sci. Soc. Am. Proc. 7: 209–214.

McSweeney, K., and I. J. Jansen, 1984. Soil structure and associated rooting behavior in minesoils. Soil Sci. Soc. Am. J. 48: 607–612.

Oades, J. M. 1984. Soil organic matter and structural stability: mechanisms and implications for management. Plant Soil 76: 319–337.

Powers, D. H., and E. L. Skidmore, 1984. Soil structure as influenced by simulated tillage. Soil Sci. Soc. Am. J. 48: 879–884.

Schafer, W. M., G. A. Nielsen, and W. D. Nettleton, 1980. Minesoil genesis and morphology in a spoil chronosequence in Montana. Soil Sci. Soc. Am. J. 44: 802–807.

Schuurman, J. J. 1965. Influence of soil density on root development and growth of oats. Plant Soil 22: 352–374.

Stevenson, F. J. 1982. Humus chemistry. John Wiley & Sons, New York, 443 pp.

Stotzky, G. 1966. Influence of clay minerals on microorganisms. III. Effect of particle size, cation exchange capacity and surface area on bacteria. Can. J. Microbiol. 12: 1235–1246.

Tate, R. L. III and D. A. Klein (eds.), 1985. Soil reclamation processes: Microbiological analyses and applications. Marcel Dekker, New York, 349 pp.

Tisdall, J. M., and J. M. Oades, 1982. Organic matter and water-stable aggregates in soils. J. Soil Sci. 33: 141–163.

van Bavel, C. H. M., and F. W. Schaller, 1951. Soil aggregation, organic matter, and yields in a long-time experiment as affected by crop management. Soil Sci. Soc. Am. Proc. 15: 399–404.

TWELVE

MATHEMATICAL MODELING OF SOIL ORGANIC MATTER TRANSFORMATIONS

As the complexity of an ecosystem and the questions asked of that system increase, it becomes essentially impossible to provide a precise picture of processes occurring therein and the environmental factors impinging on those processes. Data interpretation is obscured by system heterogeneity, whereas extrapolation between systems to solve more broadly based environmental problems is risky, to say the least. For example, with many localized agricultural or ecosystem management problems, such as the determination of the effect of a pesticide on organic matter reactions, the variable(s) of interest may be isolated, a few simple experiments conducted, and a reasonably reliable conclusion regarding the fate of the pesticide among the soil organic and mineral components developed. But, as the complexities of the ecosystem or the management problems increase, the probability of developing sufficient understanding of the biotic and abiotic interactions through a few simple experiments to propose management alternatives is greatly reduced, if not impossible.

Two examples of such difficulties may be found in studies of mineland reclamation and with management of agricultural soils to maximize or to maintain soil organic matter levels. Minelands have been and still are reclaimed with procedures based on measurement of a few chemical and physical properties of soils representative of the disturbed sites. In many cases, such management plans have been found to be inadequate when the long-term development of the ecosystem is considered, or as is common, when the system fails. With an increased understanding of the biological phenomena occurring in soils and the problems associated with toxicity of residual metals, some organic compounds, and organic and inorganic acids, failure of past reclamation management plans can now readily be attributed to our limited consideration of and/or appreciation of the basic role of the soil microbial community and soil organic matter in reclamation success (Tate and Klein, 1985).

Similar complexity of the impact of organic and mineral matter interactions on ecosystem development is noted in agricultural soils. Management of these soils to maximize steady state colloidal organic matter contents relies upon application of our basic knowledge of the degradation rate of organic matter produced in the soils as well as that supplied exogenously. The half life of soil organic matter varies with soil moisture, temperature, pH, proportions of sand, silt, and clay comprising the soil, the structure of the organic amendment itself and the variety of components contained therein, as well as the availability of a variety of microbial nutrients. Furthermore, mineralization may be precluded totally by extremes in any one of these factors. To further complicate this soil process, effects of the various environmental factors on organic matter decomposition may be additive. Therefore, decomposition activity may not occur under conditions where it logically might be anticipated. For example, pH stress could be accentuated by reduction of the soil water activity. Or, slight increases in soil moisture in heavy clay soil may greatly retard biological activity where a minimal impact of a similar change in moisture would be found in a silt loam. Thus alteration of management procedures can affect a wide variety of soil physical, chemical and biological properties which control soil organic matter decomposition.

With both the minelands and agricultural soils, a need for a quantitative picture of the soil organic matter processes that can be extrapolated to a number of sites is seen, but the complexity of the soil systems and the types of data commonly collected from various sites precludes or hampers development of such a general concept. A solution to this limited applicability of environmental data is the development of conceptual and mathematical models descriptive of the system(s) of interest. Model construction involves use of many traditional data collection techniques, but the experiments are designed in a manner to allow development of a more generalized understanding of the interactive effect of various soil and biological properties. The model derived from such studies must be sufficiently complex to be realistically descriptive of the ecosystems of interest, but not so complex that the analyses become unwieldy. To date, ecosystem models have been proven useful in research results interpretation, identification of research needs, and predicting system interactions. A successful model could be anticipated to assist the understanding of our underlying assumptions concerning ecosystem function, explain our field observations, and predict needs for further research needs; that is, provide a means of determining what information is needed to increase our knowledge of ecosystem function. The ultimate objective, whether stated or even admitted to by those developing the model, must be to produce a series of mathematical relationships which are descriptive of the processes as they occur *in situ*.

A variety of basic ecological and biological problems have been evaluated with mathematical models ranging from simple linear equations to complex multicomponent mathematical formulations. For example, Meentemeyer (1978) examined factors controlling forest litter decomposition rates to develop a general model of the process. They used stepwise multiple linear correlation regression to relate actual evapotranspiration (AET) (a measure of climate variability), lignin concentration, and AET/lignin concentration interaction to litter decomposition rates. They demonstrated that AET alone accounted for 51 percent of the variability in decay rates, AET/lignin for 19 percent, and lignin concentration for 2 percent. The effect of lignin concentration on the decomposition rate varied with climatic conditions with minimal effect being observed in low AET (but not arid) climates.

Similarly, Paranas (1975) used a model for decomposition rate of plant, animal residues, and soil organic matter to evaluate basic interactions between nitrogen nutrition and decomposition rates. Their model showed that (1) amendment of nitrogen poor soil amendments with nitrogen increased their decomposition rates, (2) the carbon : nitrogen ratio of substrates with ratios greater than a critical 20 to 30 decreased the substrate carbon : nitrogen ratio, (3) with initial carbon : nitrogen ratios below a critical value, no net change in the substrate ratios occurred with time, (4) with initial carbon : nitrogen ratios below the critical value, net mineralization of the organic nitrogen occurs, and (5) amendment of the substrate with ammonium increases the organic nitrogen mineralization rate but not necessarily the net mineralization rate. In a related study, Parnas (1976) evaluated the effect of microbial growth with two limiting substrates on the priming effect. Interactions between the microbial populations, maximal growth yield, and carbon : nitrogen rate were used to explain this interesting soil organic matter phenomenon.

Use of both conceptual and mathematical models in the study of soil organic matter transformations is rapidly increasing. Conceptual models simply involve the qualitative development of a picture or pictures of how various processes and soil factors interact to produce a functional ecosystem. This qualitative analysis may be converted to a quantitative relationship through reduction of the conceptual model to a series of mathematical equations. Conceptual models are useful in experimental design and developing a generalized understanding of ecosystem interactions. Mathematical models allow the quantitative extrapolation between sites and soil types.

Considering the prevalent use of both conceptual and mathematical models in ecological research, a basic understanding of processes involved in model development, validation, and utilization is necessary for those called upon to make decisions regarding optimal management of soil organic matter processes, in light of total ecosystem development. Thus this chapter is prepared with the objectives of providing an outline of the principles involved in developing useful mathematical models for soil organic matter transformations and of indicating the variety of soil organic matter models that have been published. Currently available models range from simple explanations of geological phenomena to complex descriptions of plant debris mineralization. Several of these will be used as examples of the current "state-of-the-art" model building. Those selected as examples were chosen for study for the principles that they exemplify, not necessarily because they are the best models available. Thus the author apologizes to those modelers who believe that their work provides a much better example of the state of the art for mathematical modeling.

12.1. DEVELOPMENT OF MODELS

The pathway for the development of a mathematical model proceeds more or less through a series of discrete steps. The process in question is first organized into a conceptual model. Once a reasonably representative conceptual model has been devised, the relationships may be quantified through conversion of the conceptual picture into mathematical relationships. The resultant mathematical formulae must be validated through comparison of model outputs with field data. In that the products of each of these developmental "steps" are scientifically useful, a given project may include all or just part of this model building process. For example, conceptual models provide valuable assistance in experimental design, whereas mathematical models which have not been validated may spur hypotheses development relating to quantification of factor interactions and general applicability of the model. Thus relationships that fail to simulate conditions generally anticipated to occur under specific environmental conditions could be rejected prior to the validation step.

Conceptual models, although they may seem new to some of us, in reality form a basis of the scientific process. Before a hypothesis can be proposed for testing, there must be some minimal understanding of the system to be studied, albeit a simple chemical reaction or a complex soil community. Whether the objective is to propose hypotheses that can be evaluated experimentally or to develop a mathematical or conceptual model, a number of specific actions are involved in conceptualization of the reactions of interest. First, the boundaries of the area of study must be delineated. These boundaries may include the physical system to be studied, the range of chemical variation to be accepted in the model, as well as time factors. Once the physical and temporal boundaries of the study are chosen, the properties of the ecosystem that are to be included in the study must be selected. Candidates for inclusion include the various physical, chemical, and biological interactions occurring in the ecosystem as effectors of the process rates as well as physical pools of the various biological and chemical entities. If the study of the system is relatively new, a number of variables that are known to change during the time frame of interest may be considered to be constant until the relationship of the process with other, perhaps more important factors, has been modeled. For example, were it necessary to evaluate catabolism of the pesticide glyphosate in soil, it might be profitable to determine its decomposition kinetics in near neutral soils under constant temperature before the effects of temperature or pH are examined. Although *a priori* it may seem reasonable to include as many variables in a model as possible, early simplification of the model could actually assist in avoiding inclusion of factors that have minimal effect on the processes in question. Once the system is defined and the important aspects of the system determined, it becomes important to evaluate the interactions of the model variables and physical and chemical entities; that is, what are the aspects of the model which accelerate or inhibit interactions

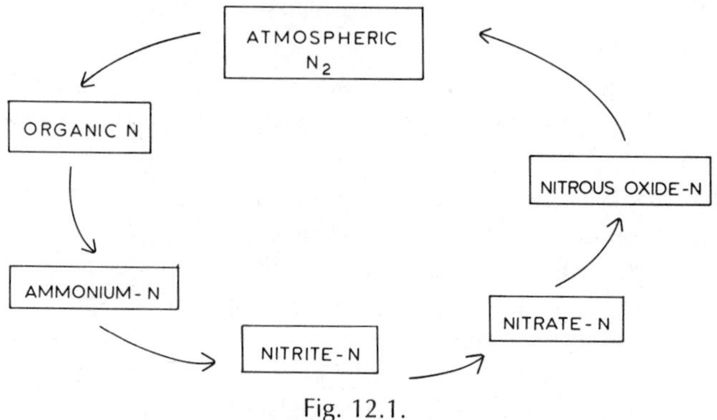

Fig. 12.1.
Simple conceptual model of the nitrogen cycle.

between the various chemical or biological entities. This determination of the interactive aspects of the model is a prelude to conversion of the conceptual model to a mathematical relationship.

A conceptual model is often depicted as a series of boxes connected by straight lines (see Fig. 12.1 as an example). The boxes generally represent various substrate pools or processes, whereas the interconnecting lines designate the interactions, rates, or directions of these processes or nutrient pools. Not all substrate pools or processes need be included in the conceptual model. With the nitrogen cycle model (Fig. 12.1), where it of interest, the organic nitrogen pool could be divided into individual components, individual reactions (including overall processes, such as nitrification or denitrification) could be designated, and inhibitory or stimulatory interactions depicted. Inclusion of any of these items would increase the complexity of the model and perhaps increase the accuracy with which it represents the ecosystem of interest. The associated danger with increased model complexity is that major concepts may be obscured by the overall model detail. Once the overall ecosystem model has been developed to the point that hypotheses may be proposed, the direction of study could involve evaluation of some of the pools (boxes) or the interactions between the pools (the lines). For example, with the model of the nitrogen cycle (Fig. 12.1), the reactions involving nitrogen mineralization could be studied or research could be concentrated on the rate of transfer of nitrogen from the ammonium pool to the nitrate pool. Generally, models of rates of interactions provide the most useful data in ecosystem studies.

Once the conceptual model has been built, if total model development is the objective, mathematical relationships which are descriptive of the processes *in situ* must be determined. Those building models of environmental processes have two options at this point—previously developed mathematical relationships may be adapted to the reactions of question or experimental data may be analyzed statistically through linear or nonlinear regression analysis or related procedures to determine the mathematical equations that fit the data in question.

Although a variety of mathematical equations have been used to represent soil biological processes, the most common mathematical formulation used to quantify the processes appears to be the Michaelis–Menten equation:

$$dS/dt = v = (V_{max}S)/(K_m + S) \quad (1)$$

where $v =$ the reaction rate, V_{max} is the maximum reaction velocity, S is the initial substrate concentration, and K_m is the Michaelis constant, which is equivalent to 1/2 the V_{max}. This equation is useful for describing those reactions that follow second-order kinetics. The numerator and the denominator of this relationship may be modified to account for conditions where the reaction is accelerated or limited by soil chemical and physical properties.

Should the process, at least in the time frame of interest, follow first-order kinetics, then the following relationship is frequently used:

$$dS/dt = -kS_0 \quad (2)$$

where S_0, t, and K represent the starting substrate concentration, time, and a metabolic rate constant specific for the substrate in question, respectively. Other known relationships have been frequently incorporated into mathematical formulations to account for affects of various soil properties on the reaction rate. For example, the Arrhenius or Q_{10} relationships are frequently used for temperature effects on reaction rates.

With the capability of measuring actual field reaction rates along with a variety of physical and chemical parameters, it is also possible to develop linear or nonlinear regression models which represent the field data collected. These equations are quite useful for the ecosystem(s) and conditions under which the supportive data were derived. Frequently, extrapolation or application of the equations to other systems is difficult. This relates in part to the fact that a number of factors affecting the reaction rate usually are combined into single kinetic constant(s) by the regression analysis procedure. Thus, the coefficients and intercepts of the regression model can be anticipated to vary between different ecosystems. Also, in many cases, climatic conditions may be such that a factor that is important in an alternative ecosystem may have minimal impact on the system for which the regression model is developed. For example, extrapolation of biodegradation data from a humid tropical soil to a temperate site would be difficult because in the tropical soil severe temperature limitations may not have been encountered and therefore accounted for in the model. This is exemplified by two simple models derived to predict subsidence of the surface of Histosols with varying water table depths. In South Florida, a subtropical region, soil subsidence is represented by the linear equation:

$$Y_f = (X_f - 2.45)/14.77 \quad (3)$$

where Y_p and X_p are the subsidence rates and water table depths, respectively.

For Histosols in Indiana, subsidence is predicted by the equation:

$$Y_i = (X_i - 9.46)/23.0 \tag{4}$$

where Y_i and X_i represent the subsidence rates and water table depths, respectively (Stephens, 1969). The difference between the two equations results from the impact of climatic variation on the biological processes whose rates are quantified in the rate constants. A more generalized equation may be derived through an increase in the complexity of the mathematical relationships by introduction of terms and/or submodels where the impact of climatic effects are accounted for. Such an model could be that developed by Browder and Volk (1978) to describe subsidence of South Florida Histosols. These workers used the Michaelis-Menten equation and other standard mathematical relationships to describe the biological reactions occurring in Histosols. Although the model was derived and validated using data from South Florida experiments, it is likely that with minor modifications of some of the values of the constants the model would become applicable to Histosols in general.

Once a mathematical approximation has been developed for the system and reactions of interest, the validity of the model in the "real" world needs to be tested. A model that only partially reflects actual field results indicates that some environmental factor(s) other than those chosen need to be included in the study, the equations developed are inappropriate for the process, or unrealistic rate constants were chosen. Also, once the model has been developed it may be possible to delete some factors from the analysis since they may be shown to have an insignificant effect on the final data ouput of the model.

Actual field data are necessary for two steps in the modeling process; assignment of values to the rate constants and validation. Of necessity, the two data pools should be independent and as representative of the actual field situation as possible. In the past, it was necessary to rely significantly on laboratory derived data. This resulted because for determination of the true effect of a single variable on a given reaction rate other factors needed to be constant. Today, it is possible through some of the more sophisticated methods for collecting field data and statistical analysis techniques to possess a data pool representative of actual field rates.

12.2. EXOGENOUS ORGANIC MATTER DECOMPOSITION

Decomposition parameters of organic matter amended to soils relate to a number of practical environmental and agricultural management questions. From the basic ecology viewpoint, the longevity of plant debris, or even decaying root biomass, controls overall ecosystem productivity in a closed system in that the bulk of the nitrogen required for biomass synthesis is derived from mineralization of plant debris. From the view of agriculture, considerable practical interest in this question results from the necessity to minimize application of nitrogenous fertilizers. Decaying crop debris may provide a significant portion of the nitrogen required in subsequent growing seasons. Some information regarding the practical impact of this nitrogen mineralization is gained from in field measurement of organic nitrogen reserves. But, considering the magnitude of world agriculture, a more generalized model capable of predicting organic matter decomposition and subsequent mineral nitrogen availability could be extremely valuable for research, if not, actual field use. As our understanding of soil organic matter transformation processes and the effect of various abiotic parameters on these processes increases, the capability of constructing such models is improving. The following discussion provides examples of some of the organic matter decomposition models that have been reported in the literature and the basic principles associated with their development.

The simplest model representative of the mineralization kinetics of organic substrates amended to soil incorporates the mathematical formulation for first order kinetics. First order kinetics are most readily observed when soil is amended with a pure substrate at enzyme saturating concentrations. The rate of disappearance of the substrate is described by Equation 2. With reactions involving first-order kinetics, there must be no change in the population density of the microbial community catalyzing the decomposition process. The total enzyme concentration must remain constant. This constancy of enzyme concentration is not unusual in soil in that decomposer population densities frequently are limited by other environmental conditions aside from the carbon substrate and many enzymes are stabilized in soil organic matter at sufficient concentrations to minimize the effect of short-term inputs due to increases of microbial population densities. With the generation times of several hours, or even days, that are common among soil microorganisms, quasi first-order kinetics may also be detected with experiments of short duration even under situations where some microbial growth occurs. We must note that in reality the reactions only appear to be first-order because the microbial populations are increasing, albeit at rates too slow for commonly used assay procedures to detect. Thus the actual variation in microbial population, thereby the enzyme concentration, is within the range of variation of experimental data collected on heterogeneous systems and thus not readily quantified. Similarly, first order kinetics are precluded when a metabolic inhibitor is produced during decomposition of the organic substrate, or if the availability of the substrate changes with time. Substrate availability is affected by substrate particle size which may be altered during the incubation period through microbial or microfaunal disruption of particle structure or due to adsorption/desorption phenomena.

If the enzymes responsible for the reactions of interest are not saturated with substrate or if microbial growth occurs, then the reaction rate is generally better represented by the Michaelis–Menten relationship (Equation 1). If enzymes levels increase during the incubation period, simple Monod kinetics may also describe the reaction rate. Monod kinetics are expressed as follows:

$$dB/dt\,(1/B) = \mu_{max} S/(K_s + S) \tag{5}$$

where B is the cell density, S is the substrate concentration, μ_{max} is the maximum specific growth rate, and K_s is the half-saturation rate.

These equations have been used successfully to evaluate mineralization of purified carbonaceous compounds amended to soil. Of more practical significance is the decomposition of a mixture of carbon compounds, such as is found in plant debris. Again, the assumption is that the substrates are added to the soil in sufficiently high concentrations to induce first- or second-order decomposition kinetics. In the former situation, the decomposition rate can be proposed to consist of the summation of the decomposition terms for each individual plant component; that is,

$$dS/dt = K_1S_1 + K_2S_2 + K_3S_3 \dots + K_nS_n \tag{6}$$

where K and S represent the rate constants and initial substrate concentrations for each plant component. This equation is reasonably representative of decomposition kinetics of most types of plant debris.

Many models which are based on this representation of organic matter decomposition contain only two terms, one for easily decomposed components and one for the more biodegradation resistant compounds (e.g., see Hunt, 1977; Jenkinson, 1977; Murayama, 1984). Murayama (1984) evaluated straw decomposition in soil with the following model:

$$Y_t = C_1e^{-K_1t} + C_2e^{-K_2t} \tag{7}$$

where Y_t is the residue remaining at time t (as percent of initial substrate) and C_1 and C_2 are the initial proportions of the substrate which decompose with rate constants of K_1 and K_2, respectively. Initially, the sum of C_1 and C_2 is 100; that is

$$Y_0 = C_1 + C_2 = 100 \tag{8}$$

The first term in Equation 8 represents the decomposition rate for the labile straw fraction and the second term is the rate for the nonlabile fraction. The constants for this equation expectedly would vary with straw source and environmental conditions, with the latter factor having the greatest impact on the decomposition rate.

Hunt (1977) divided decomposing organic matter into two components in a similar manner. The labile pool was considered to be comprised of sugars, starches, and proteins, whereas the slowly decomposing pool contained cellulose, lignins, fats, tannins, and waxes. The proportion of the organic matter remaining on day t (A_t) was defined by Equation 9:

$$A_t = Se^{-kt} + (1 - S)e^{-ht} \tag{9}$$

where S, is the initial proportion of labile material, $(1 - S)$ is the initial proportion of resistant material, and k and h are rate constants for the labile and resistant fractions, respectively. Their analysis of published data suggested that h and k may be assumed not to vary significantly among fresh substrates. Based on this assumption, they developed the equation:

$$S_0 = 0.070 + 1.11\,(N/C)^{1/2} \tag{10}$$

where S_0 is the substrate remaining assuming average values for h and k and N/C is the nitrogen:carbon ratio of the substrate. The values for h and k did vary with partial decomposed materials, such as feces, and humic materials.

Jenkinson (1977) evaluated ryegrass decomposition under field conditions over a 10-year period. He found that the decay curves were represented by a two component first-order model with 70 percent of the grass carbon decomposed with a half-life of 0.25 years and the remainder with a half life of eight years.

These studies demonstrate that within the accuracy range anticipated for complex substrates amended to soil in reasonably high concentrations, a summation of first-order terms gives a good fit for the environmental data. It is reasonable to expect that due to the large concentrations of plant residues added soil and the fact that biodegradation resistant components generally form a minor portion of the plant biomass that early in the incubation period the decomposition curve would reflect predominantly the decomposition of the readily decomposable plant components. As the readily decomposable substrates are depleted, the mineralization rate would reflect the decomposition of the more resistant plant components. Thus in the latter situation, carbon dioxide would be evolved from the soil at a rate determined by the sum of the decomposition rate for biodegradation resistant components plus the slow decomposition of any residual (or occluded when it becomes available) readily decomposable substrate.

Both the first-order and second-order models as represented by the Michaelis–Menten relationship may be complicated by catabolite repression phenomena. For this to occur, decomposition of one or a group of organic carbon substrates is precluded until supplies of more energetically favorable substrate(s) have been depleted. This situation was suggested when Tate (1984) found that amendment of newly established microbial communities in muck soils with glucose inhibited decomposition of aromatic compounds until the glucose pool had been exhausted. A further cause of variation of actual field data from model derived data may result from accumulation of intermediate decomposition products. This would be expected in soils receiving large quantities of readily decomposable substrates because aerobic metabolism could become limited by oxygen diffusion rates in pockets with high concentrations of easily decomposable substrate.

Models discussed to this point were developed under the assumption that the substrate(s) added to the soil are metabolized by the microbial community, at least in part, as a nutrient and energy source and that the substrate is added to the soil in concentrations that meet these nutritional needs. With the current use and disposal of large concentrations of xenobiotics and the occurrence of soil contaminating spills, a recent interest in decomposition kinetics of carbon compounds added to soil in low concentrations—frequently below the levels that would support microbial growth—has developed. Many of these compounds are decomposed cometabolically by the soil microbial community; that is, they are mineralized but no member of the microbial community is capable of gaining a nutrient or energy benefit from their metabolism. Schmidt et al. (1985) developed 12 kinetic models which may be used to describe catabolism of organic substrates that do not support bacterial growth. These models were found to be useful in describing growth when the responsible populations were growing logistically, logarithmically, or linearly, or not increasing in numbers. Following validation of their models, the authors concluded that the mineralization kinetics of organic compounds present at concentrations insufficient to support microbial growth were best described by either first-order models or by models which involved incorporation of growth kinetics of populations growing on alternate substrates. When metabolism of the substrate present in low concentrations occurs at the expense of a second substrate present in growth supporting concentrations, the kinetics observed reflected growth on the substrate present in high concentrations plus the concentration of the substrate of interest.

Brunner and Focht (1984) developed a deterministic kinetic model of organic carbon mineralization which is applicable to either growth or nongrowth soil conditions. The authors believe an advantage of their model is that its mixed-order nature does not require *a priori* assumptions of reaction order, discontinuity period of lag or stationary phase, or correction for endogenous mineralization rates. The model was compared to the Monod Equation 6.

12.2.1. Inclusive Organic Matter Decomposition Models

Several models have been developed in which various aspects of organic matter decomposition and the resultant contribution to plant nutrient pools have been combined. These models are generally both conceptually and mathematically complex. One such model was reported and validated by Smith (1979a; 1979b). The model, an intermediate resolution model of soil organic matter decomposition, was based upon a study of published experimental work. Organic and inorganic forms of soil nitrogen, phosphorus, and potassium were evaluated mathematically. The carbon submodel will be discussed herein as an example of the equations included in this type of model. Since the bulk of the reactions were considered to be microbially catalyzed, and polysaccharides were assumed to be the predominant carbon source, the equations are generally a combination of modified Michaelis–Menten relationships and first-order terms. The conceptual model for the carbon submodel is presented in Fig. 12.2, whereas the definition of the symbols used in the equations are listed in Table 12.1.

In that the bulk of the carbon and energy supply for the soil microbial community is provided by the soil organic matter pool and the productivity of the higher plant and animal community depends at least indirectly upon the activity of soil microorganisms, the rate of changes of this organic matter pool is an important delineator of the ecosystem. Based on the preceding assumption relating microbial activity and soil polysaccharides, the size of the soil polysaccharide pool provides a measure of the overall soil organic matter mineralization rate.

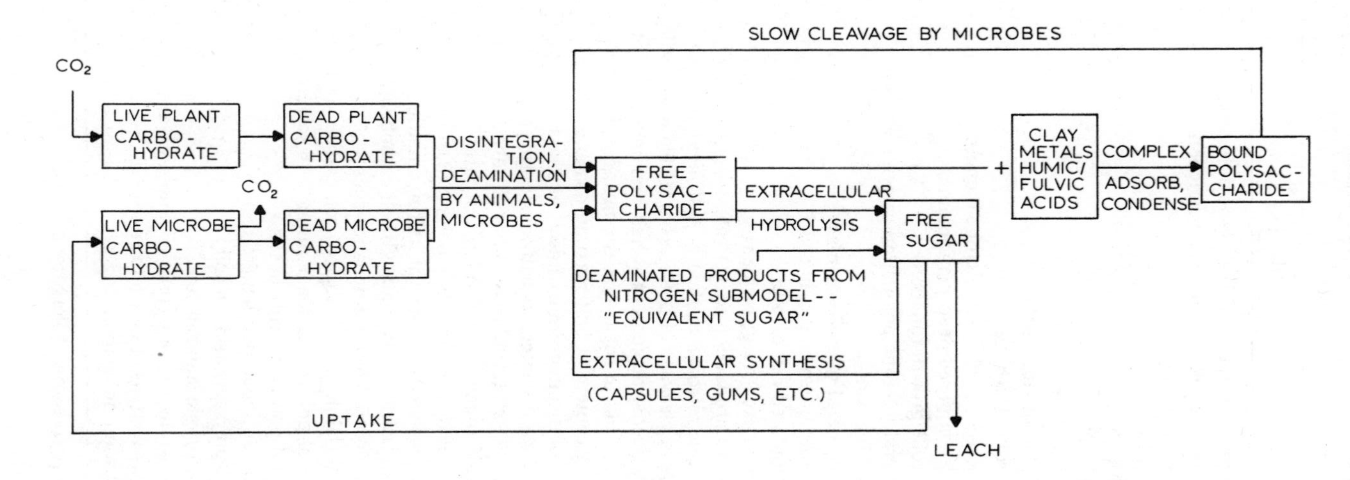

Fig. 12.2.
Conceptual model of the carbon submodel (Smith, 1979a). Reprinted with permission of Pergamon Press.

Table 12.1.
Symbols Used in the Carbon Submodel Model Developed by Smith (1979)[a]

Symbol	Definition
a_i	Parameters in the microbe growth function
b_m	Parameters in microbe death function
β_m	Microbe death rate
C_M	Labile K
C_N	Labile N
C_P	Labile P
C_S	Labile sugar
$C^m{}_s$	Maximum labile sugar
D	Microbe death function
D_m	Undecomposed dead microbe biomass
D_p	Undecomposed dead plant biomass
G	Microbe growth function
K_m	General microbe population carrying capacity
l	Leaching rate
L_m	Live general microbe biomass
N_0	Free organic nitrogen in soil
q_1	Assimilated cell carbon
q_2	Carbon used for growth energy
q_a	Rate of production of equivalent simple sugar
K_a	Organic nitrogen content at half-maximum rate
q_b	Rate of cleavage of polysaccharide from soil colloids
K_b	Polysaccharide content at half-maximum rate
q_c	Rate of use of sugar in waste metabolism
K_c	Sugar content at half-maximum rate
q_d	Rate of disintegration of dead biomass
K_d	Dead mass content at half-maximum rate
q_e	Rate of extracellular synthesis of polysaccharide
K_e	Sugar content at half-maximum rate
q_f	Rate of hydrolysis of free polysaccharide
K_f	Polysaccharide content at half-maximum rate
q_m	Total (labile + assimilated) cell carbohydrate
q_n	Carbon used for maintenance energy
q_p	Total (labile + assimilated) cell carbohydrate
q_r	Rate of polysaccharide bonding with soil colloids
S_b	Bound polysaccharide in soil
S_c	General population labile sugar
S_f	Free polysaccharide in soil
S_s	Free simple sugar in soil
V_s	Maximum rate of uptake of simple sugar
K_s	Sugar content at half-maximum rate
w	Soil solution content

[a]Reprinted with permission of Pergamon Press.

Total rate of change of free polysaccharides (dS_f/dt) was defined by Smith (1979a) to equal the sum of the polysaccharides contained in dead matter, microbially produced polysaccharides plus those polysaccharides previously adsorbed on soil minerals which are slowly released chemically minus hydrolyzed polysaccharides minus those bound to soil minerals and resistant organic matter. Mathematically this relationship is defined as follows:

$$dS_f/dt = q_d L_m(q_m L_m + q_p D_p)/(K_d + D_p + D_m) + q_e L_m S_s/(K_e + S_s) + q_b L_m S_b/(K_b + S_b) - q_f L_m S_f/(K_f + S_f) - q_r S_f \quad (11)$$

Note that all of the terms of this equation take on the form of a Michaelis–Menten equation except the last term which relates to binding of the soil polysaccharides.

Concentrations of bound polysaccharides are generally controlled in equilibrium between the adsorbed and soluble states. The rate of change of concentrations of bound polysaccharides is equal to the difference between the binding rate of the polysaccharides and their rate of release:

$$dS_b/dt = q_r S_f - q_b L_m S_b/(K_b + S_b) \quad (12)$$

Although the saccharides are found in soil organic matter as polymers, it is the simple sugars that the microbial community is mineralizing directly for carbon and energy. The rate of change of this carbon and energy form was described as follows:

$$dS_s/dt = q_f L_m S_f/(K_f + S_f) + q_a L_m N_o/(K_a + N_o) - V_s L_m(1 - C_s/C^m{}_s)S_s/(K_s + S_s) - q_e L_m S_s/(K_e + S_s) - 1S_s \quad (13)$$

The terms for this equation represent that polysaccharide formed from hydrolysis, that from deamination, microbial uptake and metabolism, microbial synthesis, and leaching, respectively. Note that the authors assume that the kinetics of decomposition of those carbon chains produced through deamination of amino acids are similar to those of the simple sugars.

A further source of carbon and energy was labile cell sugar. This soil carbon pool is depicted mathematically as follows:

$$dS_c/dt = V_s L_m(1 - C_s/C^m{}_s)S_s/(K_s + C_s) - (q_1 + q_2)GL_m - q_n L_m - q_c L_m(C_s - q_n)/(K_c + C_s - q_n) - DS_c \quad (14)$$

In this situation, the terms represent microbial uptake, growth consumption, maintenance use, waste metabolism, and death rate loss, respectively.

These reactions are dependent upon the live microbial biomass, which is equivalent to the gross production (GL_m) minus the death loss (DL_m). Since microbial growth rate is dependent upon the relative availability of essential nutrients, growth may be related to these factors as follows:

$$G = a_1 (1 - L_m/K_m)/(a_2/C_S + a_3/C_P + a_4/C_M + a_5/C_N + 1) \quad (15)$$

Through combining mortality factors into the death rate

$$D = \beta_m e^{-c_s q_n}(1 + b_m e^{-100w}) \quad (16)$$

the live microbial biomass was described as follows:

$$dL_m/dt = a_1(1 - L_m k_m)/(a_2/C_S + a_3/C_P + a_4/C_M + a_5/C_N + 1) - \beta_m e^{-c_s q_n} (1 + \beta_m e^{-100w}) L_m \quad (17)$$

These mathematical equations provide a means of quantitatively describing the processes described in the conceptual model. Combination of this submodel with submodels for nitrogen, phosphorus, and potassium plus a previous plant growth model allowed the simulation of complete element cycles within the soil-plant system.

This model developed by Smith (1979a) was modified by Knapp et al. (1983) for their evaluation of the effect of nitrogen on the disappearance of carbon from wheat straw. The conceptual model for the study of Knapp et al. (1983) is presented in Fig. 12.3. These workers examined the early stages of straw decomposition. Since this was a short-term study, lignin was assumed to be completely biodegradation resistant. Where Smith (1979a) relied on Michaelis–Menten kinetics for his model, Knapp et al. found that first order rates were adequate descriptors of their system. This resulted from the slow decomposition rates they observed. Therefore, in the Knapp et al. model, the residue pool rate of change was described as follows:

$$dR_b/dt = DL_m + y_p V_p I_s L_m/(K_p + I_s) + R_d L_m R_b \quad (18)$$

The symbols used in this model are defined in Table 12.2. The terms in this residue pool rate equation represent input dead biomass, rate of polysaccharide production, and rate of residue decomposition, respectively. Similarly, changes in the overall carbon pool are also represented by a mixture of first-order and Michaelis–Menten equations:

$$dS_s/dt = R_d L_m R_b + DI_s L_m - L_m V_s S_s (1 - I_s/S_m)/(K_s + S_s) \quad (19)$$

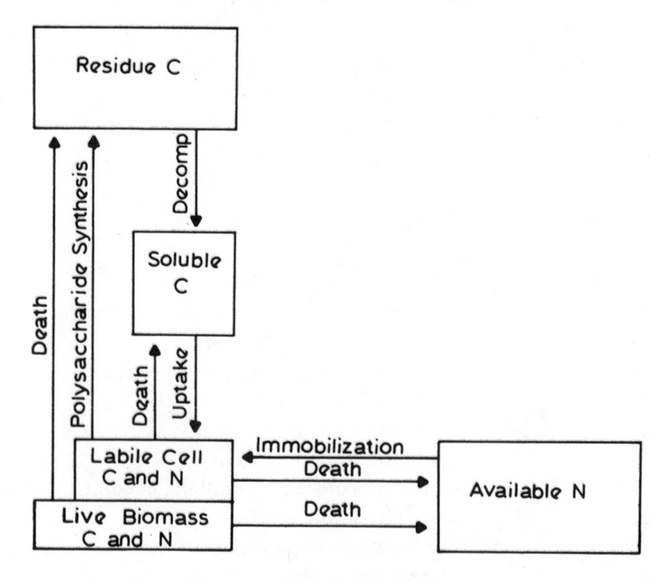

Fig. 12.3.
Carbon and nitrogen pools used for model of straw decomposition. (Knapp et al., 1983). Reprinted with permission of Pergamon Press.

The terms in this equation represent residue decomposition rate, input of labile cell carbon from dying microbes, and rate of carbon uptake by live biomass. The last term is a standard Michaelis-Menten equation with the maximum amount of carbon uptake limited by $(1 - I_s/S_m)$

In a system where the supply of plant debris has a wide carbon : nitrogen ratio, the size of the labile microbial cell pool becomes important. With a nitrogen limited system, decomposition rate of plant residues and nutrient availability for biomass production are controlled to a large part by the release of the nitrogen immobilized in microbial biomass. Knapp et al. (1983) represented the change in the microbial cell pool as follows:

$$\begin{aligned} dS_c/dt = {} & L_m V_s S_s(1 - I_s/S_m)/(K_s + S_s) \\ & + L_m V_g I_s I_n(1 - L_m/M_b)/(K_c + I_s)(K_n + I_n) \\ & - V_r I_s L_m/(K_1 + I_s) - V_p I_s L_m/(K_p + I_s) - D I_s L_m \quad (20) \end{aligned}$$

Terms in this equation represent usage of carbon for live biomass, carbon for growth, carbon for maintenance respiration, carbon for polysaccharide synthesis, and loss of labile carbon from dying microbes, respectively. Note the use of inhibitor and limitor terms in the Michaelis-Menten equations. This equation is the main departure from the Smith model discussed previously. The constants

Table 12.2.
Symbol Definitions for Wheat Straw Decomposition Model (Knapp et al., 1983)[a]

Symbol	Definition
R_b	Crop residue
D	Death rate
L_m	Live biomass
y_p	Yield for polysaccharide
V_p	Maximum uptake rate for polysaccharide
I_s	Cell labile C concentration
K_p	Saturation constant for polysaccharide
R_d	Residue decomposition rate
S_s	Cell labile C
V_s	Maximum uptake of soil C
K_s	Saturation constant for uptake of soil C
S_m	Maximum labile cell C
V_g	Maximum uptake for growth
I_n	Internal cell N concentration
M_b	Maximum live biomass
K_c	C saturation constant for growth
K_n	N saturation constant for growth
V_r	Maximum C uptake for respiration
K_r	C saturation constant for respiration
y	True growth yield
C_2	Carbon dioxide production

[a]Reprinted with permission of Pergamon Press.

for this equation are adjusted so that maintenance has highest priority and polysaccharides the lowest. Choice of values for the constants is one of the more important aspects of model development. Poorly representative constants could result in a model that deviates from *in situ* results. Also, as was done with this model, values of constants may be adjusted to mimic interprocess relationships actually observed. Alternatively, the complexity of the equations may be increased through inclusion of inhibitor or stimulator terms. Note that in equation 20 inhibitor terms are added under the assumption that growth is controlled by both carbon and nitrogen. Thus the growth rate declines when either nutrient source is limiting. Inclusion of the limitor $(1 - L_m/M_b)$ limits biomass to a value of less than M_b. Changes in the live biomass carbon were represented by this group by a mathematical description of the interactions between carbon uptake for growth, true growth yield and carbon loss to death as follows:

$$dL_m/dt = yL_m V_g I_s I_n(1 - L_m/M_b)/(K_c + I_s)(K_n + I_n) - DL_m \qquad (21)$$

Change in live biomass carbon equals the product of the carbon uptake for growth and true growth yield minus the carbon loss to death. This live biomass carbon does not include labile cell carbon or storage carbohydrates. Finally carbon dioxide evolution from the soil was proposed to be described by the following kinetic relationship:

$$dC_2/dt = (1 - y)L_m V_g I_s I_n (1 - L_m/M_b)/(K_c + I_s)(K_n + I_n) + V_r I_s L_m/(K_r + I_s) + (1 - y_p)L_m I_s V_p/K_p + I_s) \qquad (22)$$

where Y_p is equal to the true growth yield for storage carbohydrate. Knapp et al. concluded from this model that the residue decomposition rate was controlled by the microbial biomass produced during the early decomposition phases. The magnitude of this biomass pool appeared to be limited primarily by the initial quantities of readily available carbon and nitrogen in the straw residue.

Both the Smith and the Knapp et al. models are good examples of the complexities associated with just the carbon submodels of these ecosystem inclusive models and the relatively simplicities of the mathematical relationships to describe the reactions kinetics. Although the processes are complex, they are relatively closely approximated by a combination of first-order and Michaelis-Menten relationships. The systems are fine-tuned through the addition of inhibitor or control terms as was pointed out with the Knapp et al. model.

12.2.2. Impact of Abiotic Soil Properties on Model Development

The models discussed thus far involve simple kinetic equations which are not modified to account for effects of such soil factors as pH, temperature, and moisture. These abiotic soil properties could be "ignored" in a simple system or for work involving closely related ecosystems, but if an objective of the study is to produce a generally applicable model, kinetic equations must accommodate the impact of abiotic factors on reaction rates. As with the preceding biochemical process models, general mathematical formulations, such as the Q_{10} or Arrhenius relationships, may be adapted to fit the model conditions or equations may be derived statistically. To exemplify the complexities associated with inclusion of abiotic parameters as rate limitors, several models will again be used to demonstrate how these factors may readily be combined to provide a more general model of biogeochemical cycles.

A good example of modeling the impact of abiotic factors on soil organic matter processes is provided by Bunnel et al. (1977) in their study of the effect of moisture and temperature on microbial respiration and substrate weight loss. Moisture interactions with microbial respiration was expressed by two mathematical relationships. In the simplest situation—one in which microbial respiration varies with soil water content, the changes in respiration was described by mathematical equations which incorporate the following term for moisture effect:

$$M/(a_1 + M) \qquad (23)$$

where M is the percent water content on a dry weight basis, and a_1 is the percent water content at which activity is at half its optimal value. As the soil moisture level reaches saturation, water availability ceases to limit respiration. Under water saturated conditions atmospheric gas exchange controls microbial respiratory processes. Thus the interaction of the microbial respiration with soil moisture is related to the diffusion pathway for gaseous microbial nutrients, which is affected directly by the geometry of the system. Thus the mathematical representation of the interaction varies with the system of study. Bunnell et al. (1977) present a simple relationship involving respiration in a highly organic substrate. In this situation, the Michaelis–Menten equation provides a good approximation of the reaction kinetics as follows:

$$1 - M/(a_2 + M) = a_2/(a_2 + M) \tag{24}$$

where a_2 is the percent water content at which gas exchange is one half maximum.

Bunnell et al. (1977) found that a substrate specific Q_{10} relationship adequately described the temperature effect on microbial respiration. This factor entered their equations with the following formulation:

$$a_3 \times a_4^{(T-10)/10} \tag{25}$$

where a_3 is the respiration rate that would occur at 10°C if neither moisture or free oxygen were limiting, and a_4 is the Q_{10} coefficient. They concluded that although the Arrhenius relationship is commonly used in ecosystem models, it is more applicable to systems involving activity of a single microbial species.

These moisture and temperature relationships were combined into a formulation for microbial respiration as follows:

$$R(T,M) = M/(a_1 + M) \times a_2/(a_2 + M) \times a_3 \times a_4^{(T-10)/10} \tag{26}$$

Each model is only applicable under the specific conditions delimited by the system boundaries selected during initial model conceptualization. Use of the equations beyond this range could lead to erroneous conclusions. For example, it must be noted that this model does not account for an upper lethal temperature or a freezing effect. Similarly, the use of the model is not appropriate at both the high and low moisture extremes.

Hunt (1977) developed the following relationship to fit previously published data on the effect of temperature on decomposition rate:

$$E = e^{(-5.66+0.240\,T-0.00239\,T^2)} \tag{27}$$

where T is the temperature in °C. They assumed, based on published data, that the respiration rate was insignificant near 0°C and that the rate dropped linearly between 38 and 45°C.

Smith (1979a) in the soil organic matter decomposition model discussed previously used a temperature relationship derived by Wildung et al. (1975) combined with some temperature related parameter equations to describe the temperature effect on this important soil degradation process. The relevant equations are as follows:

$$Y = a(1 - e^{bw}) \tag{28}$$

where Y represents soil respiration, w is the moisture content, and a and b are defined as follows:

$$a = 0.47(T - 5)^{1/3} \tag{29}$$

and

$$b = 0.3e^{-0.27T} \tag{30}$$

where T is the temperature in °C.

Cassman and Munns (1980) and Clark and Gilmour (1983) used regression analysis to determine the interactive effects of soil temperature and moisture on organic matter decomposition processes. In their studies, mineralization was evaluated in laboratory studies with controlled temperature and moisture levels. Data were subsequently analyzed to generate mathematical equations predictive of mineralization as a function of temperature and moisture. Clark and Gilmour (1983) used their model to conclude that a single equation relating temperature with the first-order rate constant for organic matter decomposition Equation 2 should not be applied to the decomposition of organic matter in both unsaturated and water saturated soils.

These studies again demonstrate the availability of mathematical relationships, such as the Q_{10} and Arrhenius equations, that may be used for analysis of the effect of abiotic factors on soil biological reactions. Complementing these standard formulae with results from regression analyses provides a significant array of equations which may be used to model basic soil organic matter reactions.

12.3. CONCLUSIONS

Mathematical and conceptual models are useful, at times, even essential, for the evaluation of complex biological and chemical processes occurring in native ecosystems. These models aid in interpretation of experimental results, in identification of addition research needs, and in the prediction of ecosystem interactions. Whereas many current and past models yield results which only partly mimic actual field observations, as the complexity of the mathematical formulations and the basic knowledge pool of processes occurring in the field increase, more representative models are being developed. Keys to improvement of models reside to a large degree with the team assembled. Ideally, a modeling group should include specialists in statistical and mathematical formulation development and use, ecologists or biologists familiar with the ecosystem of interest, plus environmental scientists versed in the physical and chemical reactions occurring therein, as well as meteorologists to assist in assessment of microclimatic interactions with biotic and abiotic properties of the ecosystem. Finally, reliable data must be available for both model development and validation.

REFERENCES

Browder, J. A., and B. G. Volk, 1978. Systems model of carbon transformations in soil subsidence. Ecol. Modelling 5: 269-292.

Bunnell, F. L., D. E. N. Tait, P. W. Flanagan, and K. van Cleve, 1977. Microbial respiration and substrate weight loss. I. A general model of the influences of abiotic variables. Soil Biol. Biochem. 9: 33-40.

Brunner, W., and D. D. Focht, 1984. Deterministic three-half-order kinetic model for microbial degradation of added carbon substrates in soil. Appl. Environ. Microbiol. 47: 167-172.

Cassman, K. G., and D. N. Munns, 1980. Nitrogen mineralization as affected by soil moisture, temperature, and depth. Soil Sci. Soc. Am. J. 44: 1233-1237.

Clark, M. D., and J. T. Gilmour, 1983. The effect of temperature on decomposition at optimum and saturated soil water contents. Soil Sci. Soc. Am. J. 47: 927-929.

Hunt, H. W. 1977. A simulation model for decomposition in grasslands. Ecology 58: 469-484.

Jenkinson, D. S. 1977. Studies on the decomposition of plant material in soil. V. The effects of plant cover and soil type on the loss of carbon from ^{14}C labelled ryegrass decomposing under field conditions. J. Soil Sci. 28: 424-434.

Knapp, E. B., L. F. Elliott, and G. S. Campbell, 1983. Carbon, nitrogen, and microbial biomass interrelationships during the decomposition of wheat straw: A mechanistic simulation model. Soil Biol. Biochem. 15: 455-461.

Meentemeyer, V. 1968. Macroclimate and lignin control of litter decomposition rates. Ecology 59: 465-472.

Murayama, S. 1984. Decomposition kinetics of straw saccharides and synthesis of microbial saccharides under field conditions. J. Soil Sci. 35: 231-242.

Parnas, H. 1975. Model for decomposition of organic material by microorganisms. Soil Biol. Biochem. 7: 161-169.

Parnas, H. 1976. A theoretical explanation of the priming effect based on microbial growth with two limiting substrates. Soil Biol. Biochem. 8: 139-144.

Schmidt, S. K., S. Simkins, and M. Alexander, 1985. Models for the kinetics of biodegradation of organic compounds not supporting growth. Appl. Environ. Microbiol. 50: 323-331.

Smith. O. L. 1979a. An analytical model of the decomposition of soil organic matter. Soil Biol. Biochem. 11: 585-606.

Smith, O. L. 1979b. Application of a model of the decomposition of soil organic matter. Soil Biol. Biochem. 11: 607-618.

Stephens, J. C. 1969. Peat and muck drainage problems. J. Irrig. Drainage Div. Am. Soc. Civ. Eng. 95: 285–305.

Tate, R. L. III 1984. Function of protease and phosphatase activities in subsidence of Pahokee muck. Soil Sci. 138: 271–278.

Tate, R. L., and D. A. Klein (eds.), Soil reclamation processes: Microbiological analyses and applications. Marcel Dekker, New York, 349 pp.

Wildung, R. E., T. R. Garland, and R. L. Buschbom, 1975. The interdependent effects of soil temperature and water content on soil respiration rate and plant root decomposition in arid grassland soils. Soil Biol. Biochem. 7: 373–378.

THIRTEEN

ECOSYSTEM MANAGEMENT AND SOIL ORGANIC MATTER LEVELS

With our increasing awareness of the fragile nature of the ecosystems in which we reside, the need is felt to develop management plans and/or regulations concerning their development and/or recreational utilization. Best management and land use plans are now common topics of discussion for governmental bodies. Unfortunately, many examples exist where such plans have been enthusiastically prepared and implemented only to be shown in subsequent years to be abject failures. Frequently, the inability to achieve project objectives is derived primarily from an earlier dwelling on aboveground processes nearly to the total exclusion of consideration of basic soil reactions and components. The soil organic matter fraction is frequently one of the least appreciated, most severely abused aspects of ecosystem management plans. As our understanding of the role of this soil component in the total ecosystem increases, it becomes apparent that failure to consider soil organic matter as a controlling aspect of the total ecosystem severely reduces the prospects of management or regulatory success.

Ecosystem management is required in sites ranging from intensively cultivated farmlands to the less intrusive modification of more pristine sites, such as national forests and reserves. In parks and forests, soil and plant interactions are frequently controlled with a goal of preserving the aesthetic and/or unique aspects of the ecosystems. This contrasts the highly managed agricultural soils where the overall goal is to maximize crop yields. Along this gradient of degree (and perhaps even desirability) of intervention into the ecosystem is slash and burn agriculture as is practiced in less developed regions of the world. With this form of agriculture, plant nutrients accumulated in soil organic matter under the native vegetation are "mined" for crop production. This plant nutrient source is adequate to support crop yields at the subsistence level. Once the nitrogen reserves of the soil are depleted or reduced below levels sufficient for good crop production, the site is abandoned and a new area cleared. In each of these examples, the soil organic matter pool is manipulated to provide the maximum aesthetic and/or economic benefit.

Whether the project involves preservation of natural sites or increased agricultural production, accomplishment of the objectives requires optimization of soil organic matter levels. This requires integration of the current knowledge regarding this soil fraction into a coherent management plan. Accordingly, this chapter is prepared with the overall objective of evaluating the impact of various soil management procedures on soil organic matter to develop a coherent picture of the role of such procedures in ecosystem preservation, albeit a national forest or farmland. In that most biological, chemical and physical benefits derived from soil organic matter are derived from increasing soil organic matter, the topic will be evaluated by determining the impact of various management practices on the quantities of organic matter in the soil and how the prevailing trend of decreasing soil organic matter reserves in developed soils may be reversed.

13.1. IMPACT OF ECOSYSTEM DISRUPTION AND AGRICULTURAL SOILS

In all cases, the quantities of organic matter present in any given soil are equivalent to the difference between inputs and losses from the soil through biological decomposition and leaching. This may be expressed, from perhaps a somewhat naive but simple viewpoint, mathematically as follows:

$$dS/dt = k_1 I - k_2 D - k_3 L \tag{1}$$

where I, D, and L represent organic matter input, degradation, and leaching, respectively, and k_1, k_2, and k_3 are specific rate constants for the processes specified. In most cases, leaching effects are sufficiently low that this term may be disregarded. Leaching is most significant where soil erosion is a major problem. Control or reduction of this loss of organic matter with top soil reduces, in most cases, the leaching term to insignificant levels. Thus organic matter levels in most well managed sites may be said to be the difference between input and decomposition. As described in Chapter 12, it is obvious that both synthesis and decomposition reactions are controlled by a variety of biological, chemical, and physical soil properties. It is reasonable to conclude that under conditions where biomass synthesis is encouraged by favorable soil and climatic conditions, biological degradation is also maximized. This is seen frequently in tropical and subtropical soils where climatic conditions are highly favorable for biologically catalyzed processes. Thus abundant organic matter input into tropical soils is frequently seen, yet the quantities of organic matter contained therein may not be greater or even as great as that found in temperate regions. This results from the increased decomposition rate of organic matters present in the soils. Similarly, biomass production is reduced under arctic conditions, but so are decomposition processes. Thus the level of organic matter present in the soils reaches an equilibrium controlled by the combination of physical and biological controllers of organic matter transformations characteristic of the specific ecosystem. But, this parallel stimulation or retarding of synthesis and decomposition does not imply that the degradation and input terms of the preceding equation do not vary independently. Whereas the terms may both be affected by a given soil factor, the actual change in the rates may differ. Thus it is reasonable to assume that in the comparison of tropical and temperate soils, the actual final steady-state organic matter level in the soils of the two regions would differ even if all other factors exclusive of climate were identical. Basically, the overall conclusion from these observations is that the combination of positive and negative controllers of soil organic matter reactions define the system rather than, in most cases, the overriding impact of a single factor. Exceptions to this rule involve a few special environments. For example, in highly acidic soils or soils around hot springs, the effect of the extremes of pH and temperature are obvious. In less drastically impacted systems, one or only a few factors may be the major limitation of was shown with Mollisols of the Southern Great Plains (U.S.A.) (Nichols, 1984). In a survey of 65 pedons, organic matter levels were found to relate significantly to soil clay contents and to a lesser extent with precipitation.

This interactive effect of abiotic soil properties was discussed in Chapter 1 from the view of demonstrating that the soil organic matter accumulated in a given soil reached a steady-state level characteristic of that ecosystem. Of interest to this discussion is the impact of intervention into the soil ecosystem on the steady-state level of organic matter. The most commonly discussed observation is the generally, but not universally, observed decline in soil organic matter levels as virgin soil is cleared and brought into agricultural production. Laws and Evans (1949) provide data that are a good example of this phenomenon. Cultivation of a Houston black clay soil for 50 to 90 years resulted in a disruption of the soil physical structure as was shown by decreased soil aggregation, air space and water holding capacity. Changes in these physical properties were accompanied by a 50 percent reduction of organic matter in surface soil samples. Similarly, cultivation of forest soils in Georgia was shown to result in a mean decrease from 3.29 to 1.43 percent organic matter after 25 or more years of cultivation (Giddons, 1957). Also, declines in soil nutrients resulted in Nigerian forest soils following clear cutting and cropping. As would be expected, return of crop residues (maize) to the soil as a mulch reduced the organic matter loss (Ayanaba et al., 1976).

Although there is a decline in the positive aspects of soil physical structure with loss of this colloidal organic matter, in the short term, some positive benefits can be derived from the net mineralization of the soil organic matter reserves. Reinhorn and Avnimelech (1974) found that up to nine tons of nitrogen/ha could be released during the first few years of cropping of a newly cultivated soil. Smith and Young (1975) found that the potential for mineralization of soil organic nitrogen was greater in virgin than in cultivated soils. This could be explained by the greater tendency for the accumulated organic matter to contain larger quantities of less humified organic matter than generally are found in extensively cultivated soils.

This nitrogen liberation following development of virgin forest soils is essential in slash and burn agricultural systems. Thus whereas the drastic reduction of soil organic matter levels which most commonly accompanies development for agriculture is detrimental to overall soil structure, these mineralization processes are essential for this agricultural practice which is predominant in many regions of the world. Whereas mineralization of the soil organic matter reserves allows limited crop production, this type of agriculture can truly be referred to as a mining operation in that the nutrient reserves are rapidly depleted. Sanchez et al. (1983) found in Peruvian that 25 percent of the soil organic carbon and nitrogen were mineralized during the first year of slash and burn agriculture. The mineralization rate approached an equilibrium each year thereafter. Phosphorus and magnesium became deficient during the second year. Calcium became limiting during the first 30 months, and zinc and manganese during the fourth year. Other mineral deficiencies occurred sporadically during the eight-year study.

Fortunately, the reduction of the organic matter of the slash and burn soils, or any other developed soil, is not irreversible. The traditional return of deforested slash and burn sites to forest after crop yields decline allows the gradual return or "resupply" *per se* of the soil "bank" of plant nutrients. Compared with the rapid rate of loss of the nutrient pool, this accumulation of organic matter in the absence of any exogenously supplied organic matter (see Section 13.3) is exceedingly slow. For example, White el a. (1976) found in soils of South Dakota (U.S.A.) that total nitrogen and organic matter increased about 0.001 and 0.02 percent annually in agricultural soils established in pasture. With the commonly observed declines in soil organic matter of 50 or more percent, several decades would be required to return the soil to the original state.

Because of the need to provide food for the world's population, some reduction of the world's soil organic matter reserves is inevitable. Therefore, the best management plan for agricultural soils generally becomes not to avoid reduction of the soil organic matter pool, but rather to minimize the decline. From a more positive view, the objective of the management plan might be to maximize the quantities of organic matter retained as the soils are managed for optimal crop production. Maximization of the steady-state organic matter levels results in both maximizing of soil structural development and in achievement of the ultimate objective of the development project, sustained economical crop yields.

13.2. MANAGEMENT OF ORGANIC MATTER IN FOREST SOILS

The objectives of the management of more pristine systems, such as forests, are quite similar to those encountered with agricultural soils. Because of the normal changes in organic matter involved with ecosystem succession, intervention into the ecosystem is frequently desirable so that the forest can be maintained in a more juvenile state for a longer period than would normally occur. Practical aspects of forest management also relate to the need to increase wood production—a goal similar to traditional agriculture, and to reduce the danger of fire damage to surrounding residential communities. Frequently, reduction of forest litter accumulations lowers the potential for uncontrolled fire damage to both the forest and its surroundings. Since if left undisturbed this litter layer eventually matures into soil organic matter, management of the forest litter layer has a direct impact on soil organic matter and associated biological processes.

Reduction of the organic matter pool through controlled burning can be substantial. Barnette and Hester (1930) quantified the organic matter losses due to prescribed burning in forested lands near Gainesville, Florida (U.S.A.). Comparison with organic matter accumulated in soils that had not been burned over a 42 year period suggested that burning results in annual losses of approximately 5.5×10^4 kg organic matter from the forests. Vance and Henderson (1984) evaluated nitrogen availability following long-term burning in an oak-hickory (*Quercus* spp.-*Carya* spp.) forest. Nitrogen mineralization rates were determined with laboratory incubated soil samples. Burning caused significantly reduced quantities of extractable ammonium. The lower extractable nitrogen concentrations appeared to result from an adverse effect of the long-term burning on substrate quality; that is, mineralization susceptibility of the substrates was reduced by heat induced reactions of the mineralization substrates.

The most dramatic management procedure associated with forests from the view of maintaining the soil ecosystem could be concluded to be those aspects of wood harvest—especially with clear cutting. In this situation, although the final plan may be to replant the tree community, prior to canopy closure and development of those plant and animal communities generally associated with forest evolution, major changes occur in the soil organic matter and microbial populations. Generally, the soil organic matter level associated with the previous existing forest declines and microbial populations involved in the decomposition of this organic matter decomposition increase several fold. Due to the decreased evapotranspiration of the ecosystem and the destruction of the balance between nutrient regeneration and biomass production, excess plant nutrients are lost from the site in drainage waters—frequently to the detriment of surrounding ecosystems. This scenario has been replicated in both tropical and temperate forests.

Cunningham (1963) evaluated the loss of organic carbon, total nitrogen, and total phosphorus from a clear cut forested tropical soil in Ghana over a three-year period. The rate of organic matter decomposition decreased with time. Greatest impact of harvesting of the wood was found in the fully exposed (clear cut) plots. After three years, the fully exposed plots contained less mineral nitrogen, had a lower cation exchange capacity and pH, and contained less exchangeable potassium than did the shaded soil. The clear cut soil also was more compacted. Thus drainage was impeded and soil erosion accelerated.

Evaluation of the overall changes in soil organic matter following clear cutting leads to the *a priori* conclusion that the microbial population densities initially should increase. This would be followed by a decline following exhaustion of the pool of readily decomposable organic matter. Prior to establishment of the precutting conditions in reestablished forests, the microbial populations would be anticipated to decrease to levels significantly lower than existed prior to cutting. As the inputs of organic matter derived from new tree growth increase, these microbial populations will also increase until they approach the preharvest levels. Data collected by Lundegren (1982) on a scots pine stand reflect these population dynamics. Bacterial populations three years after harvesting were lowest in the clear cut plots compared to control plots. Return of the slash to the clear cut plots increased the quantities of organic matter available for microbial nutrition thereby allowing augmentation of the populations compared to those receiving no organic matter inputs.

Likens et al. (1978) studied nutrient dynamics in a northern hardwood forest following clear cutting over a number of years. As a result of the changes in soil physical properties due to removal of the aboveground biomass and the associated reduction in evapotranspiration, leaching of plant nutrients increased several fold following harvest. During the 10 years following cutting, approximately 499 kg nitrate-nitrogen, 450 kg calcium and 166 kg of potassium per hectare were lost in stream waters. This compared to 43, 131, and 21.7 kg per hectare of nitrate-nitrogen, calcium, and potassium, respectively, in an adjacent forested ecosystem. The concentrations and net loss of these nutrients in stream water peaked during the second year of the study, before the onset of regrowth. This suggests that quantities of nutrients lost from the ecosystem were controlled primarily by soil processes rather than by increased incorporation of the nutrients into new plant biomass. Most probably, the decline in nutrient loss from the sites coincided with depletion of the readily decomposed organic matter pools. The study site was an experimentally deforested ecosystem where the wood products were not removed and the forest floor was essentially undisturbed. Also, vegetation regrowth was suppressed for three years. These data were compared to a commercially clear cut northern hardwood forest. As previously stated, maximum nutrient losses in stream waters occurred during the first or second year following cutting. This nutrient leaching was not prevented by natural vegetation regrowth. The authors concluded that the biogeochemical behavior of the commercially cut watershed paralleled the experimentally cut system but was more subdued. This reduced leaching from the commercial system resulted from nutrient assimilation by new biomass. But, due to the reduced organic matter inputs compared to the preharvest condition, a net mineralization of the soil organic matter still occurred. In both the commercial and experimental systems, once the net excess of readily decomposable organic matter was depleted a new steady-state relationship between organic matter inputs and decomposition was established.

The data from both the managed forest ecosystem and the agricultural soils support the premise that each steady-state ecosystem has an organic matter level balanced between inputs (synthesis and amendments) and outputs (decomposition). Disturbance of the system, necessitates an obligatory re-establishment of the steady-state organic matter level. With the ecosystems discussed, this steady-state level is generally less than was found in the system prior to disruption of the native flora. Thus a net production of mineral plant nutrients occurs. These nutrients may be harvested as occurs with subsistence level agriculture or they may be lost in drainage waters. The question thus becomes one of managing the system in a manner to minimize the organic matter loss and to maximize the exploiting of the obligatory nutrient mineralization for subsequent biomass production—albeit reestablishment of the forest site or for agricultural production.

13.3. MANAGEMENT OF SOILS FOR OPTIMAL ORGANIC MATTER ACCUMULATION

Soil organic matter levels may be controlled in developed soils through maximization of native organic matter inputs, minimizing decomposition, or through amendment of the soil with exogenous supplies of organic residues. In each situation, different final steady state levels of organic matter will be achieved. We can conclude that a gradient exists with least colloidal organic matter contained in poorly managed, poorly yielding agricultural soils, and maximal concentrations in those soils where native inputs are augmented with animal or plant residues. Slight benefits in colloidal organic matter levels are gained through maximization of crop yields and significant gains have been achieved through reduction of decomposition processes through minimizing tillage operations. Each of these processes and the origin of the ecosystem benefits will be examined in detail in this section.

13.3.1. Cropping Benefits

As indicated previously, the steady-state level of colloidal organic matter in any soil depends upon the difference between the organic matter input and outputs. These two rates are in turn controlled by the physical and chemical status of the soil. Thus based on the assumptions that (1) plant growth contributes organic carbon to the soil reserves (the minimum quantity being root material plus exudates, and the maximum being total incorporation of a fallow crop), and (2) a standard cropping procedure is associated with crop types and agricultural regions (different cropping practices may be used between fields, but it is reasonable to assume a common management plan within a given field site), it is concluded that optimization of crop yields allows for greater organic matter accumulation than would occur with tillage and limited crop production. The decline of organic matter levels in cultivated soils as compared to undisturbed native ecosystems results from the combined impact of tillage and the change in organic matter inputs. Thus, with a single tillage practice this organic matter supply varies with crop yield and harvest practice (i.e., the quantity of crop residue returned to the soil). Thus it must be concluded that optimal utilization of a tilled agricultural soil must include maximization of crop yields. (This is not to say that the cropping practice which gives the greatest crop yield is necessarily going to be that which results in the highest steady-state soil organic matter levels.) This conclusion has been supported by comparison of soil organic matter concentrations between a variety of conventional management procedures.

Moldenhauer et al. (1967) in a long-term study of interactions between management procedures, infiltration, erosion, and crop yield, found that erosion varied inversely with high nitrogen fertility which was accompanied by trends towards increased colloidal organic matter levels. Morachan et al. (1972) found that return of crop residues to soil could maintain soil organic matter levels. They evaluated organic matter in soils cropped to monocultures of corn (*Zea mays* L.). Yields were maximized through optimization of fertilization. The soil carbon content, wet aggregate stability and water retention rate progressively increased with the rate of addition of organic material over the 13-year duration of the study. Similarly, van Cleve and Moore (1978) found that maximization of soil fertility on aspen (*Populus tremuloides* Michx.) forest soils increased soil biological activity up to 42 and 33 percent for the nitrogen and phosphorus amendments, respectively. Higher soil organic matter levels were proposed to result from the greater aspen growth response to the nitrogen fertilization. The augmentation of the soil biological activity observed in the fertilized soils was attributed to the combined effect of the increased soil organic matter content and the improved soil nitrogen and phosphorus fertility. Odell et al. (1984) also noted in their study of the Morrow Plot soils (Urbana, Illinois U.S.A.) that augmentation of crop yields retarded the loss of soil nitrogen and organic carbon. Each of these studies leads to the conclusion that good farm management, whether for field crops or forest products, can reduce the decline in native soil organic matter levels. The loss of organic matter is not prevented in most soils. Thus if the soil organic matter levels are to be maintained at the highest levels possible consistent with economic crop yields, the effect of tillage on soil organic matter must be reduced and/or the input of organic matter must be increased. Both of these alternatives have been incorporated into current agricultural procedures through no-till agriculture or the utilization of sludge, crop residues, mulches, or manures. Benefits are accrued from both increased organic matter inputs and reduced decomposition losses (Equation 1). Biological decomposition of the soil organic reserves is reduced as the physical disruption of the soil declines through the practice of minimizing tillage, whereas the input term can be greatly increased through amendment of the soil with residues.

13.3.2. Tillage Alternatives

Tillage practices, aside from their association with provision of an adequate medium for crop development and the aesthetic aspects of a clean cultivated field, impact significantly on the quantities of colloidal organic matter retained within the soil matrix. Excessive tillage encourages organic matter decomposition through destruction of natural physical barriers to organic matter decomposition, such as physical occlusion, oxygen limitation due to reduced diffusion into the soil pores, and reduction of soil moisture levels and water holding capacity. Current agricultural practices range from intensive tillage of the soil prior to planting to no-till agriculture. A compromise between the two alternatives involves no-till soil management interspersed with occasional (3-to-5 year intervals) working of the soil to alleviate the soil compaction commonly associated with no-till operations.

The effect of various tillage practices on soil structural properties, especially those associated with crop yields, has been intensively examined. Recent work has been concentrated on optimizing crop cultural conditions while utilizing cropping procedures which minimize disruption of the soil structure. Because of the breadth of the topic and the extent of the effort currently expended to evaluate no-till agricultural management procedures for various crops, the objective of this discussion will be to summarize some of the biological and physical aspects of no-till agriculture and their impact on the quantities of colloidal organic matter contained in the ecosystem. Although variation in some soil physical properties between the extreme of soil management—no tillage and intensive tillage—will be discussed, the primary emphasis of this chapter will involve evaluation of the impact of tillage on soil organic matter transformations—primarily those reactions involved in biogeochemical cycles. The rates of mineralization of major plant nutrients through these cycles provide an indication of the turn-over rates of readily decomposable organic matter. For a more detailed analysis of this agricultural system than can be presented herein, the reader is referred to a treatise edited by Phillips and Phillips (1984). Also, Blevins et al. (1984) have provided an excellent review of the changes in soil properties in no-till agriculture.

13.3.2.1. Soil Physical Properties and No-Till Agriculture

Two major soil properties which strongly influence biological processes, especially overall biomass production—thereby supplies of fixed carbon entering the soil ecosystem—strongly affected by the intensity of soil tillage are soil temperature and soil moisture. Soil temperature and moisture values are the result of both climatic factors and soil surface and profile properties. Increased soil shading through retention or use of mulches as well as increases in the total soil organic matter, which indirectly increase the water holding capacity, dampens the rate of change of soil temperature. The impact of mulches on soil moisture and temperature must be considered in evaluating no-till operations in that this management procedure involves both reduction of breaking up of the soil structure and retention of the soil cover. The soil surface in no-till operations is generally covered with herbicide treated plant populations. Thus in no-till systems, the early spring temperatures are generally reduced and the fall temperatures elevated (e.g., see Unger, 1978).

Similarly, soil moisture increases with the higher water holding capacity of the higher organic matter containing no-till soils. Thus rain water would be anticipated to be less subject to overland flow under no-till management than traditionally is detected in conventional systems. Jones et al. (1969) in a study of tillage and mulching effects on corn yield, soil moisture, and runoff, found that the mulched soils had the lowest runoff and highest crop yields and soil moisture of the soil treatments. Tillage treatments alone had little effect on these variables. Most of the benefit they observed was derived from the mulch. Soil water conserved by the mulched resulted in an average increase in grain yield of 1932 kg/ha. In a related study, Angle et al. (1984) in a study designed to compare the quantities of nutrients and sediments in runoff from conventional and no-till corn (*Zea mays* L.) watersheds observed major effects of soil tillage practices environmental problems. They found over nine times more runoff originated from the conventional till watershed as compared to the no-till operation. This increased overland flow resulted in approximately 40 fold more suspended sediment in the runoff waters of the conventional till compared to the no-till soils. This increased erosion would necessarily be associated with greater regional problems with movement of nutrients from the agricultural soils into surface waters. Increased soil moisture retention in the no-till systems may be either beneficial or detrimental. This increased soil moisture could stimulate crop production and microbial activity during times of droughts, but with excessive rainfall (especially in soils with limited drainage) augmented soil moisture could retard biologically catalyzed processes. For a more detailed review of the interaction of soil moisture and tillage procedures, see Phillips (1984).

An indirect effect of extreme significance relates to the interactive effect of increased soil moisture on soil temperature. The greater the soil moisture, the more the soil temperature changes are buffered. Thus in temperate regions, no-till soils are anticipated to reach temperatures conducive to plant growth (seed germination for annuals) later in the spring than occurs with conventionally managed soils. The magnitude of this temperature effect on biomass production was shown by Rennie and Heimo (1984) when they found that cool soil temperature regimes with initial soil temperature of 5°C rising to 20°C at the heading stage reduced barley growth by approximately one third compared to temperatures in the 15 to 25°C range. Final crop yield was not affected. Gupta et al. (1983) concluded that soil surface temperature variations between different residue and tillage treatements were primarily the result of variation in surface residue covers. The found 12 and 19°C maximum temperature differences during the fall and spring, respectively, for no residue and surface residue treatments for the same tillage condition.

Potter et al. (1985) measured surface soil thermal properties in conventional till, chisel plow, and no-till operations. Thermal diffusivity was significantly greater in no-till systems than in either of the other management systems. This indicated that the thermal conductivity was also greater in the no-till operation. Direct measurement indicated that this value was more than 20 percent greater in the no-till operation. The soil residue cover associated with the no-till operation had a greater impact on soil temperature and heat flux than it did on the soil thermal properties.

13.3.2.2. Nutrient Cycling and Tillage

These examples of the effect of tillage on soil physical properties suggest that major changes in the associated biological properties may also be anticipated to result from reduced soil mixing. It may be concluded *a priori* that reduction of tillage results in augmented colloidal organic matter accumulations in that the soil properties approach those characteristic of the system prior to development. Achievement of the original soil organic matter levels through reduced tillage can not be expected since the fixed carbon inputs from the crop most likely differ from those of the native grassland or forest and working of the soil encourages compaction. Were the system initially a desert, then the combination of water management and soil husbandry may actually increase colloidal soil organic matter pools. Of prime concern to agriculturalists is the effect of reduced soil tillage on plant nutrient availability. Considerable nutrient benefit is derived from the mineralization of soil organic matter residues. But, mineralization, immobilization, and denitrification rates are all modified by tillage practices. With the expense associated with plant nutrient supplies, ideally gains in soil structure through minimization of tillage should be coupled with at least maintenance of plant nutrient dynamics efficiencies associated with more traditional agriculture.

Most of the soil nitrogen pool exists as organic nitrogen (Chapter 1). Variation of the degree of soil tillage alters the size of this pool and the total quantities of nutrients mineralized. Lamb et al. (1985) provided a good example of soil nitrogen losses in a variety of wheat (*Triticum aestivum* L.) tillage systems. The study conducted in western Nebraska on a Duroc loam (fine silty, mixed, mesic, Pachic Haplustolls) was designed to evaluate changes in soil nitrogen in the surface 30 cm of soil for no-till soils, stubble mulch soils, and plow tillages. After 12 years of cultivation soil nitrogen losses compared with the native grassland site was 3, 8, and 19 percent for no-till, stubble mulch, and plow tillages, respectively. The losses of surface soil and the organic matter contained therein due to erosion was minimized by the surrounding grasslands. Stinner et al. (1984) compared nutrient budgets and nutrient cycling in conventional tillage, no-tillage, and old-field ecosystems on the Georgia Piedmont (U.S.A.). Nutrient incorporation into crops, weeds, and old-field plants was the largest flow in each of the systems. Leaching losses were greatest in the conventional tilled systems and least in the old field. No significant differences in leaching losses for potassium, phosphorus, and magnesium were observed between the conventional and the no-tillage systems. Leaching of these nutrients was reduced in the old field ecosystem. As would be anticipated, both the no-till system and the old-field systems retained larger surface litter fractions than did the conventional till. Their data suggested accelerated litter decomposition and subsequent nutrient mineralization occurred in the conventional till soils compared to the other ecosystems. The results of these studies dramatically demonstrate the savings of soil organic reserves as a result of minimizing soil mixing through cultivation.

Keeney and Bremner (1964) evaluated the effect of cultivation on nitrogen distribution in 10 virgin soils and their cultivated analogs. Total nitrogen, nonexchangeable ammonium nitrogen, nonhydrolyzable nitrogen, and hydrolyzable (total, ammonium, hexosamine, amino acid, and unidentified) pools were assessed. They found that cultivation led to significant decreases in all nitrogen forms except nonexchangeable ammonium nitrogen, but had little affect on the percentage distribution of nitrogen amongst the various forms. The average percent loss of total nitrogen due to cultivation was 36.2 percent with average losses of 70.5, 28.9, 29.5, 20.0, 17.1, 3.5, and 0.5 percent for the hydrolyzable nitrogen, amino acid nitrogen, nonhydrolyzable nitrogen, unidentified hydrolyzable nitrogen, hydrolyzable ammonium nitrogen, hexosamine nitrogen, and nonexchangeable ammonium nitrogen, respectively. Lamb et al. (1985) in their study found that the nonexchangeable and exchangeable ammonium nitrogen were not affected by cultivation. They found that the nonhydrolyzable nitrogen fraction of their soils was reduced by all forms of tillage. This fraction accounted for a substantial proportion of the total soil nitrogen loss.

Although tillage practice does have a major effect on root distribution and morphology (Barber, 1971) and the resultant organic inputs from these plant structures, the most studied interaction with soil organic matter affected by tillage practice appears to be the stimulation and/or inhibition of various microbial processes associated with pool size of mineral nitrogen. Due to the economic problems associated with fixed nitrogen supplies, considerable work has been concentrated on evaluation of various aspects of the nitrogen cycle in no-till or minimally tilled soils. Several changes in microbial activities can be predicted to result from the differences in soil physical structure between tilled and non-tilled soils. With the increased inputs of surface mulches and crop debris in the no-till systems, both the organic matter supply term and the decomposition term of Equation 1 would be increased. Thus the steady-state levels degradative microbial activities and associated enzymes are increased in the no-till systems as compared to conventional tilled sites. This is not to say that the peak in activity in conventional tilled sites following incorporation or crop debris may not exceed the overall activity in the no-tilled soils, but the mean activity for the growing season would of necessity be elevated in the absence of tillage. Doran (1980a) in a study of surface soils from long-term no-till and conventional tillage plots from seven locations (U.S.A.) found that microbial counts of aerobic microorganisms, facultative anaerobes, and denitrifiers increased in surface soils of no-till plots as compared to the parallel plowed soils. Phosphatase and dehydrogenase activities also increased as did the water contents and organic carbon and nitrogen contents of the no-till soils. These changes were noted specifically in the surface soils where the impact of cultivation would be maximized. In the 7.5 to 15 cm and 15 to 30 cm depths, the trends were reversed. All the measured activities were actually higher in the conventional tilled soils. This would be anticipated in that reduction of the tillage causes changes in soil air diffusion rates and organic matter distributions such that these factors would be expected to become limiting in the deeper layers of the no-till soil profiles.

In a related study, Linn and Doran (1984) measured denitrification potential by the acetylene block method and found that this process was also accelerated in the no-till soils. Dick (1984) in a study of a the impact of tillage on enzymatic activities in Hoytville silty clay loam (Mollic Ochraqualfs) and Wooster silt loam (Typic Fragiudalfs) found that acid phosphatase, alkaline phosphatase, arylsulfatase, invertase, amidase, and urease activities in the surface profile of no-till soil were significantly elevated over levels detected in conventional tilled soils. The species of crop also affected the enzyme activities.

With the increased cycling of plant carbon in the no-till sites, associated nitrogen transformations may also be elevated. Thus such processes as mineralization, immobilization, nitrification, and denitrification are anticipated to be stimulated. These hypotheses have been supported with data from a variety of agricultural systems. Moldboard plowing was found to result in uniform nitrogen mineralization potentials throughout the top 15 cm of a Palouse silt loam (fine-silty, mixed, mesic Pachic Ultic Haploxerolls) in the fall sampling but this potential was greater for chisel plowed, and no-till plots at the 0 to 5 cm depth and less at the 5 to 20 cm and 10 to 15 cm depths (El-Haris et al., 1983). Thus the average nitrogen mineralization potential for the top 15 cm of the soil profile was unaffected by tillage or crop rotation in the fall sampling. In the spring, the average nitrogen mineralization potentials for the chisel plowing or no-till plots were significantly higher than was observed for the moldboard plowing. Crop rotation also had a significant impact on the nitrogen mineralization potential. The reverse of the mineralization reaction, immobilization, is also stimulated in no-till plots (Rice and Smith, 1984). In each of three soil types, the immobilization in the no-till plots was stimulated 1.6 to 2.2 fold in the no-till plots over conventional tillage systems. In contrast, Carter and Rennie (1984) did not find any differences in mineralization and immobilization rates between no-tillage and conventional tillage systems. Denitrification and nitrification rates have also been shown to be stimulated in no-till soils (Rice and Smith, 1983; Aulakh et al., 1984a; Aulakh et al., 1984b; Broder et al., 1984; Clay et al., 1985; Groffman, 1985). As was detected with nitrogen mineralization, these reactions may be stimulated in the surface portions of the soil as compared with conventional tilled soils, but reductions deeper in the soil profile may negate the increases.

With the input of organic carbon with a wide carbon : nitrogen ratio, it could be anticipated that immobilization would be a problem in no-till systems. Indeed it is reasonably common to observe data demonstrating this fact. Similarly, the energy supplied by the carbon, the greater soil moisture, and the limitations on oxygen diffusion predict increased denitrification rates in no-tilled soils compared to conventionally tilled soils. This too has been shown readily. These effects on the rate of nitrogen cycle processes raises the concern about the expense of nitrogenous fertilizer utilization in no-till systems in the long run. Adequate data is now available to show that in "young" no-till soils, more nitrogen must be added to produce the same crop yields commonly observed in conventionally tilled soils. But, examination of the dynamics of organic matter transformation processes in soil suggests that any reduction of utilization of no-till procedures to reduce expenses of supplying this fixed nitrogen may be short sited. It was shown in Chapter 1, that many decades are required for many soil systems to reach equilibrium. The optimal management of no-till sites suggests that this time will be shortened, but perhaps not into the range of times currently suggested as being long-term research experiments in today's literature. Thus the probability is high that as the length of the average no-till study is increased, the quantities of nitrogen that must be added to provide comparable yields in no-till systems compared to conventional practices may decrease. This would result from the gross quantities of nitrogen transferred between the various nitrogen pools in the soil being higher in the no-till system compared to conventional sites, but the actual quantities available to the plant biomass being at least equivalent in the two systems. Support for this hypothesis is derived from observation of the biomass yields of native grasslands and the rich quantities of organic matter accumulated therein.

13.3.3. Soil Organic Amendments

Amendment of agricultural soils with organic nutrients has most probably been commonly practiced since soils were first tilled for crop production. Prior to development of industrial techniques for nitrogen fixation, the most obvious benefits of such practices were associated with improved soil and crop nutrient status. A more long-term and less appreciated—at least until organic matter levels becoming limiting—benefit of soil amendment with organic materials involves the gradual but substantial improvement of soil structure. Today, materials such as green manures, a variety of animal feces, and sewage and other waste sludge are added to soils. At least superficially, these materials are said to be used to improve the soils, but considering the problems which may be associated with disposal of some waste sludge, the societal benefit from their admixture with soil may be derived primarily if not solely from the disposal of the troublesome waste. Although there are political, environmental, economic, and scientific aspects of land disposal of waste organics, the discussion contained herein will be limited to analysis of the impact of organic amendments on soil organic matter levels, microbial activities, soil plant nutrient status. Some of the environmental problems associated with use of organic waste materials in agricultural production will also be summarized. Primary emphasis of this analysis will involve the role of these amendments in promoting ecosystem stability and long term productivity of the soils.

The benefits from soil organic amendments have been shown in a variety of soil types with a number of different organic "wastes." The long term advantage of use of farmyard amendment was demonstrated in the Rothamsted classical experiments (Jenkinson and Rayner, 1977). Farmyard manure was added to a plot annually over the 20-year period from 1852 through 1871. None was added thereafter. But, after 100 years, the manured plot still retained more organic carbon and nitrogen than the unamended control plot. The amount and duration of the elevated organic matter levels and the derived benefits varies with the type of amendment used. This benefit of an organic amendment is associated with its degree of humification. Kolenbrander (1974) noted that green plant foliage was least effective and peat was most effective as soil amendments. Farmyard manure was found to be twice as an effective soil amendment as cereal straw, whereas peat was about 2.5 fold as effective as farmyard manure.

This is not to say that return of crop residue to soil is not beneficial. We must recall that the quantities of organic matter retained in a soil results from the equilibrium between synthesis and decomposition. Thus any increase in organic matter inputs has a positive impact on total soil organic matter. Barber (1979) in a study of corn residue management and the level of soil organic matter found that about 11 percent of the carbon in the cornstalk residues was synthesized into new soil organic matter. The actual percent did vary, in that in a parallel fertility experiment, about 8 percent of the residue carbon was found in colloidal soil organic matter after 12 years. Similarly they concluded that at least 18 percent of the carbon in roots entered the soil organic matter pool. This study was conducted in Iowa on Raub silt loam (Aquic Arquidoll). Larson et al. (1972) in a study of the effect of corn (*Zea mays* L.) on soil organic matter levels of Marshall silty clay loam (Typic Hapludoll) after 11 years cultivation estimated that to prevent the loss of organic carbon through the return of cornstalk residue approximately 6 tons of carbon/h/yr needed to be added. Whereas the crop residues do contribute positively to the soil organic matter pool, it must be stressed that more humified residues encourage greater organic matter accumulations and associated physical, chemical, and biological benefits. Hofman and van Ruymbeke (1980) in a 12-year comparison of soil humus content with management practices concluded that the incorporation of the total crop residues could not replace the incorporation of farmyard manure. These data support the conclusion that incorporation of crop residues are useful primarily from the view of reducing the rate of decline in soil organic matter of the cultivated soils, but applications of major quantities of humified material is necessary to increase the steady-state organic matter level.

For maximum benefit to be derived from organic amendments, the microbial community must be stimulated. This allows increased humification of the amendment, production of plant nutrients, as well as a variety of other chemical and physical reactions necessary for maintenance and improvement of soil structure and crop production. As the humification state of the amendment increases, the rapidity of microbial attack on the substrate declines. Augmentation of microbial populations occurs with crop residues (e.g., see Doran, 1980b), waste sludge (see, Stroo and Jencks, 1985), and so on. The question associated with these benefits is involved more with the dynamics of the reactions than with the degree of benefit. With more biologically decomposable substrates such as crop residues, the microbial populations increase rapidly with the majority of the organic substances mineralized during the first year. This is to be compared to the more resistant substances where the increase in microbial populations is less dramatic—in fact, if the substance is sufficiently humified, a statistically significant increase in microbial activity may be difficult to demonstrate. Thus with the crop residues with a large carbon : nitrogen ratio, available plant nutrients would be anticipated to decline rapidly due to immobilization and the majority of the fixed carbon would be totally mineralized. With the more resistant amendments such as farmyard manure, mineralization reactions are more subdued. Therefore, nutrient demands of the soil microbial populations are reduced, resulting in occurrence of less nutrient immobilization, if any. With this difference in decomposition kinetics, greater long-term benefits would be associated with the greater humified substrates. With the use of the more resistant products, the microbial community would be active, but the processes would be sufficiently slow to encourage greater humification of the fixed carbon.

13.3.3.1. Physical Benefits of Soil Amendment

Any increase in soil organic matter or decrease in the rate of loss of soil organic matter results in improvement of overall ecosystem productivity. Obvious benefits of soil amendment are associated with decreased bulk density and improved water infiltration rates and water holding capacity. Amendment of manure to field plots of Holtville silty clay (Typic Torrifluvents) over nine years resulted in increased water infiltration rates during the growing season (Meek et al., 1982). Manure amendment had minimal effect on the water intake rate when measured between cropping periods of the soils that had been recently tilled. The increased water infiltration rates during the growing season would be helpful in areas where salinity affects crop yields. Scholl and Pase (1984) found that straw and pine bark amendments to New Mexico (U.S.A.) coal mine spoils improved the soil water content at the end of the growing season. Combination of soil amendment with pine bark and contour furrowing increased water contents in one of the two years of the study. Sewage sludge amendment to Beltsville silt loam increased the saturated hydraulic conductivity initially. This improved conductivity lasted 50 to 80 days before returning to that of the original soils (Epstein, 1975). In a related study, sewage sludge compost was also shown to improve the soil water content and retention capacity of a silt loam soil (Epstein et al., 1976). Comparison of the moisture content verses water potential indicated that more water was available to the plant community in sludge amended soils. Similarly, amendment of soil with methane generator sludge increased the available water held between -10 and -100 kPa matrix potential increased as the amount of sludge added to the soil increased (Atalay and Blanchard, 1984).

Within the soil matrix, these moisture benefits are derived from the improved soil structure—bulk densities are reduced by the soil amendment (Laverdiere and DeKimpe, 1984; Summerfeldt and Chang, 1985), and the capability of the organic matter itself to retain water. Mulches, as are commonly associated with no-till agriculture (see Smika, 1983), also improve the soil moisture status. In a comparison of the position of wheat straw (standing or flat on the soil surface) compared with bare soil, maximum water retention was found with a mixture of one half the straw standing with the remainder flat. Correlation of soil water losses with wind movement indicates that the effect of this surface mulch is to reduce evaporative water losses.

13.3.3.2. Nutrient Benefits

The major benefit from soil incorporation of organic residues results from its positive impact on soil structure. Since these materials also contained fixed plant nutrients, the mineralization kinetics of these potential nitrogen, phosphorus, and sulfur sources must be examined to evaluate the effect of these soil amendments on plant nutrition. In native sites, such as virgin forests, total ecosystem productivity is closely coupled with the mineralization rate of the aboveground plant debris and subsurface root products. In agricultural or managed soils, this coupling is frequently of little significance in that industrially produced plant nutrient utilization is also included in the management plan. Thus the question associated with nutrient aspects of organic soil amendments is related to whether they become a net source or sink for mineral nutrients. With succulent green plant debris, mineralization of the fixed carbon reserves generally leads to at least a transient immobilization of available mineral nitrogen. With subsequent cycling of the organic matter in this system, the immobilized nutrients do eventually become available to the plant community. With more humified substrates, although they have a wide carbon : nitrogen ratio, they may actually become a net source of nutrients. The desirability of the various types of amendment, albeit green manure or sewage sludge, relies not only upon the immediate effects on plant nutrition but also upon the long-term implications of their utilization on overall ecosystem stability and productivity. The necessity of utilization of larger quantities of fixed nitrogen with amendment is less of an economic problem if it has been demonstrated that this initial sacrifice will be recovered with greater future production or more sustained productivity of fragile sites.

Because of urbanization, disposal of municipal wastes has become a major governmental and environmental concern. Sewage sludge comprises a major portion of these waste materials. The agricultural benefits, aside from those accrued by the municipality with disposal of the troublesome material, have been examined in a wide variety of field, greenhouse and laboratory studies. In many instances, especially when the problems associated with toxic and/or health threatening microorganisms are considered, the agricultural benefits from the use of municipal sludge is minimal. This is certainly true in situations when the transportation costs from the sewage plant to the land spreading site are considered. Notable examples exist where sewage sludge has been dried, bagged and sold commercially. This has certainly been a feasible disposal means that met a limited market. The economic market for these materials is limited and has most certainly been saturated. Thus interest returns to demonstrating a driving need for use of sewage and related sludges as agricultural soil conditioners. If the sludge contains significant quantities of readily decomposable carbon compounds and is added in sufficient concentrations, nitrogen deficiencies may be induced in the crop (Terman et al., 1972). Control of the quality of the sludge, the application rate, and associated agricultural practices can lead to a positive effect on crop yields (e.g., see Mays et al., 1973) with minimal impacts on regional water sheds (see King, 1973). Mineralization of the organic nitrogen compounds in the sludge is one of the principle factors controlling the quantities of sludge that can be added to agricultural lands. Parker and Sommers (1983) in their study of the mineralization of organic nitrogen in municipal sludge found that approximately 25 percent of the total organic nitrogen was mineralizable in raw and primary sludge, 40 percent in waste activated, 15 percent in anaerobically digested, and 8 percent in composted sludge. They developed the following equation through regression analysis to describe the mineralization rate of sludge organic nitrogen in soil:

$$\text{Sludge Mineralized (\%)} = 6.37\ (\%\ \text{sludge organic N}) + 4.63$$

Thus, with proper management, municipal sludges have been shown to be useful for increased agricultural crop production. Similarly, McIntosh et al. (1984) noted that sludge amendment stimulated growth of hybrid poplar (*Populus deltoides* spp., *P. angulata* × *P. trichocarpa*) but not white pine (*Pinus strobus* L.). Other sludge materials, such as industrial wastes, may also be added to soil, but care must be taken to avoid difficulties associated with any toxic materials contained therein and to complement the nutrients contained in the waste materials with any other necessary fertilizer amendments.

13.3.3.3. Environmental Problems of Sludge Amendment

There are most certainly soil structural benefits and to a certain degree plant nutrient gains to be accrued from the agricultural utilization of high organic matter soil amendments, including a variety of industrial and municipal sludges. These benefits are strongly offset, if not negated in some instances, by a variety of potential environmental problems which must be examined before implementation of management plans involving amendment of waste materials to productive soil systems. Probably the first difficulty that comes to mind when municipal sewage products are considered involves the potential for inclusion of human viruses (see Bitton et al., 1984) and mutagens or carcinogens (see Angle and Baudler, 1984). The feasibility of the disposal plan depends totally upon the survival of the pathogen or the biodegradation stability of the toxicant under the prevailing field environmental conditions. In situations where the longevity of the pathogenic organisms is short or the human contact with the field or products derived therefrom is minimal, complications arising from the sludge use are likely minor. Care must be taken to avoid contamination of mature crops, especially those consumed without processing.

Of probably greater concern is the concentrations of metals contained in sludges. The concentration per unit volume of the sludge may be low, but with repeated applications to the soils, significant concentrations may accumulate in the soil. Also the potential for bio-accumulation must be considered. Factors that must be considered to determine the utility of metal contaminated sludge materials include the soil organic matter content and the soil pH (see Chapter 10).

3.4. CONCLUSIONS

Sustaining highly productive agricultural systems is encouraged by the maintenance and/or improvement of the soil organic matter status. This objective can be achieved by increasing organic matter inputs or reducing losses through biodegradation or leaching. Erosion control is obligatory for the solution of the latter problem. Organic matter inputs may be augmented through maximization of crop yields and through soil amendment. Use of partially humified organic products, such as manures or sludges, increases the longevity of organic matter derived benefits. Because of associated problems of metal contamination, potential public health problems, and transportation and application costs, residue application must be carefully managed. Tillage reduction leads to a parallel decrease in organic matter decomposition rates which in turn encourages augmented or at least sustained colloidal soil organic matter levels. Proper soil husbandry must include organic matter conservation plans, such as residue management and tillage control, to assure the long-term productivity of the ecosystem.

REFERENCES

Angle, J. S., and D. M. Baudler, 1984. Persistence and degradation of mutagens in sludge-amended soil. J. Environ. Qual. 13: 143-146.

Angle, J. S., G. McClung, M. S. McIntosh, P. M. Thomas, and D. C. Wolf, 1984. Nutrient losses in runoff from conventional and no-till corn watersheds. J. Environ. Qual. 13: 431-435.

Atalay, A., and R. W. Blanchard, 1984. Evaluation of methane generator sludge as a soil amendment. J. Environ. Qual. 13: 341-344.

Aulakh, M. W., D. A. Rennie, and E. A. Paul, 1984a. Gaseous nitrogen losses from soils under zero-till as compared with conventional-till management systems. J. Environ. Qual. 13: 130-136.

Aulakh, M. S., D. A. Rennie, and E. A. Paul, 1984b. The influences of plant residues on denitrification rates in conventional and zero tilled soils. Soil Sci. Soc. Am. J. 48: 790-794.

Ayanaba, A., B. Tuckwell, and D. S. Jenkinson, 1976. The effects of clearing and cropping on the organic reserves and biomass of tropical forest soils. Soil Biol. Biochem. 8: 519-525.

Barber, S. A. 1971. Effect of tillage practice on corn (*Zea mays* L.) root distribution and morphology. Agron. J. 63: 724-726.

Barber, S. A. 1979. Corn residue management and soil organic matter. Agron. J. 71: 625-627.

Barnette, R. M., and J. B. Hester, 1930. Effect of burning upon the accumulation of organic matter in forest soils. Soil Sci. 29: 281-284.

Bitton, G., O. C. Pancorbo, and S. R. Farrah, 1984. Virus transport and survival after land application of sewage sludge. Appl. Environ. Microbiol. 47: 905-909.

Belvins, R. L., M. S. Smith, and G. W. Thomas. 1984. Changes in soil properties under no-tillage. In R. E. Phillips, and S. H. Phillips (eds.), No-Tillage agriculture. van Nostrand Reinhold, New York, pp. 190-230.

Broder, M. W., J. W. Doran, G. A. Peterson, and C. R. Fenster, 1984. Fallow tillage influence on spring populations of soil nitrifiers, denitrifiers, and available nitrogen. Soil Sci. Soc. Am. J. 48: 1060-1067.

Carter, M. R., and D. A. Rennie, 1984. Nitrogen transformations under zero and shallow tillage. Soil Sci. Soc. Am. J. 48: 1077-1081.

Cunningham, R. K. 1963. The effect of clearing a tropical forest soil. J. Soil Sci. 14: 334-345.

Dick, W. A. 1984. Influence of long-term tillage and crop rotation combinations on soil enzyme activities. Soil Sci. Soc. Am. J. 48: 569-574.

Doran, J. W. 1980a. Soil microbial and biochemical changes associated with reduced tillage. Soil Sci. Soc. Am. J. 44: 765-771.

Doran, J. W. 1980b. Microbial changes associated with residue management with reduced tillage. Soil Sci. Soc. Am. J. 44: 518-524.

El-Haris, M. K., V. L. Cochran, L. F. Elliott, and D. F. Bezdicek, 1983. Effect of tillage, cropping, and fertilizer management on soil nitrogen mineralization potential. Soil Sci. Soc. Am. J. 47: 1157-1161

Epstein, E. 1975. Effect of sewage sludge on some soil physical properties. J. Environ. Qual. 4: 139-142.

Epstein, E., J. M. Taylor, and R. L. Chaney, 1976. Effects of sewage sludge compost applied to soil on some soil physical and chemical properties. J. Environ. Qual. 5: 422-426.

Giddens, J. 1957. Rate of loss of carbon from Georgia soils. Soil Sci. Soc. Am. J. 21: 513-515.

Groffman, P. M. 1985. Nitrification and denitrification in conventional and no-tillage soils. Soil Sci. Soc. Am. J. 49: 329-334.

Gupta, S. C., W. E. Larson, and D. R. Linden, 1983. Tillage and surface residue effects on soil upper boundary temperatures. Soil Sci. Soc. Am. J. 47: 1212-1218.

Hofman, G., and M. van Ruymbeke, 1980. Evolution of soil humus content and calculation of global humification coefficients on different organic matter treatments during a 12-year experiment with Belgian silt soils. Soil Sci. 129: 92-94.

Jenkinson, D. S., and J. H. Rayner, 1977. The turnover of soil organic matter in some of the Rothamsted classical experiments. Soil Sci. 123: 298-305.

Jones, J. N., Jr., J. E. Moody, and J. H. Lillard, 1969. Effects of tillage, and mulch on soil water and plant growth. Agron. J. 61: 719-721.

Keeney, D. R., and J. M. Bremner, 1964. Effect of cultivation on the nitrogen distribution in soils. Soil Sci. Soc. Am. Proc. 28: 653-656.

King, L. D. 1973. Mineralization and gaseous loss of nitrogen in soil-applied liquid sewage sludge. J. Environ. Qual. 2: 356-358.

Kolenbrander, G. J. 1974. Efficiency of organic manure in increasing soil organic matter content. Trans. Int. Congr. Soil Sci. 10th. 2: 129-136.

Lamb, J. A., G. A. Peterson, and C. R. Fenster, 1985. Wheat fallow tillage systems' effect on a newly cultivated grassland soils' nitrogen budget. Soil Sci. Soc. Am. J. 49: 352-356.

Larson, W. E., C. E. Clapp, W. H. Pierre, and Y. B. Morachan, 1972. Effects of organic residues on continuous corn: II. organic carbon, nitrogen, phosphorus and sulfur. Agron. J. 64: 204-208.

Laverdiere, M. R., and C. R. DeKimpe, 1984. Agronomic use of clay soils from Abitibi, Quebec: 2. Effects of organic amendments and cultivation on crop production. Soil Sci. 137: 128-133.

Laws, W. D., and D. D. Evans, 1949. The effects of long-time cultivation on some physical and chemical properties of two Rendzina soils. Soil Sci. Soc. Am. Proc. 14: 15-19.

Likens, G. E., F. H. Bormann, R. S. Pierce, and W. A. Reiners, 1978. Recovery of a deforested ecosystem Science (Washington, D.C.) 199: 492-496.

Linn, D. M., and J. W. Doran, 1984. Aerobic and anaerobic microbial populations in no-till and plowed soils. Soil Sci. Soc. Am. J. 48: 794-799.

Lundgren, B. 1982. Bacteria in a pine forest soil as affected by clear-cutting. Soil Biol. Biochem 14: 537-542.

Mays, D. A., G. L. Terman, and J. C. Duggan, 1973. Municipal compost: effects on crop yields and soil properties. J. Environ. Qual. 2: 89-92.

McIntosh, M. S., J. E. Foss, D. C. Wolf, K. R. Brandt, and R. Darmody, 1984. Effect of composted municipal sewage sludge on growth and elemental composition on white pine and hybrid popular. J. Environ. Qual. 13: 60-62.

Meek, B., L. Graham, and T. Donovan, 1982. Long term effects of manure on soil nitrogen, potassium, sodium, organic matter, and water infiltration rates. Soil Sci. Soc. Am. J. 46: 1014-1019.

Moldenhauer, W. C., W. H. Wischmeier, and D. T. Parker, 1967. The influence of crop management on runoff, erosion, and soil properties of a Marshall silty clay loam. Soil Sci. Soc. Am. Proc. 31: 541-546.

Morachan, Y. B., W. C. Moldenhauer, and W. E. Larson, 1972. Effects of increasing amounts of organic residues on continuous corn: I. Yields and soil physical properties. Agron. J. 64: 199-203.

Nichols, J. D. 1984. Relation of organic carbon to soil properties and climate in Southern Great Plains. Soil Sci. Soc. Am. J. 48: 1382-1384.

Odell, R. T., S. W. Melsted, and W. M. Walker, 1984. Changes in organic carbon and nitrogen of Morrow Plot soils under different treatments, 1904-1973. Soil Sci. 137: 160-171.

Parker, C. F., and L. E. Sommers, 1983. Mineralization of nitrogen in sewage sludges. J. Environ. Qual. 12: 150-156.

Phillips, R. E. 1984. Soil moisture. In R. E. Phillips and S. H. Phillips (eds.), No-tillage agriculture. van Nostrand Reinhold, New York, pp. 66-86.

Phillips, R. E., and S. H. Phillips (eds.), No-tillage agriculture. van Nostrand Reinhold, New York, 306 pp.

Potter, K. N., R. M. Cruse, and R. Horton, 1985. Tillage effects on soil thermal properties. Soil Sci. Soc. Am. J. 49: 968-973.

Reinhorn, T., and Y. Avnimelech, 1974. Nitrogen release associated with the decrease in soil organic matter in newly cultivated soils. J. Environ. Qual. 3: 118-121.

Reennie, D. A., and M. Heimo, 1984. Soil and fertilizer-N transformations under simulated zero till: Effect of temperature regimes. Can. J. Soil Sci. 64: 1-8.

Rice, C. W., and M. S. Smith, 1983. Nitrification of fertilizer and mineralized ammonium in no-till and plowed soil. Soil Sci. Soc. Am. J. 47: 1125-1129.

Rice, C. W., and M. S. Smith, 1984. Short-term immobilization of fertilizer nitrogen at the surface of no-till and plowed soils. Soil Sci. Soc. Am. J. 48: 295-297.

Sanchez, P. A., J. H. Villachica, and D. E. Bandy, 1983. Soil fertility dynamics after clearing a tropical rainforest in Peru. Soil Sci. Soc. Am. J. 47: 1171-1178.

Scholl, D. G., and C. P. Pase, 1984. Wheatgrass response to organic amendments and contour furrowing on coal mine spoil. J. Environ. Qual. 13: 479-482.

Smika, D. E. 1983. Soil water change as related to position of wheat straw mulch on the soil surface. Soil Sci. Soc. Am. J. 47: 988-991.

Smith, S. J., and L. B. Young, 1975. Distribution of nitrogen forms in virgin and cultivated soils Soil Sci. 120: 354-360.

Stinner, B. R., D. A. Crossley, Jr., E. P. Odum, and R. L. Todd, 1984. Nutrient budgets and internal cycling of N, P, K, Ca, and Mg in conventional tillage, no-tillage, and old-field ecosystems on the Georgia Piedmont. Ecology 65: 354-369.

Stroo, H. F., and E. M. Jencks, 1985. Effect of sewage sludge on microbial activity in an old, abandoned minesoil. J. Environ. Qual. 14: 301-304.

Summerfeldt, T. G., and C. Chang, 1985. Changes in soil properties under annual applications of feedlot manure and different tillage practices. Soil Sci. Soc. Am. J. 49: 983-987.

Terman, G. L., J. M. Soileau, and S. E. Allen, 1972. Municipal waste compost: Effects on crop yields and nutrient content in greenhouse pot experiments. J. Environ. Qual. 2: 84-89.

Unger, P. W. 1978. Straw mulch effects on soil temperatures and sorghum germination and growth. Agron. J. 70: 858-864.

Vance, E. D., and G. S. Henderson, 1984. Soil nitrogen availability following long-term burning in an oak-hickory forest. Soil Sci. Soc. Am. J. 48: 184-190.

van Cleve, K., and T. A. Moore, 1978. Cumulative effects of nitrogen, phosphorus and potassium fertilizer additions on soil respiration, pH, and organic matter content. Soil Sci. Soc. Am. J. 42: 121-124.

White, E. M., C. R. Krueger, and R. A. Moore, 1976. Changes in total N, organic matter, available P, and bulk densities of a cultivated soil 8 years after tame pastures were established. Agron. J. 68: 581-583.

FOURTEEN

SOIL ORGANIC MATTER: A CURRENT AND FUTURE CONCERN

In the absence of societal pressures arising from high population densities, urbanization, and industrialization, changes in the world's soil organic matter reserves and the resultant impact on land use were of little concern. The plant nutrients in the organic matter pool could be mined at will and the recreational sites enjoyed in what today would be considered solitude. But, there are few regions of the world today where this is possible. In the past, a slight decline in crop yields or even environmental quality would perhaps necessitate field and/or village relocation, but would fall short of the severe famine and environmental disruptions commonly reported in scientific literature and the popular press today. There are now severe burdens on our soil reserves. Soil ecosystems are called on to provide more food and fiber than ever before, to serve as a waste receptacles, to mitigate any toxic impacts of disposal of these wastes, and to withstand the repeated incursions associated with recreational development of our more aesthetic ecosystems. Thus land management plans which allow for the maximum utilization (some would say exploiting) of the ecosystem while preserving as closely as possible the essential nature of the system in a quasi pristine form for essentially all ecosystems must be developed. Since changes in soil organic matter reactions have strong implications on total ecosystem function and stability, this development of best management procedures calls for increased investigation of soil organic matter cycles plus application of the volumes of information that have been collected and as yet not incorporated into ecosystem concepts relating to these processes.

The importance of the soil organic fraction in future ecosystem management plans is exemplified by listing a few of the current environmental concerns that rely upon our understanding of this soil fraction. If a top priority list of environmental concerns relating to organic matter was developed, there would most probably be a few differences of opinion of order and perhaps even inclusion of specific topics. But, most lists would likely include these current and long-term environmental problems (these topics are obviously not listed according to priority.):

1. **Maintenance and improvement of soil structure through use of agricultural procedures that reverse the currently observed decline in soil organic matter reserves.** As the world's population increases, greater acreages of land are required to meet housing and recreational needs. Thus quantities of soils available for cultivation are declining throughout the world. Unfortunately, the land that is being incorporated into growing cities, towns, and villages is in most cases prime agricultural land. Thus the farmers are asked to provide increasing quantities of food and fiber on land that is either declining in productivity because of diminishing soil structure resulting from loss of organic matter reserves or land that never contained the proper soil structure for intensive agriculture. In either case, management schemes must be developed and implemented that allow for the conservation and where possible, augmentation of the colloidal soil organic matter pool.
2. **Restoring site safety and/or productivity following industrial spills or improper use of xenobiotics.** Industrially produced chemicals, be they pesticides for agricultural or household use, precursors for the variety of plastic or synthetic polymers incorporated into our daily life, or products or byproducts of the biomedical profession, are being incorporated into the soil ecosystem through inadvertent spills, disposal processes, and even, in many cases, proper usage. The soil microbial community is called upon to detoxify, decompose, or otherwise mitigate the adverse properties and reactions of these industrial products in a variety of soil ecosystems. In some situations, these xenobiotics are totally mineralized in the absence of adverse environmental reactions. But, the soil microbial community is not infallible. Some of the chemicals may not be decomposed by the microbial community. Thus they remain to react chemically with various soil components, enter the food chain, or even migrate from the spill site. Others may be incorporated intact into humic substances never to be released as free products again. Some may react with the humic polymers through reactions which allow subsequent hydrolysis and release of the chemical into the ecosystem. In the latter situation, soil previously felt to be safe for agricultural or even recreational use may later be found to contain a toxicant.
3. **Urban waste management**. As urban populations increase, the regional concentration of organic waste products requiring disposal becomes nearly unmanageable. These materials have been burned, buried in land fills, and towed out to sea. Wastes that tend to accumulate include the common household garbage generally trucked to landfills, low level radioactive wastes also buried in landfill type sites, and sewage sludges. Biodegradable organic matter comprises a major portion of each of these materials. Their incorporation into landfill sites makes this a quasi soil organic matter problem because the substances behave as would most any other exogenously supplied organic material. Unfortunately these materials contain components which may lead to toxicity problems with crops produced on the soil, contaminate ground water, or even preclude aboveground plant community

development. Although many of the reactions involved in the biodecomposition of land disposed wastes are common to natural soil organic matter reaction sequences, the high concentrations of organic matter incorporated into the soil at waste disposal sites create special conditions that must be dealt with by site management. These include the potential for anaerobic methane production, organic acid production and subsequent mobilization of metallic ions including the radioisotopes contained in low level radioactive wastes, and onsite synthesis of toxicants. Many of the principles discussed in this treatise will be useful in developing improved organic waste disposal plans.

4. **Recreation.** The demand for recreational sites is increasing in most parts of the world. This is in opposition to the desire to preserve many ecosystem types in "pristine" conditions. To protect the ecosystem from the anthropogenic problems associated with various tourist incursions and to manage the sites to maintain their desirable aspects, as complete understanding as possible of nutrient cycling in these "nondeveloped" areas is needed. This includes the rates of incorporation of plant nutrients into soil organic matter, the subsequent mineralization of this plant nutrient pool, and then assimilation into aboveground biomass. Common ecosystems where such knowledge is in great demand include both forests and desert sites; both are fragile. Recreational damage may appear more quickly with the desert floor compared to forests in that disruption of the desert crust changes the stable ecosystem to a mass of shifting sand. This type of damage is long lasting and detrimental to all undeveloped parks and recreational areas where basic ecological principles are not considered in the development plan.
5. **Reclamation.** Following the dramatic disturbance associated with a wide variety of mining operations, a generally unaesthetic site remains. With due consideration of the ecological, biological, and soil science principles associated with the "remains" of the mining operation, a reasonably stable, productive ecosystem can be developed. The-long term stability of the created ecosystem is strongly dependent upon the predevelopment concern for encouragement of a productive soil ecosystem; that is, soil organic matter and associated microbiological processes must established. Return of the original top soil and management of the waste to remove obvious plant toxicants (e.g., acid with acid mine wastes) may allow for immediate plant community growth. But the ecosystem is continually changing. If the stability of the site is to be encouraged, the predominant soil properties must be those that allow for continued positive development of the plant community, not gradual decline until essentially a desert exists in what was probably a productive ecosystem prior to intervention for mining.
6. **Desertification.** The world's desert regions are expanding. Although a variety of reasons, perhaps excuses, have been proposed to account for this observation, a major cause must be considered to be poor soil husbandry. Many of these soils are associated with some of the oldest of the world's

societies. There has been great demand on the productivity of the soils in the past, but with the current increased populations of the regions the demands far exceed the capability of the soil ecosystem to produce. In many cases, the soils have been robbed of any natural inputs of organic matter while being maintained in conditions that encourage rapid depletion of any pools of native colloidal organic matter. Any hope for reversing this encroachment of the desert must include sound management of the soil organic reserves.

7. **Land Use Planning.** Soils are a natural resource. They must be treated as such. It is a prime maxim of soil science that not all soils are adequate for all uses that society can propose for them. The properties of the soil as well as the demands of the intended use must be considered in land use planning—not just the generally discussed sociological and economic implications. Questions to be considered should include the capability of the soil to deal with any by products of the development and the impact of the current utilization on future soil development.

Each of these areas of concern will become increasingly important as societal pressure on the ecosystem intensifies. All involve to a large degree management of the soil organic matter, including the biotic communities. Thus there is an increasing need to apply newly developed chemical and biological techniques to provide field data as well as to continue increasing the sophistication of soil organic matter models so that more prudent resource use plans can be developed. This necessitates greater application of currently available data plus intensified efforts to increase our understanding of this often neglected soil component—colloidal organic matter. It is the author's hope that this treatise will at least in part spur the needed research and provide a sound background of the basic information required to meet these current and future needs.

INDEX